Introductory Statistical Thermodynamics

Introductory Statistical Thermodynamics

by

Nils Dalarsson and Mariana Dalarsson
Royal Institute of Technology, Stockholm, Sweden

Leonardo Golubović
West Virginia University, Morgantown, USA

AMSTERDAM • BOSTON • HEIDELBERG • LONDON
NEW YORK • OXFORD • PARIS • SAN DIEGO
SAN FRANCISCO • SINGAPORE • SYDNEY • TOKYO
Academic Press is an imprint of Elsevier

Academic Press is an imprint of Elsevier
30 Corporate Drive, Suite 400, Burlington, MA 01803, USA
525 B Street, Suite 1800, San Diego, California 92101-4495, USA
84 Theobald's Road, London WC1X 8RR, UK

Notices
Knowledge and best practice in this field are constantly changing. As new research and experience broaden our
understanding, changes in research methods, professional practices, or medical treatment may become necessary.

Practitioners and researchers must always rely on their own experience and knowledge in evaluating and using
any information, methods, compounds, or experiments described herein. In using such information or methods
they should be mindful of their own safety and the safety of others, including parties for whom they have a
professional responsibility.

To the fullest extent of the law, neither the Publisher nor the authors, contributors, or editors, assume any liability
for any injury and/or damage to persons or property as a matter of products liability, negligence or otherwise, or
from any use or operation of any methods, products, instructions, or ideas contained in the material herein.

Library of Congress Cataloging-in-Publication Data
Introductory statistical thermodynamics / edited by N. Dalarsson, M. Dalarsson [and] L. Golubović.
    p. cm.
  Includes bibliographical references and index.
  ISBN 978-0-12-384956-4 (hardback)
1. Statistical thermodynamics. I. Dalarsson, M. (Mirjana) II. Dalarsson, N. (Nils)
III. Golubović, L. (Leonardo)
  QC311.5.E44 2011
  536'.7–dc22

                        2010042085

British Library Cataloguing-in-Publication Data
A catalogue record for this book is available from the British Library.

ISBN: 978-0-12-384956-4

For information on all Academic Press publications
visit our Web site at *www.elsevierdirect.com*

Typeset by: diacriTech, India

Printed and bound by CPI Group (UK) Ltd, Croydon, CR0 4YY

Contents

Some sections are labeled by an asterisk *) as explained in the preface.

Preface

The aim of this book is to discuss basic theoretical results of the statistical physics and thermodynamics, which cover the fundamental physical concepts used for the macroscopic description of systems with very large numbers of constituent particles. The macroscopic concepts used in classical thermodynamics are derived from the microscopic theories of constituent particles, i.e., quantum mechanics and statistical mechanics. However, in literature, the subject of classical thermodynamics is frequently introduced from a purely macroscopic point of view. Such an approach is conceptually more difficult, and its reasoning is often hard to follow by many physics and engineering students. In particular, the significance and the real physical meaning of the fundamental concept of entropy are difficult, if not impossible, to explain on purely macroscopic grounds.

Therefore, we chose to approach the subject from the microscopic point of view. It is based on the fact that all macroscopic systems consist of microscopic constituent particles (molecules, atoms, or elementary particles) obeying the laws of quantum mechanics. Using basic postulates of statistical mechanics and elementary results from quantum mechanics, we derive the general thermodynamic description of physical systems at macroscopic level. This description is largely independent on the details of the microscopic models describing the interactions of the particles in various physical systems.

Despite the initial microscopic approach to the subject, the central part of the book consists of physical considerations on a purely macroscopic level. By adopting some simple quantum-mechanical models of the constituent particles of a system, the book shows how one can calculate macroscopic thermodynamic quantities on the basis of the relevant microscopic results. The book is focused on the study of systems in thermodynamic equilibrium. However, the statistical approach to the discussion of various equilibrium situations provides the students with the needed preparation for the extension to the discussion of nonequilibrium systems.

The basic plan of the book is the following: The first part of the book covers the microscopic models of constituent particles. It discusses three basic models from quantum mechanics. Some sections of the first part of the book are of a purely quantum-mechanical and mathematical significance. Although not directly related to the traditional subjects of thermodynamics, they are considered in some detail in order to establish a complete understanding of the quantum description of macroscopic systems of particles.

The second part of the book is devoted to the detailed derivation of the basic notions of classical statistical mechanics and the general laws of macroscopic thermodynamics that relate various thermodynamic quantities. The third part of the book covers the application of the general macroscopic laws, derived in the second part, to the physically interesting cases of the ideal and nonideal gases. Finally, in the fourth part of the book, a discussion of quantum statistical mechanics and some relativistic phenomena is presented. The quantum relativistic thermodynamics includes a study of some macroscopic phenomena in the expanding early universe.

Some sections of the book cover purely mathematical and/or quantum-mechanical concepts that are not directly related to what is traditionally the scope of courses of Statistical Physics and Thermodynamics. These sections are marked by an asterisk and can be omitted by the instructors adopting a less theoretical and mathematical approach to the subject. These sections can also be omitted by readers who have already taken a Quantum Mechanics course prior to taking the course in Statistical Physics and Thermodynamics, and are familiar with the quantum-mechanical and mathematical concepts covered in these sections.

The book is intended as a text for an introductory one-semester course in thermodynamics for undergraduate students of physical sciences or engineering. However, it can also be used as material for an introductory graduate course in thermodynamics. The book has evolved from a set of lecture notes originally prepared by one of the authors, N. Dalarsson, but has been subsequently expanded with a number of new topics over the last fifteen years.

It is the intention of the book to provide the readers with a high level of detail in derivations of all equations and results. All algebraic manipulations are outlined in great detail, such that they can be followed by an interested college student with very little or no risk of ever getting lost. It is our experience that a common show stopper for a young college student, trying to master a subject, are the phrases in the literature claiming that something can be derived from something else by some "straightforward although somewhat tedious algebra." If a student cannot readily reproduce that "straightforward" algebra, which often is the case, the usual reaction under the time pressure of the studies is to accept the claim as a fact. And from that point on, throughout the rest of the course, the deeper understanding of the subject is lost.

The book contains a number of solved problems. They have been selected to illustrate the theoretical concepts discussed throughout the book. The solutions are given in detail to help master the theoretical concepts from the book. It is common in the literature to include the problems at the end of each chapter and to outline the solutions or just provide answers at the end of the book. In the present book, we choose a different approach and add a number of problems with detailed solutions at the end of each chapter. Furthermore, a number of detailed theoretical derivations throughout the book can also be used as homework or exam problems.

There is a number of advanced books on statistical thermodynamics, and we have benefited from some of those. These sources are listed in the bibliography at the end of the book, as well as a few other books recommended as suitable further reading. Regretfully, in an introductory book such as this one, it was neither possible to include an extensive list of all original references and major textbooks and monographs on the subject nor to mention all the people that have contributed to our understanding of the subject.

We hope that our readers will find that we have fulfilled the objective of providing a self-contained and self-explanatory book, which provides a smooth introduction to this important subject and that they will enjoy reading this book as much as we enjoyed writing it.

N. Dalarsson, M. Dalarsson, and L. Golubović
Stockholm, Sweden, and Morgantown, WV, USA, June 2010

1 Introduction

The subject of this book is a discussion of systems consisting of many particles. Good examples for this are atomic and molecular gases; however, the theoretical concepts we will develop can also be applied to the studies of liquids, solids, electromagnetic radiation (photons), crystal lattice vibrations (phonons), etc. Essentially, all physical, engineering, chemical, or biological systems of practical interest consist of many constituent particles. Thus, the concepts developed in this book are useful for the description of a large number of natural phenomena.

The first systematic approach to the investigation of macroscopic systems from a macroscopic empirical point of view began in the nineteenth century. The subject of empirical thermodynamics was developed before the discovery of the atomic nature of matter. The proposal that heat is a form of energy was first put forward by Rumford (1798) and Davy (1799). It was generally accepted after the experimental work of Joule (1843–1849). The first theoretical explanation of the principles of heat engines was given by the French engineer Carnot (1824). A consistent form of the theoretical thermodynamics was formulated by Clausius and Kelvin (1850), and it was greatly developed and generalized by the American physicist Gibbs (1876–1878).

The microscopic approach to thermodynamics started with the work of Clausius, Maxwell, and Boltzmann. Maxwell formulated the distribution law of molecular velocities (1859), while Ludwig Boltzmann derived his fundamental transport equation (1872). Thereafter, Chapman and Enskog (1916–1917) developed systematic methods for solving the Boltzmann equation. The modern equilibrium statistical mechanics was also initiated in the works of Boltzmann (1872). It was then significantly generalized by the work of Gibbs (1902). The discovery of quantum mechanics introduced some changes; however, the basic framework of the Gibbs theory survived the advent of quantum mechanics.

In an attempt to discuss common systems involving many particles, such as gases, one realizes that the laws of quantum mechanics are needed to adequately describe the motions of the individual constituent particles of these systems. On the other hand, the atomic nuclei, often being the part of these constituent particles (such as atoms or molecules), are normally not affected by their motion. So, the nuclear forces play a secondary role. Likewise, the gravitational forces between the constituent particles are also generally negligible. Thus, the major forces affecting the constituent particles in the common systems such as gases, liquids, and solids are only the well-understood electromagnetic interactions, such as the familiar electrostatic forces.

Introductory Statistical Thermodynamics

In view of this, the fundamental physical laws governing these systems are understood well. Unfortunately, however, there are some major difficulties of using the microscopic laws to the systems with a large number of constituent particles. Due to the presence of an enormous number of degrees of freedom (such as all particle positions), such systems are analytically and computationally intractable by using the ordinary methods of classical and quantum mechanics. The major goal of statistical mechanics is thus to relate this tantalizing quantitative complexity to the apparent simplicity of the macroscopic description typically involving a small number of data such as temperature, pressure, particle density, etc. We would like to understand, for example, how the ordinary concept of temperature emerges from the complexity of many particle systems.

Let us now define the important concepts of "microscopic" ("small scale") and "macroscopic" ("large scale"). A system is called microscopic, if it is of atomic dimensions or smaller ($\leq 10^{-10}$ m). Individual atoms and molecules are examples of such systems. On the other hand, a system is called macroscopic, if it is visible by, for example, a microscope using ordinary light ($\geq 10^{-6}$ m). A macroscopic system consists typically of a large number of constituent particles. Such a system is a visible (by the microscope) amount of a gas or a liquid. When studying a macroscopic system, we are not concerned with the detailed behavior of each of the individual particles constituting the system. Instead, we usually study macroscopic quantities describing the system like volume, pressure, temperature, etc.

If the macroscopic quantities describing an isolated system are constant in time, we say that the system is in equilibrium. If an isolated system is not in equilibrium, the time-dependent macroscopic quantities, describing the system, will change until the system reaches an equilibrium.

By applying the statistical physics to the systems consisting of a large number of constituent particles, it is possible to gain a considerable understanding of these systems by standard mathematical methods. This success of statistical physics may seem miraculous, since macroscopic systems consist of such a huge number of particles. Yet, as we are not concerned with the detailed mechanical behavior of every individual constituent particle, we can apply the statistical laws to the system. In macroscopic systems such as gases, due to having a very large number of identical particles, the statistical laws become highly effective and many important problems become tractable.

Part One

Quantum Description of Systems

2 Introduction and Basic Concepts

2.1 Systems of Identical Particles

The subject of the present discussion is the description of systems consisting of a large number of identical particles. One such system is a gas consisting of a large number of constituent molecules. Let us assume that the system under consideration occupies the volume V and contains N noninteracting particles. The system is surrounded by some medium and in general it is allowed to exchange both energy and particles with the surrounding medium.

As the constituent particles of the system are microscopic objects, e.g., gas molecules, their motion is ruled by the laws of quantum mechanics and they can only occupy certain discrete energy levels. Let n_i be the number of particles, which occupy the energy levels within the energy interval $(E_i, E_i + \Delta E_i)$.

If the system is in an equilibrium state, the total number of particles in the system is given by

$$N = \sum_i n_i, \tag{2.1}$$

where the sum goes over all energy intervals $(E_i, E_i + \Delta E_i)$. The energy of each particle in the system is the sum of its kinetic and potential energies

$$E_i = T_i + \vartheta_i, \tag{2.2}$$

and the total internal energy of the system can be written in the form

$$U = \sum_i n_i E_i. \tag{2.3}$$

Above, we consider the so-called ideal systems that consist of mutually noninteracting particles. For the ideal systems, the potential energies $\vartheta_i = \vartheta_i(\vec{r})$ of the particles are due to the influence of some external fields. For example, a gas under consideration is normally under the influence of the Earth' gravity. If the potential energies due to external fields are nearly the same for all constituent particles, then such a system is nearly homogeneous, that is, it has nearly uniform density. Under this condition, the potential energies due to external fields are inessential for the description of the

Introductory Statistical Thermodynamics

system's macroscopic behavior. Thus, the macroscopic thermodynamic properties of the system are determined by the kinetic energies of its constituent particles.

The simplest realizations of an ideal system are the gases of elementary particles that have no internal structure or monatomic gases of inert elements (such as helium, argon, neon, etc.), which internal structure can be ignored to a good approximation. In such ideal gases, the constituent particles of the system can be treated as point-like particles without an internal structure. Because of this, the kinetic energy of each constituent particle is only due to the translational motion of the particle with velocity $\vec{v} = \vec{v}(t)$ and momentum $\vec{p} = m\vec{v}$. In general, the motion of the particles in an ideal gas is a three-dimensional motion, which can be described by the Descartes co-ordinates (x, y, z). Thus, an ideal gas has three translational degrees of freedom.

Sometimes the constituent molecules of a gas do have an internal structure and cannot be treated as ideal point particles. For example, the molecules of common gases in nature are often diatomic like hydrogen (H_2), oxygen (O_2), etc. Such molecules can rotate and vibrate about their center of mass. If we analyze an individual molecule in a co-ordinate system where its center of mass is at rest, i.e., in its center-of-mass frame, then it has zero translational kinetic energy. In this frame, the energy of a molecule includes only the energies of rotations and vibrations about its center of mass. In other words, the gas has a number of additional degrees of freedom due to the rotations and vibrations of the constituent molecules.

2.2   Quantum Description of Particles *)

In order to study the properties of the individual constituent particles and calculate their kinetic energies, used to determine the macroscopic thermodynamic properties of the system, an introduction to some basic concepts of quantum mechanics is needed.

According to the basic quantum-mechanical assumption, a particle of energy E and momentum vector $\vec{p}$ is associated with a wave of frequency ω and the wave vector $\vec{\kappa}$. In other words, material particles have a dual wave-particle nature. The quantum-mechanical relations between the particle quantities E and $\vec{p}$ and the wave parameters ω and $\vec{\kappa}$, respectively, are given by

$$E = \hbar\omega, \quad \vec{p} = \hbar\vec{\kappa}, \tag{2.4}$$

where $\hbar = h/2\pi$ and $h = 6.626068 \times 10^{-34}$ Js is the Planck constant. The motion of such a "matter wave" is described by a complex wave function

$$\psi = \psi(\vec{r}, t). \tag{2.5}$$

The squared magnitude of the wave function (2.5) is then interpreted as the probability density of finding the particle in the space point $\vec{r} = (x, y, z)$ at the time t. Since the particle is located somewhere in the three-dimensional space, the absolute square of the wave function (2.5) satisfies the following normalization condition

$$\int_V |\psi(\vec{r}, t)|^2 \; d^3\vec{r} = 1, \tag{2.6}$$

where the integration is performed over the entire available three-dimensional space. Unlike the classical mechanics, quantum mechanics can provide only the information about the probabilities and expectation values of various physical quantities such as the particle position, momentum, energy, etc. Thus, the expectation value of the energy E is given by

$$< E >= \int_V \psi^*(\vec{r}, t)\hat{H}\psi(\vec{r}, t)\mathrm{d}^3\vec{r}, \tag{2.7}$$

where $\psi^*(\vec{r}, t)$ is the complex conjugate of the wave function $\psi(\vec{r}, t)$, and $\hat{H}$ is the energy operator or the *Hamiltonian operator*. The form of the operator $\hat{H}$ will be determined later. Likewise, the expectation value of the momentum vector $\vec{p}$ is given by

$$<\vec{p}>= \int_V \psi^*(\vec{r}, t)\hat{\vec{p}}\psi(\vec{r}, t)\mathrm{d}^3\vec{r}, \tag{2.8}$$

where $\hat{\vec{p}}$ is the *Momentum operator*. In order to determine the forms of the energy and momentum operators, we can use the wave function of a suitable plane wave, given by

$$\psi(\vec{r}, t) = N \exp(i\vec{\kappa} \cdot \vec{r} - i\omega t) = N \exp\left[\frac{i}{\hbar}(\vec{p} \cdot \vec{r} - Et)\right]. \tag{2.9}$$

Using the result (2.9), we see that

$$i\hbar \frac{\partial}{\partial t}\psi(\vec{r}, t) = E\psi(\vec{r}, t) = \hat{H}\psi(\vec{r}, t), \tag{2.10}$$

and

$$-i\hbar \frac{\partial}{\partial x}\psi(\vec{r}, t) = p_x\psi(\vec{r}, t) = \hat{p}_x\psi(\vec{r}, t)$$

$$-i\hbar \frac{\partial}{\partial y}\psi(\vec{r}, t) = p_y\psi(\vec{r}, t) = \hat{p}_y\psi(\vec{r}, t) \tag{2.11}$$

$$-i\hbar \frac{\partial}{\partial z}\psi(\vec{r}, t) = p_z\psi(\vec{r}, t) = \hat{p}_z\psi(\vec{r}, t).$$

The results (2.11) can be summarized in the following vector expression

$$-i\hbar\nabla\psi(\vec{r}, t) = \vec{p}\psi(\vec{r}, t) = \hat{\vec{p}}\psi(\vec{r}, t). \tag{2.12}$$

Substituting (2.10) into (2.7) and (2.12) into (2.8), respectively, and using the normalization condition (2.6), we obtain

$$< E >= E \int_V |\psi(\vec{r}, t)|^2\mathrm{d}^3\vec{r} = E, \tag{2.13}$$

and

$$<\vec{p}> = \vec{p} \int_V |\psi(\vec{r},t)|^2 \; d^3\vec{r} = \vec{p}, \tag{2.14}$$

as expected. In the simple case of the plane wave, the expectation values of the energy and the momentum of the particle are equal to their exact classical values. From (2.10) and (2.12), we can now identify the operators $\hat{H}$ and $\hat{\vec{p}}$ as follows:

$$\hat{H} = i\hbar \frac{\partial}{\partial t}, \quad \hat{\vec{p}} = -i\hbar \, \nabla. \tag{2.15}$$

Since the wave functions $\psi(\vec{r},t)$ are the functions of $\vec{r}$ and t, i.e., we are working in the ordinary three-dimensional representation, the operators of the coordinates and time are simply equal to the coordinates and time. Thus, we have

$$\hat{x} = x, \quad \hat{y} = y, \quad \hat{z} = z, \quad \hat{t} = t. \tag{2.16}$$

Let us now recall the classical expression for the total energy of a single nonrelativistic particle moving in the field of some general potential $V = V(\vec{r},t)$, i.e.,

$$E = \frac{p^2}{2m} + V(\vec{r},t). \tag{2.17}$$

Multiplying both sides with the wave function $\psi(\vec{r},t)$, we obtain

$$E\psi(\vec{r},t) = \frac{p^2}{2m}\psi(\vec{r},t) + V(\vec{r},t)\psi(\vec{r},t). \tag{2.18}$$

Substituting the dynamical variables with their quantum-mechanical operators, we obtain the following operator equation

$$\hat{H} \, \psi(\vec{r},t) = \frac{\hat{p}^2}{2m} \, \psi(\vec{r},t) + \hat{V}(\hat{\vec{r}},\hat{t})\psi(\vec{r},t). \tag{2.19}$$

Substituting the operator definitions (2.15) and (2.16), we obtain the most general time-dependent *Schrödinger equation*, in the following form

$$i\hbar\frac{\partial\psi}{\partial t} = -\frac{\hbar^2}{2m}\nabla^2\psi + V(\vec{r},t)\psi, \tag{2.20}$$

where $\psi = \psi(\vec{r},t)$. The Schrödinger equation (2.20) is the fundamental equation of quantum mechanics. In case of a stationary potential $V(\vec{r},t) = V(\vec{r})$, i.e., a potential that is constant in time, the wave function can be written in the form

$$\psi(\vec{r},t) = \psi(\vec{r})\exp\left(\frac{iEt}{\hbar}\right). \tag{2.21}$$

Substituting (2.21) into (2.20), we obtain the following time-independent Schrödinger equation

$$-\frac{\hbar^2}{2m}\nabla^2\psi(\vec{r}) + V(\vec{r})\,\psi(\vec{r}) = E\,\psi(\vec{r}). \tag{2.22}$$

In this book, we will only analyze the stationary phenomena, and we will, therefore, use the time-independent Schrödinger equation (2.22) only.

2.3 Problems with Solutions *)

Problem

A commutator of two quantum-mechanical operators $\hat{A}$ and $\hat{B}$ is defined by

$$\left[\hat{A},\hat{B}\right] = \hat{A}\hat{B} - \hat{B}\hat{A}.$$

Calculate the commutators $\left[x,\,\hat{p}_x\right]$ and $\left[\hat{H},\,t\right]$.

Solution

The commutator $\left[x,\,\hat{p}_x\right]$ applied to an arbitrary wave function ψ gives

$$\left[x,\,\hat{p}_x\right]\psi = -i\hbar\left[x,\,\frac{\partial}{\partial x}\right]\psi, \tag{2.23}$$

or

$$\left[x,\,\hat{p}_x\right]\psi = -i\hbar\left[x\frac{\partial\psi}{\partial x} - \frac{\partial}{\partial x}(x\psi)\right] = i\hbar\,\psi. \tag{2.24}$$

Thus, we finally obtain the commutator

$$\left[x,\,\hat{p}_x\right] = i\hbar. \tag{2.25}$$

Similarly, the commutator $\left[\hat{H},\,t\right]$ applied to an arbitrary wave function ψ gives

$$\left[\hat{H},\,t\right]\psi = i\hbar\left[\frac{\partial}{\partial t},\,t\right]\psi, \tag{2.26}$$

or

$$\left[\hat{H},\,t\right]\psi = i\hbar\left[\frac{\partial}{\partial t}(t\psi) - t\frac{\partial\psi}{\partial t}\right] = i\hbar\,\psi. \tag{2.27}$$

Thus, we finally obtain the commutator

$$\left[\hat{H},\,t\right] = i\hbar. \tag{2.28}$$

3 Kinetic Energy of Translational Motion

3.1 Hamiltonian of Translational Motion

In order to calculate the translational kinetic energy of the constituent particles of a system, let us assume that a system consisting of N noninteracting point particles is enclosed within a cubic box with impenetrable walls. The side of the box is equal to a and its volume is thereby equal to $V = a^3$, as indicated in Fig. 3.1.

In classical mechanics, the kinetic energy of a nonrelativistic particle of mass m, moving with a velocity $\vec{v}_i$, is given by

$$E_i = \frac{1}{2}mv_i^2. \tag{3.1}$$

If we introduce the momentum of a particle $\vec{p}_i = m\vec{v}_i$, we obtain

$$E_i = \frac{p_i^2}{2m} = \frac{1}{2m}\left(p_{ix}^2 + p_{iy}^2 + p_{iz}^2\right). \tag{3.2}$$

For the sake of simplicity, at this point, we can drop the label of the particle and write

$$E = \frac{p^2}{2m} = \frac{1}{2m}\left(p_x^2 + p_y^2 + p_z^2\right). \tag{3.3}$$

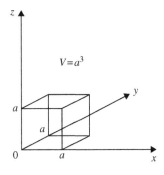

Figure 3.1 A system in a cubic box.

In quantum mechanics, the kinetic energy (3.3) is replaced by the Hamiltonian operator $\hat{H}$ as follows:

$$\hat{H} = \frac{\hat{p}^2}{2m} = \frac{1}{2m}\left(\hat{p}_x^2 + \hat{p}_y^2 + \hat{p}_z^2\right).$$
(3.4)

3.2 Schrödinger Equation for Translational Motion *)

The motion of a constituent particle is described by its wave function $\psi(x, y, z)$, which satisfies the Schrödinger equation

$$\hat{H}\psi = E\psi.$$
(3.5)

In order to specify the operator $\hat{H}$, we use the quantum-mechanical prescriptions

$$\hat{p}_x = -i\hbar\frac{\partial}{\partial x}, \quad \hat{p}_y = -i\hbar\frac{\partial}{\partial y}, \quad \hat{p}_z = -i\hbar\frac{\partial}{\partial z}.$$
(3.6)

Thus, the Hamiltonian operator (3.4) becomes

$$\hat{H} = -\frac{\hbar^2}{2m}\left(\frac{\partial^2}{\partial x^2} + \frac{\partial^2}{\partial y^2} + \frac{\partial^2}{\partial z^2}\right) = -\frac{\hbar^2}{2m}\nabla^2.$$
(3.7)

The Schrödinger equation (3.5) then reads

$$-\frac{\hbar^2}{2m}\nabla^2\psi = E\psi,$$
(3.8)

or using (3.7)

$$\frac{1}{\psi}\frac{\partial^2\psi}{\partial x^2} + \frac{1}{\psi}\frac{\partial^2\psi}{\partial y^2} + \frac{1}{\psi}\frac{\partial^2\psi}{\partial z^2} = -\frac{2mE}{\hbar^2}.$$
(3.9)

3.3 Solution of the Schrödinger Equation *)

In order to find a solution of the equation (3.9), we assume that the solution $\psi(x, y, z)$ can be written in the form

$$\psi(x, y, z) = X(x)Y(y)Z(z),$$
(3.10)

where the functions $X(x)$, $Y(y)$, and $Z(z)$ are the functions of a single variable x, y, and z, respectively. The equation (3.9) then reduces to

$$\frac{1}{X}\frac{d^2X}{dx^2} + \frac{1}{Y}\frac{d^2Y}{dy^2} + \frac{1}{Z}\frac{d^2Z}{dz^2} = -\frac{2mE}{\hbar^2}.$$
(3.11)

By examination of the equation (3.11), we see that the first term on the left-hand side of the equation is a function of the co-ordinate x only. Analogously, the second term on the left-hand side of the equation (3.11) is a function of the co-ordinate y only, and the third term on the left-hand side of the equation (3.11) is a function of the co-ordinate z only. Thus, the equation (3.11) has the form

$$f(x) + g(y) + h(z) = \text{Constant.} \tag{3.12}$$

The equality (3.12) is possible only if each of the three functions $f(x)$, $g(y)$, and $h(z)$ is a constant. Otherwise, since x, y, and z are independent variables, by their arbitrary change, we could always spoil the equality (3.12). Thus, we can write

$$\frac{1}{X}\frac{d^2X}{dx^2} = -k_x^2, \quad \frac{1}{Y}\frac{d^2Y}{dy^2} = -k_y^2, \quad \frac{1}{Z}\frac{d^2Z}{dz^2} = -k_z^2, \tag{3.13}$$

where, in order to satisfy the equation (3.11), we require

$$k_x^2 + k_y^2 + k_z^2 = \frac{2mE}{\hbar^2}. \tag{3.14}$$

From the equation (3.14), we see that if we know the three quantities k_x, k_y, and k_z, we can easily calculate the translational energy of a constituent particle in our system as follows:

$$E = \frac{\hbar^2}{2m}\left(k_x^2 + k_y^2 + k_z^2\right). \tag{3.15}$$

The three differential equations (3.13) are ordinary differential equations which can be written in the form

$$\frac{d^2X}{dx^2} + k_x^2 X = 0, \tag{3.16}$$

$$\frac{d^2Y}{dy^2} + k_y^2 Y = 0, \tag{3.17}$$

$$\frac{d^2Z}{dz^2} + k_z^2 Z = 0. \tag{3.18}$$

In order to solve these three differential equations, we need to specify the boundary conditions for our system in the cubic box. In quantum mechanics, the quantity $\pi(x, y, z) = |\psi(x, y, z)|^2$ is interpreted as the probability of finding a constituent particle at the point in space specified by the co-ordinates (x, y, z). According to our initial assumption, all of the constituent particles of the system are confined to the interior of the cubic box with impenetrable walls, and no particle can be found outside the box. Thus, the probability $\pi(x, y, z)$ and thereby the complex function $\psi(x, y, z)$ must be equal to zero for any point in space which lies outside the cubic box, i.e., where x, y, or z are less than zero or larger than the side of the cube a. The boundary conditions for our problem are, therefore, specified as follows:

$$\psi(x, y, z) = 0 \quad \text{if} \quad x < 0 \quad \text{or} \quad x > a, \tag{3.19}$$

$$\psi(x, y, z) = 0 \quad \text{if} \quad y < 0 \quad \text{or} \quad y > a, \tag{3.20}$$

$$\psi(x, y, z) = 0 \quad \text{if} \quad z < 0 \quad \text{or} \quad z > a. \tag{3.21}$$

But since the function $\psi(x, y, z)$ must be a continuous function, we must have

$$\psi(0, y, z) = \psi(a, y, z) = 0, \tag{3.22}$$

$$\psi(x, 0, z) = \psi(x, a, z) = 0, \tag{3.23}$$

$$\psi(x, y, 0) = \psi(x, y, a) = 0, \tag{3.24}$$

or using the equation (3.10)

$$X(0) = 0, \quad X(a) = 0, \tag{3.25}$$

$$Y(0) = 0, \quad Y(a) = 0, \tag{3.26}$$

$$Z(0) = 0, \quad Z(a) = 0. \tag{3.27}$$

The set of conditions (3.25), (3.26), and (3.27) is the set of the required boundary conditions for the three differential equations (3.16), (3.17), and (3.18), respectively. They are called the boundary conditions, since they specify the behavior of our solutions at the boundaries of the cubic box. The three differential equations, given by the equations (3.16), (3.17), and (3.18), have the following general solutions

$$X(x) = A_1 \sin k_x x + B_1 \cos k_x x, \tag{3.28}$$

$$Y(y) = A_2 \sin k_y y + B_2 \cos k_y y, \tag{3.29}$$

$$Z(z) = A_3 \sin k_z z + B_3 \cos k_z z. \tag{3.30}$$

From the above boundary conditions, we readily obtain

$$X(0) = B_1 = 0, \quad Y(0) = B_2 = 0, \quad Z(0) = B_3 = 0, \tag{3.31}$$

and

$$X(a) = A_1 \sin k_x a = 0 \quad \Rightarrow \quad k_x a = n_x \pi, \tag{3.32}$$

$$Y(a) = A_2 \sin k_y a = 0 \quad \Rightarrow \quad k_y a = n_y \pi, \tag{3.33}$$

$$Z(a) = A_3 \sin k_z a = 0 \quad \Rightarrow \quad k_z a = n_z \pi, \tag{3.34}$$

where $n_x = 1, 2, 3, \ldots$, $n_y = 1, 2, 3, \ldots$, and $n_z = 1, 2, 3, \ldots$ are three arbitrary nonzero integers, which label the quantum state of the particle in the box. The solution to the Schrödinger equation (3.8) then reads

$$\psi(x, y, z) = A \sin\left(\frac{n_x \pi}{a} x\right) \sin\left(\frac{n_y \pi}{a} y\right) \sin\left(\frac{n_z \pi}{a} z\right), \qquad (3.35)$$

where $A = A_1 \cdot A_2 \cdot A_3$ is the normalization constant.

3.4 Normalization of the Wave Function *)

The normalization constant A can be determined using the initial assumption that the particle must be somewhere in the box. Thus, the total probability of finding the particle inside the box must be equal to unity, i.e.,

$$\int_0^a \int_0^a \int_0^a |\psi(x, y, z)|^2 dx\,dy\,dz = 1. \qquad (3.36)$$

Substituting the solution (3.35) into (3.36), we obtain

$$A^2 \int_0^a \sin^2\left(\frac{n_x \pi}{a} x\right) dx \int_0^a \sin^2\left(\frac{n_y \pi}{a} y\right) dy \int_0^a \sin^2\left(\frac{n_z \pi}{a} z\right) dz = 1. \qquad (3.37)$$

If we now use the general result

$$\int \sin^2(kx) dx = \frac{1}{2}\left(x - \frac{1}{2k} \sin 2kx\right) + \text{Constant}, \qquad (3.38)$$

we obtain from (3.37)

$$A^2 \left(\frac{a}{2}\right)\left(\frac{a}{2}\right)\left(\frac{a}{2}\right) = 1 \quad \Rightarrow \quad A = \left\{\frac{2}{a}\right\}^{3/2}. \qquad (3.39)$$

Finally, the normalized solution of the Schrödinger equation (3.8) is given by

$$\psi(x, y, z) = \left\{\frac{2}{a}\right\}^{3/2} \sin\left(\frac{n_x \pi}{a} x\right) \sin\left(\frac{n_y \pi}{a} y\right) \sin\left(\frac{n_z \pi}{a} z\right). \qquad (3.40)$$

3.5   Quantized Energy of Translational Motion

In thermodynamics, we are primarily interested in the result for the translational energy of the constituent particles in the system (3.15). Substituting the results (3.32), (3.33), and (3.34) into (3.15), we obtain

$$E = \frac{\hbar^2 \pi^2}{2ma^2}\left(n_x^2 + n_y^2 + n_z^2\right), \tag{3.41}$$

or using $\hbar = h/2\pi$

$$E = \frac{h^2}{8ma^2}\left(n_x^2 + n_y^2 + n_z^2\right). \tag{3.42}$$

The volume of the cubic box is equal to $V = a^3$, and we can finally write

$$E = \frac{h^2}{8mV^{2/3}}\left(n_x^2 + n_y^2 + n_z^2\right) = E_0\left(n_x^2 + n_y^2 + n_z^2\right), \tag{3.43}$$

where the elementary quantum of the energy of a constituent particle is defined by

$$E_0 = \frac{h^2}{8mV^{2/3}}. \tag{3.44}$$

The order of magnitude of E_0 can be calculated using the result for the Planck constant $h = 6.626068 \times 10^{-34}$ Js. A typical microscopic gas particle has a mass of the order $m \sim 10^{-27}$ kg, and our system is assumed to be contained in a cubic box of a macroscopic size $V \sim 1\,\mathrm{m}^3$. Thus, we obtain

$$E_0 \sim \frac{10^{-67}}{10^{-27} \times 1} \sim 10^{-40}\,\mathrm{J}. \tag{3.45}$$

The result for the energy of translation of constituent particles (3.43) will be used to calculate the thermodynamic properties of an ideal gas. This result shows that the energy levels of a particle are spherically distributed and that the particles can only occupy a discrete spectrum of energy eigenvalues, i.e.,

$$E = E\left(n_x, n_y, n_z\right) = 3E_0, 6E_0, 9E_0, 11E_0, 12E_0 \ldots. \tag{3.46}$$

A constituent particle cannot have an energy of translation, which is not a multiple of E_0. For example, there is no particle with energy $E = 3.5E_0$. However, from the estimate (3.45), we see that the spacing between the adjacent energy levels E_0 is extremely small, and it is often justified to use the limit of a continuous energy distribution in practical calculations.

3.6   Problems with Solutions *)

Problem

An electron of mass $m_e = 9.11 \times 10^{-31}$ kg is confined in a cube of a volume
$V = 10^{-6}$ m^3.

(a) Calculate the energy eigenvalues of the electron in the quantum states with
$n_x^2 + n_y^2 + n_z^2 \leq 14$.
(b) How degenerate is each of these quantum states?

Solution

(a) The energy eigenvalues are given by

$$E = \frac{h^2}{8mV^{2/3}} \left(n_x^2 + n_y^2 + n_z^2 \right) = E_0 n^2, \tag{3.47}$$

where

$$E_0 = \frac{h^2}{8mV^{2/3}} \approx 6 \times 10^{-34} \text{ J}, \tag{3.48}$$

and

$$n^2 = n_x^2 + n_y^2 + n_z^2. \tag{3.49}$$

The first six energy eigenvalues (with $n^2 \leq 14$) are given by

(n_x, n_y, n_z)	n^2	E
(1, 1, 1)	3	1.8×10^{-33} J
(1, 1, 2)	6	3.6×10^{-33} J
(1, 2, 2)	9	5.4×10^{-33} J
(1, 1, 3)	11	6.6×10^{-33} J
(2, 2, 2)	12	7.2×10^{-33} J
(1, 2, 3)	14	8.4×10^{-33} J

(b) The degeneracy of the above quantum states is given by the number of arrangements of the numbers (n_x, n_y, n_z) corresponding to the same value of E (or n^2).
For example, the quantum state with $n^2 = 6$ can be realized by three different

arrangements of the numbers (n_x, n_y, n_z), i.e., $(1, 1, 2)$, $(1, 2, 1)$, and $(2, 1, 1)$. Thus, the degeneracy of the first six quantum states is

(n_x, n_y, n_z)	n^2	Degeneracy
$(1, 1, 1)$	3	1
$(1, 1, 2)$	6	3
$(1, 2, 2)$	9	3
$(1, 1, 3)$	11	3
$(2, 2, 2)$	12	1
$(1, 2, 3)$	14	6

For example, the quantum state with $n^2 = 14$ can be realized by six different arrangements of the numbers (n_x, n_y, n_z), i.e., $(1, 2, 3)$, $(1, 3, 2)$, $(2, 1, 3)$, $(2, 3, 1)$, $(3, 1, 2)$, and $(3, 2, 1)$. From the above table, we see that the energy eigenvalues of an electron in a cubic box are not evenly distributed.

4 Energy of Vibrations

4.1 Hamiltonian of Vibrations

In reality, the constituent particles of systems cannot be exactly modeled as point particles. They typically have an internal structure. A good example is the common diatomic molecules such as H_2, O_2, N_2, CO, etc. Such a molecule can not only translate (with its center of mass moving within a confining box) but also execute internal motions such as rotation around its center of mass (discussed in Chapter 5 "Kinetic Energy of Rotations") as well a vibrational motion in which the length of the chemical bond connecting the two atoms oscillates with angular frequency ω. In this chapter, we will describe this vibrational motion of a diatomic molecule by means of the simple harmonic oscillator model with the total energy (i.e., classical Hamiltonian) of the form

$$E_V = \frac{p^2}{2m} + \frac{1}{2}m\omega^2 x^2, \tag{4.1}$$

where $m = (1/m_1 + 1/m_2)^{-1}$ is the so-called reduced mass of the molecule (with m_1 and m_2 being the masses of the two atoms), and x is the deviation of the molecule's chemical bond length away from its mechanical equilibrium value. In quantum mechanics, the energy of the harmonic oscillator (4.1) is replaced by the Hamiltonian operator $\hat{H}_V$ as follows:

$$\hat{H}_V = \frac{\hat{p}^2}{2m} + \frac{1}{2}m\omega^2 x^2. \tag{4.2}$$

The vibrations of a constituent particle are described by the wave function $\psi(x)$, which satisfies the Schrödinger equation

$$\hat{H}_V \psi(x) = E_V \psi(x). \tag{4.3}$$

In order to define the operator $\hat{H}_V$, we use the quantum-mechanical prescription

$$\hat{p}_x = -i\hbar \frac{\partial}{\partial x}. \tag{4.4}$$

Introductory Statistical Thermodynamics

Substituting (4.4) into (4.3) with (4.2), we obtain the equation

$$\left[-\frac{\hbar^2}{2m}\frac{d^2}{dx^2} + \frac{1}{2}m\omega^2 x^2\right]\psi(x) = E_V\psi(x). \tag{4.5}$$

4.2 Solution of the Schrödinger Equation *)

In order to solve the equation (4.5), it can be rewritten as follows:

$$\frac{d^2\psi}{dx^2} + \left(\lambda - \alpha^2 x^2\right)\psi = 0, \tag{4.6}$$

where

$$\lambda = \frac{2mE_V}{\hbar^2}, \quad \alpha = \frac{m\omega}{\hbar}. \tag{4.7}$$

Introducing here a new variable $u = \sqrt{\alpha}\,x$, we obtain

$$\frac{d^2\psi}{du^2} + \left(\frac{\lambda}{\alpha} - u^2\right)\psi = 0. \tag{4.8}$$

In order to solve the equation (4.8), we introduce a new function $H = H(u)$ as follows:

$$\psi(u) = e^{-u^2/2}H(u). \tag{4.9}$$

The second derivative of the wave function (4.9) is given by

$$\frac{d^2\psi}{du^2} = e^{-u^2/2}\left[\frac{d^2H}{du^2} - 2u\frac{dH}{du} + \left(u^2 - 1\right)H\right]. \tag{4.10}$$

Substituting (4.9) and (4.10) into (4.8), we obtain

$$\frac{d^2H}{du^2} - 2u\frac{dH}{du} + \left(\frac{\lambda}{\alpha} - 1\right)H = 0. \tag{4.11}$$

The equation (4.11) is solved by expanding the function $H = H(u)$ into the power series as follows:

$$H = \sum_{k=0}^{\infty} a_k u^k = \sum_{j=0}^{\infty} a_j u^j. \tag{4.12}$$

From (4.12), we obtain

$$\frac{dH}{du} = \sum_{k=1}^{\infty} k a_k u^{k-1} \Rightarrow 2u\frac{dH}{du} = \sum_{k=0}^{\infty} 2k a_k u^k. \tag{4.13}$$

$$\frac{d^2H}{du^2} = \sum_{j=2}^{\infty} j(j-1) a_j u^{j-2} = \sum_{k=0}^{\infty} (k+2)(k+1) a_{k+2} u^k. \tag{4.14}$$

Substituting (4.12)–(4.14) into (4.11), we obtain

$$\sum_{k=0}^{\infty} \left[(k+1)(k+2)\, a_{k+2} - \left(2k - \frac{\lambda}{\alpha} + 1 \right) a_k \right] u^k = 0. \tag{4.15}$$

From the equation (4.15), we obtain the recurrence formula for the coefficients of the power series (4.12) as follows:

$$a_{k+2} = \frac{2k + 1 - \lambda/\alpha}{(k+1)(k+2)}\, a_k. \tag{4.16}$$

If we know the boundary conditions $a_0 = H(0)$ and $a_1 = \dot{H}(0)$, the formula (4.16) allows us to calculate the coefficients of the entire infinite series (4.12). As the squared magnitude of the wave function $\psi = \psi(x)$ gives the probability density of finding the vibrating particle at a certain distance x, the appropriate boundary condition for the wave function is that it must approach zero as $x \rightarrow \infty$. However, the wave function (4.9) with an infinite series (4.12) is divergent for $u \rightarrow \infty$ (or $x \rightarrow \infty$). The only way to obtain a proper bound-state wave function is to choose the parameter λ in such a way to terminate the power series and reduce it to a polynomial. Let us, therefore, assume that for some integer $k = n$, we have

$$\frac{\lambda_n}{\alpha} = 2n + 1, \quad n = 0, 1, 2 \dots \tag{4.17}$$

Substituting (4.17) into (4.16), we obtain

$$a_{k+2} = \frac{(2k+1) - (2n+1)}{(k+1)(k+2)}\, a_k = \frac{2(k-n)}{(k+1)(k+2)}\, a_k. \tag{4.18}$$

From (4.18), it is clear that $a_k \equiv 0$ for any $k > n$, and the power series is reduced to a polynomial. The polynomials obtained in this way are known as Hermite polynomials and they satisfy the Hermite differential equation

$$\frac{d^2 H_n}{du^2} - 2u \frac{dH_n}{du} + 2n H_n = 0. \tag{4.19}$$

4.3 Quantized Energy of Vibrations

Let us now use the definition of the parameter λ in (4.7), where we replace the energy E_V by the quantized energy of vibrations E_n. We then obtain from (4.17)

$$\frac{\lambda_n}{\alpha} = \frac{2E_n}{\hbar \omega} = 2n + 1, \quad n = 0, 1, 2 \dots, \tag{4.20}$$

or

$$E_n = \hbar \omega \left(n + \frac{1}{2} \right), \quad n = 0, 1, 2 \dots . \tag{4.21}$$

The equation (4.21) is the celebrated result for the quantized energy levels of the simple harmonic oscillator used here to model the vibrational motion of diatomic molecules. In thermodynamics, we are primarily interested in this result for the energy of vibrations of constituent particles in the system (4.21). However, for the sake of completeness, we will derive the result for the normalized wave function of the vibrating constituent particle as well.

4.4   Hermite Polynomials *)

Using the formula (4.18), we can easily calculate the first few Hermite polynomials. They are determined up to an arbitrary multiplicative constant, which is determined by the boundary conditions $a_0 = H(0)$ and $a_1 = \dot{H}(0)$. In particular, the first two polynomials $H_0(u)$ and $H_1(u)$ have only one term and are determined by an arbitrary choice of the above boundary conditions. If we choose $a_0 = 1$ for $H_0(u)$ and $a_1 = 2$ for $H_1(u)$, we have

$$H_0(u) = 1, \quad H_1(u) = 2u. \tag{4.22}$$

Further, if we choose $a_0 = -2$ for $H_2(u)$, $a_1 = -12$ for $H_3(u)$, and $a_0 = 12$ for $H_4(u)$, respectively, we obtain

$$H_2(u) = -2 + 4u^2$$

$$H_3(u) = -12u + 8u^3 \tag{4.23}$$

$$H_4(u) = 12 - 48u^2 + 16u^4.$$

The particular choices of the above boundary conditions were made to fit a practical formula, which can also be used for calculation of the Hermite polynomials. It is known as *Rodrigues' formula*, and it reads

$$H_n(u) = (-1)^n e^{u^2} \frac{d^n}{du^n}\left(e^{-u^2}\right). \tag{4.24}$$

It is easily shown, by direct calculation, that the *Rodrigues' formula* reproduces the results (4.22) and (4.23). The Hermite polynomials can be generated using the following generating function

$$G(u,t) = e^{2tu-t^2} = e^{u^2-(t-u)^2} = \sum_{k=0}^{\infty} \frac{H_k(u)}{k!} t^k. \tag{4.25}$$

In order to prove that the polynomials $H_k(u)$ in (4.25) are indeed the Hermite polynomials, we calculate the n-th derivative with respect to t of both of its sides. The n-th derivative of the generating function in the exponential form is given by

$$\frac{\partial^n}{\partial t^n} G(u,t) = e^{u^2} \frac{\partial^n}{\partial t^n} e^{-(t-u)^2} = (-1)^n e^{u^2} \frac{\partial^n}{\partial u^n} e^{-(t-u)^2}, \tag{4.26}$$

while the n-th derivative of the series expansion of the generating function is obtained as follows:

$$\frac{\partial^n}{\partial t^n} G(u,t) = \frac{\partial^n}{\partial t^n} \sum_{k=0}^{\infty} \frac{H_k(u)}{k!} t^k = \sum_{k=0}^{\infty} \frac{H_k(u)}{k!} \frac{d^n}{dt^n} \left(t^k \right). \tag{4.27}$$

Using now the elementary result

$$\frac{d^n}{dt^n} \left(t^k \right) = \frac{k!}{(k-n)!} t^{k-n}, \tag{4.28}$$

we obtain from (4.27)

$$\frac{\partial^n}{\partial t^n} G(u,t) = \sum_{k=n}^{\infty} \frac{H_k(u)}{(k-n)!} t^{k-n} = \sum_{j=0}^{\infty} \frac{H_{j+n}(u)}{j!} t^j. \tag{4.29}$$

Using now the results (4.26) and (4.29), we obtain

$$(-1)^n e^{u^2} \frac{\partial^n}{\partial u^n} e^{-(t-u)^2} = \sum_{j=0}^{\infty} \frac{H_{j+n}(u)}{j!} t^j. \tag{4.30}$$

The equality (4.30) is valid for any value of t. In particular, for $t = 0$, we obtain

$$H_n(u) = (-1)^n e^{u^2} \frac{\partial^n}{\partial u^n} e^{-u^2}, \tag{4.31}$$

and it is identical with the *Rodrigues' formula* (4.24) above. Thus, we have shown that the polynomials $H_k(u)$ in (4.25) are the Hermite polynomials.

4.5   Normalization of the Wave Function *)

Using now (4.9), the wave function of the vibrating constituent particle of the system, which occupies the eigenstate with energy E_n given by (4.21), becomes

$$\psi_n(u) = A_n e^{-u^2/2} H_n(u), \tag{4.32}$$

where A_n is a normalization constant. Using now the generating function for Hermite polynomials (4.25), we can obtain the normalization condition for the harmonic oscillator wave functions (4.32) and calculate the normalization constant A_n. Let us begin with the definition of the generating function (4.25), in the following form

$$e^{2tu-t^2} = \sum_{n=0}^{\infty} \frac{H_n(u)}{n!} t^n, \quad e^{2su-s^2} = \sum_{m=0}^{\infty} \frac{H_m(u)}{m!} s^m. \tag{4.33}$$

Multiplying the two generating functions in (4.33) with each other, we obtain

$$e^{2tu-t^2+2su-s^2} = \sum_{m=0}^{\infty}\sum_{n=0}^{\infty} \frac{s^m t^n}{m!n!} H_m(u)H_n(u). \tag{4.34}$$

Multiplying now both sides of the equality (4.34) by $\exp(-u^2)$ and integrating from $-\infty$ to $+\infty$, we obtain the following equation

$$I = \int_{-\infty}^{+\infty} du\, e^{-u^2+2tu-t^2+2su-s^2}$$

$$= \sum_{m=0}^{\infty}\sum_{n=0}^{\infty} \frac{s^m t^n}{m!n!} \int_{-\infty}^{+\infty} du\, e^{-u^2} H_m(u)H_n(u). \tag{4.35}$$

Using now the equality

$$-u^2 + 2tu - t^2 + 2su - s^2 = -(u - s - t)^2 + 2st, \tag{4.36}$$

we can calculate the integral on the left-hand side of (4.35)

$$I = e^{2st} \int_{-\infty}^{+\infty} du\, e^{-(u-s-t)^2} = \sqrt{\pi}\, e^{2st} = \sqrt{\pi} \sum_{m=0}^{\infty} \frac{2^m s^m t^m}{m!}. \tag{4.37}$$

Substituting (4.37) into (4.35), we obtain

$$\sqrt{\pi} \sum_{m=0}^{\infty} \frac{2^m s^m t^m}{m!} = \sum_{m=0}^{\infty}\sum_{n=0}^{\infty} \frac{s^m t^n}{m!n!} \int_{-\infty}^{+\infty} du\, e^{-u^2} H_m(u)H_n(u), \tag{4.38}$$

or

$$\sqrt{\pi}\, 2^m t^m = \sum_{n=0}^{\infty} \frac{t^n}{n!} \int_{-\infty}^{+\infty} du\, e^{-u^2} H_m(u)H_n(u). \tag{4.39}$$

From the equation (4.39), we immediately see that the only nonzero term in the sum on the right-hand side is the one with $n = m$. Thus, we obtain the following orthonormality condition for the Hermite polynomials

$$\int\limits_{-\infty}^{+\infty} du\, e^{-u^2} H_m(u) H_n(u) = \delta_{mn} \sqrt{\pi}\, 2^n\, n!. \tag{4.40}$$

Using now the results (4.32) and (4.40), as well as the orthonormality of the oscillator wave functions $\psi_n(x)$, we can now calculate

$$\delta_{mn} = \int\limits_{-\infty}^{+\infty} dx\, \psi_m(x) \psi_n(x)$$

$$= A_m A_n \frac{1}{\sqrt{\alpha}} \int\limits_{-\infty}^{+\infty} du\, e^{-u^2} H_m(u) H_n(u)$$

$$= \delta_{mn} A_m A_n \sqrt{\frac{\pi}{\alpha}}\, 2^n\, n!. \tag{4.41}$$

For $m = n$, we thus obtain

$$A_n^2 = \frac{1}{2^n\, n!} \sqrt{\frac{\alpha}{\pi}}, \tag{4.42}$$

or using (4.7)

$$A_n = \sqrt{\frac{1}{2^n\, n!}} \left(\frac{\alpha}{\pi}\right)^{1/4} = \sqrt{\frac{1}{2^n\, n!}} \left(\frac{m\omega}{\pi\hbar}\right)^{1/4}. \tag{4.43}$$

Substituting (4.43) into (4.32), we obtain the final result for the normalized wave function of a vibrating constituent particle of our system, which occupies the quantum state with energy E_n, in the following form

$$\psi_n(x) = \sqrt{\frac{1}{2^n\, n!}} \left(\frac{m\omega}{\pi\hbar}\right)^{1/4} \exp\left(-\frac{m\omega x^2}{2\hbar}\right) H_n\left(\sqrt{\frac{m\omega}{\hbar}}\, x\right). \tag{4.44}$$

4.6 Problems with Solutions *)

Problem

Consider a very small (microscopic) superconductor series LC-circuit with the inductance L and the capacitance C. The capacitor has a charge $q = q(t)$, and there are no external sources connected to the circuit. Identifying the conserved total electromagnetic energy of the circuit with the classical Hamiltonian, determine the quantum states of the LC-circuit.

Solution

Since there are no external sources, the total voltage over the closed LC-circuit is equal to zero, i.e.,

$$u_L + u_C = L\frac{di(t)}{dt} + \frac{q(t)}{C} = L\frac{d^2q(t)}{dt^2} + \frac{q(t)}{C} = 0. \tag{4.45}$$

In the above result, we have used the definition of the electric current $i(t) = dq(t)/dt$ and the characteristic relations for inductors and capacitors

$$u_L(t) = L\frac{di(t)}{dt}, \quad u_C(t) = \frac{q(t)}{C}. \tag{4.46}$$

Multiplying the equation (4.46) by $dq(t)/dt$, we obtain

$$L\frac{dq(t)}{dt}\frac{d}{dt}\left[\frac{dq(t)}{dt}\right] + \frac{1}{C}q(t)\frac{dq(t)}{dt} = 0, \tag{4.47}$$

or

$$\frac{d}{dt}\left[\frac{L}{2}\left(\frac{dq}{dt}\right)^2 + \frac{q^2}{2C}\right] = 0. \tag{4.48}$$

Thus, the conserved Hamiltonian of the LC-circuit is given by

$$H = \frac{L}{2}\left(\frac{dq}{dt}\right)^2 + \frac{q^2}{2C}, \tag{4.49}$$

or

$$H = \frac{p_q^2}{2L} + \frac{1}{2}L\omega^2 q^2, \quad \omega = \frac{1}{\sqrt{LC}}, \tag{4.50}$$

where we introduce the "charge momentum" p_q by

$$p_q = L\frac{dq}{dt}. \tag{4.51}$$

The equation (4.50) is the well-known harmonic oscillator equation, where the inductance L plays the role of the mass m and the capacitor charge $q(t)$ plays the role of the co-ordinate $x(t)$. Now, we can use the quantum-mechanical prescription to define the momentum operator

$$\hat{p}_q = -i\hbar\frac{\partial}{\partial q}. \tag{4.52}$$

Thus, we obtain the Schrödinger equation for the LC-circuit

$$\left[-\frac{\hbar^2}{2L}\frac{d^2}{dq^2} + \frac{1}{2}L\omega^2 q^2 \right] \psi(q) = W\psi(q). \tag{4.53}$$

The quantum states are then defined by the eigenvalue equation

$$W_n = \hbar\omega\left(n+\frac{1}{2}\right) = \frac{\hbar}{\sqrt{LC}}\left(n+\frac{1}{2}\right), \quad n = 0,1,2\ldots. \tag{4.54}$$

The wave functions of the LC-circuit are given by

$$\psi_n(q) = \sqrt{\frac{1}{2^n\,n!}}\left(\frac{L\omega}{\pi\hbar}\right)^{1/4} \exp\left(-\frac{L\omega q^2}{2\hbar}\right) H_n\left(\sqrt{\frac{L\omega}{\hbar}}\,q\right). \tag{4.55}$$

5 Kinetic Energy of Rotations

5.1 Hamiltonian of Rotations

5.1.1 Kinetic Energy and Hamiltonian Operator

In addition to the vibrational motion discussed in Chapter 4, "Energy of Vibrations," a diatomic molecule can execute yet another form of internal motion, the rotational motion around its center of mass, discussed in this chapter. A diatomic molecule can be modeled as a rigid rotator with a moment of inertia I, as shown in the Fig. 5.1.

The energy of a free rigid rotator, i.e., a rigid rotator subject to no external forces, is equal to its kinetic energy of rotation

$$E_R = T_R = \frac{1}{2} I \dot{\varphi}^2. \tag{5.1}$$

The angular momentum of a rigid rotator is defined by

$$L = \frac{\mathrm{d}E}{\mathrm{d}\dot{\varphi}} = I \dot{\varphi}. \tag{5.2}$$

Substituting (5.2) into (5.1), we obtain

$$E_R = \frac{L^2}{2I}. \tag{5.3}$$

The three components of the angular momentum vector $\vec{L} = (L_x, L_y, L_z)$ in Descartes co-ordinates are given by

$$L_x = yp_z - zp_y,$$
$$L_y = zp_x - xp_z, \tag{5.4}$$
$$L_z = xp_y - yp_x.$$

Figure 5.1 Diatomic molecule as a rigid rotator.

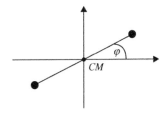

In (5.4), we use the components of the linear momentum vector $\vec{p} = m\vec{v} = (p_x, p_y, p_z)$ in Descartes co-ordinates. In quantum mechanics, the kinetic energy of the rigid rotator (5.3) is replaced by the Hamiltonian operator $\hat{H}_R$ as follows:

$$\hat{H}_R = \frac{\hat{L}^2}{2I} = \frac{1}{2I}\left(\hat{L}_x^2 + \hat{L}_y^2 + \hat{L}_z^2\right). \tag{5.5}$$

The rotation of the particle is then described by its wave function $\psi(\theta, \phi)$, which satisfies the Schrödinger equation

$$\hat{H}_R \psi(\theta, \phi) = E_R \psi(\theta, \phi), \tag{5.6}$$

where (θ, ϕ) are the two spherical angular co-ordinates, which are suitable for the description of the rotational motion of the constituent particles. In order to specify the operator $\hat{H}_R$, we use the quantum-mechanical prescriptions

$$\hat{p}_x = -i\hbar\frac{\partial}{\partial x}, \quad \hat{p}_y = -i\hbar\frac{\partial}{\partial y}, \quad \hat{p}_z = -i\hbar\frac{\partial}{\partial z}. \tag{5.7}$$

Substituting (5.7) into (5.4), we obtain the three components of the angular momentum operator in Descartes co-ordinates

$$\hat{L}_x = -i\hbar\left(y\frac{\partial}{\partial z} - z\frac{\partial}{\partial y}\right), \tag{5.8}$$

$$\hat{L}_y = -i\hbar\left(z\frac{\partial}{\partial x} - x\frac{\partial}{\partial z}\right), \tag{5.9}$$

$$\hat{L}_z = -i\hbar\left(x\frac{\partial}{\partial y} - y\frac{\partial}{\partial x}\right). \tag{5.10}$$

5.1.2 Angular Momentum Operator *)

In order to solve the Schrödinger equation (5.5), we need to express the components of the angular momentum operator (5.8–5.10) in terms of the spherical angles (θ, ϕ). To begin with, we can use expressions giving the Descartes co-ordinates (x, y, z) in terms of the spherical co-ordinates (r, θ, ϕ), as shown in Fig. 5.2.

From Fig. 5.2, we obtain

$$x = r\sin\theta\cos\phi, \quad y = r\sin\theta\sin\phi, \quad z = r\cos\theta. \tag{5.11}$$

The inverse expressions for spherical co-ordinates in terms of the Descartes co-ordinates are then given by

$$r = \sqrt{x^2 + y^2 + z^2}, \tag{5.12}$$

$$\theta = \arctan\left(\frac{\sqrt{x^2 + y^2}}{z}\right), \tag{5.13}$$

$$\phi = \arctan\left(\frac{y}{x}\right). \tag{5.14}$$

Figure 5.2 Spherical co-ordinates.

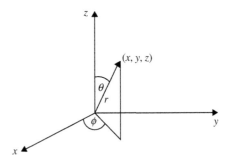

We can now write the following general expressions

$$\frac{\partial}{\partial x} = \frac{\partial r}{\partial x}\frac{\partial}{\partial r} + \frac{\partial \theta}{\partial x}\frac{\partial}{\partial \theta} + \frac{\partial \phi}{\partial x}\frac{\partial}{\partial \phi}, \tag{5.15}$$

$$\frac{\partial}{\partial y} = \frac{\partial r}{\partial y}\frac{\partial}{\partial r} + \frac{\partial \theta}{\partial y}\frac{\partial}{\partial \theta} + \frac{\partial \phi}{\partial y}\frac{\partial}{\partial \phi}, \tag{5.16}$$

$$\frac{\partial}{\partial z} = \frac{\partial r}{\partial z}\frac{\partial}{\partial r} + \frac{\partial \theta}{\partial z}\frac{\partial}{\partial \theta} + \frac{\partial \phi}{\partial z}\frac{\partial}{\partial \phi}. \tag{5.17}$$

After the explicit calculation of the derivatives, using the expressions (5.12–5.14), we obtain

$$\frac{\partial}{\partial x} = \sin\theta\cos\phi\,\frac{\partial}{\partial r} + \frac{\cos\theta\cos\phi}{r}\frac{\partial}{\partial \theta} - \frac{\sin\phi}{r\sin\theta}\frac{\partial}{\partial \phi}, \tag{5.18}$$

$$\frac{\partial}{\partial y} = \sin\theta\sin\phi\,\frac{\partial}{\partial r} + \frac{\cos\theta\sin\phi}{r}\frac{\partial}{\partial \theta} + \frac{\cos\phi}{r\sin\theta}\frac{\partial}{\partial \phi}, \tag{5.19}$$

$$\frac{\partial}{\partial z} = \cos\theta\,\frac{\partial}{\partial r} - \frac{\sin\theta}{r}\frac{\partial}{\partial \theta}. \tag{5.20}$$

Substituting (5.18–5.20) into (5.8–5.10), we can now derive the results for the components of the angular momentum operator in terms of spherical angles (θ, ϕ). It turns out that all the dependence on the radial co-ordinate r is canceled and we obtain the following expressions

$$\hat{L}_x = i\hbar\left(\sin\phi\,\frac{\partial}{\partial \theta} + \frac{\cos\phi}{\tan\theta}\frac{\partial}{\partial \phi}\right), \tag{5.21}$$

$$\hat{L}_y = i\hbar\left(-\cos\phi\,\frac{\partial}{\partial \theta} + \frac{\sin\phi}{\tan\theta}\frac{\partial}{\partial \phi}\right), \tag{5.22}$$

$$\hat{L}_z = -i\hbar\frac{\partial}{\partial \phi}. \tag{5.23}$$

Using the results (5.21–5.23), we can now calculate

$$\hat{L}_z^2 \psi(\theta, \phi) = -\hbar^2 \frac{\partial^2 \psi}{\partial \phi^2}, \tag{5.24}$$

and

$$
\begin{aligned}
&\left(\hat{L}_x^2 + \hat{L}_y^2\right) \psi(\theta, \phi) \\
&= -\hbar^2 \left[\left(\sin\phi \frac{\partial}{\partial\theta} + \frac{\cos\phi}{\tan\theta} \frac{\partial}{\partial\phi} \right) \left(\sin\phi \frac{\partial\psi}{\partial\theta} + \frac{\cos\phi}{\tan\theta} \frac{\partial\psi}{\partial\phi} \right) \right. \\
&\quad \left. + \left(-\cos\phi \frac{\partial}{\partial\theta} + \frac{\sin\phi}{\tan\theta} \frac{\partial}{\partial\phi} \right) \left(-\cos\phi \frac{\partial\psi}{\partial\theta} + \frac{\sin\phi}{\tan\theta} \frac{\partial\psi}{\partial\phi} \right) \right] \\
&= -\hbar^2 \left(\frac{\partial^2\psi}{\partial\theta^2} + \frac{1}{\tan\theta} \frac{\partial\psi}{\partial\theta} + \frac{1}{\tan^2\theta} \frac{\partial^2\psi}{\partial\phi^2} \right) \\
&= -\hbar^2 \left[\frac{1}{\sin\theta} \frac{\partial}{\partial\theta} \left(\sin\theta \frac{\partial\psi}{\partial\theta} \right) + \frac{1}{\tan^2\theta} \frac{\partial^2\psi}{\partial\phi^2} \right]. \tag{5.25}
\end{aligned}
$$

Putting together the results (5.24) and (5.25), we finally obtain the square of the angular momentum operator as follows:

$$\hat{L}^2 = \hat{L}_x^2 + \hat{L}_y^2 + \hat{L}_z^2 = -\hbar^2 \left[\frac{1}{\sin\theta} \frac{\partial}{\partial\theta} \left(\sin\theta \frac{\partial}{\partial\theta} \right) + \frac{1}{\sin^2\theta} \frac{\partial^2}{\partial\phi^2} \right]. \tag{5.26}$$

Substituting the result (5.26) into (5.5), we finally obtain the Hamiltonian operator of rotations $\hat{H}_R$ as follows:

$$\hat{H}_R = -\frac{\hbar^2}{2I} \left[\frac{1}{\sin\theta} \frac{\partial}{\partial\theta} \left(\sin\theta \frac{\partial}{\partial\theta} \right) + \frac{1}{\sin^2\theta} \frac{\partial^2}{\partial\phi^2} \right]. \tag{5.27}$$

5.2   Solution of the Schrödinger Equation *)

According to the result (5.6), the rotations of a constituent particle are described by its wave function $\psi(\theta, \phi)$, which satisfies the Schrödinger equation

$$\hat{H}_R \psi(\theta, \phi) = E_R \psi(\theta, \phi). \tag{5.28}$$

Substituting the result (5.27) into the result (5.28), we obtain the equation

$$\frac{1}{\sin\theta} \frac{\partial}{\partial\theta} \left(\sin\theta \frac{\partial\psi}{\partial\theta} \right) + \frac{1}{\sin^2\theta} \frac{\partial^2\psi}{\partial\phi^2} + \lambda\psi = 0, \tag{5.29}$$

where we introduce the parameter λ as follows:

$$\lambda = \frac{2IE_R}{\hbar^2}. \tag{5.30}$$

Multiplying the equation (5.29) by $\sin^2 \theta$ and separating the terms which depend on θ and ϕ, we obtain

$$\sin\theta \frac{\partial}{\partial\theta}\left(\sin\theta \frac{\partial\psi}{\partial\theta}\right) + \lambda\sin^2\theta\,\psi = -\frac{\partial^2\psi}{\partial\phi^2}. \tag{5.31}$$

In order to find a solution of the equation (5.31), we assume that the solution $\psi(\theta, \phi)$ can be written in the form

$$\psi(\theta, \phi) = P(\theta)\Phi(\phi), \tag{5.32}$$

where the functions $P(\theta)$ and $\Phi(\phi)$ are the functions of a single variable θ and ϕ, respectively. The equation (5.31) then reduces to

$$\frac{1}{P}\sin\theta \frac{d}{d\theta}\left(\sin\theta \frac{dP}{d\theta}\right) + \lambda\sin^2\theta = -\frac{1}{\Phi}\frac{d^2\Phi}{d\phi^2}. \tag{5.33}$$

By examination of the equation (5.33), we see that its left-hand side is a function of the co-ordinate θ only, while its right-hand side is a function of the co-ordinate ϕ only. Thus, the equation (5.33) has the form $f(\theta) = g(\phi)$. Such an equality is possible only if both of the functions $f(\theta)$ and $g(\phi)$ are equal to the same constant. Otherwise, since θ and ϕ are independent variables, by their arbitrary change, we could always spoil the equality. Thus, we can write

$$\frac{1}{P}\sin\theta \frac{d}{d\theta}\left(\sin\theta \frac{dP}{d\theta}\right) + \lambda\sin^2\theta = -\frac{1}{\Phi}\frac{d^2\Phi}{d\phi^2} = m^2, \tag{5.34}$$

where m^2 is an arbitrary positive constant. The equation for $\Phi(\phi)$ then becomes

$$\frac{d^2\Phi}{d\phi^2} + m^2\Phi = 0. \tag{5.35}$$

The normalized solutions of the equation (5.35) have the following general form

$$\Phi(\phi) = \frac{1}{\sqrt{2\pi}}\exp(im\phi). \tag{5.36}$$

To proceed, we note that (θ, ϕ) and $(\theta, \phi + 2\pi)$ correspond to the same spatial orientation in Fig. 5.2. Because of this, the condition $\Phi(\phi + 2\pi) = \Phi(\phi)$ must be required. It implies that $\exp(i2m\pi) = 1$, or that the parameter m is an integer ($m = 0, \pm 1, \pm 2, \ldots$).

The physical meaning of the quantization of the parameter m is easily clarified by using the results (5.23) and (5.27). From these results, we see that the operators $\hat{H}_R$ and $\hat{L}_z$ commute with each other, i.e., for an arbitrary wave function $\psi(\theta, \phi)$, we have

$$\hat{H}_R \hat{L}_z \psi(\theta, \phi) - \hat{L}_z \hat{H}_R \psi(\theta, \phi) = 0, \tag{5.37}$$

or in the operator form

$$[\hat{H}_R, \hat{L}_z] = \hat{H}_R \hat{L}_z - \hat{L}_z \hat{H}_R = 0. \tag{5.38}$$

This means that the eigenstates of the Hamiltonian $\hat{H}_R$, described by the wave function (5.32), are also the eigenstates of the operator $\hat{L}_z$ and that we can use the result (5.36) to calculate the eigenvalues of the operator $\hat{L}_z$. Thus, we obtain from (5.23)

$$\hat{L}_z \Phi(\phi) = -i\hbar \frac{d}{d\phi} \frac{1}{\sqrt{2\pi}} \exp(im\phi) = m\hbar \Phi(\phi). \tag{5.39}$$

Thus, the rotating particle can only have certain quantized values of the angular momentum vector component in z-direction, i.e., $L_z = m\hbar$. The elementary quantum of the angular momentum component L_z is the Planck constant $\hbar$. The equation for $P(\theta)$ is given by (5.34)

$$\frac{1}{P} \sin\theta \frac{d}{d\theta} \left(\sin\theta \frac{dP}{d\theta} \right) + \lambda \sin^2\theta - m^2 = 0, \tag{5.40}$$

or after some rearranging

$$\frac{1}{\sin\theta} \frac{d}{d\theta} \left(\sin\theta \frac{dP}{d\theta} \right) + \left(\lambda - \frac{m^2}{\sin^2\theta} \right) P = 0. \tag{5.41}$$

Introducing here a new variable $\mu = \cos\theta$ with $d\mu = -\sin\theta d\theta$, we obtain

$$\frac{d}{-\sin\theta d\theta} \left(\sin^2\theta \frac{dP}{-\sin\theta d\theta} \right) + \left(\lambda - \frac{m^2}{1 - \cos^2\theta} \right) P = 0, \tag{5.42}$$

or

$$\frac{d}{d\mu} \left[\left(1 - \mu^2 \right) \frac{dP}{d\mu} \right] + \left(\lambda - \frac{m^2}{1 - \mu^2} \right) P = 0, \tag{5.43}$$

or

$$\left(1 - \mu^2 \right) \frac{d^2P}{d\mu^2} - 2\mu \frac{dP}{d\mu} + \left(\lambda - \frac{m^2}{1 - \mu^2} \right) P = 0. \tag{5.44}$$

In order to solve the equation (5.44), let us first denote the solution of the complete equation with $m \neq 0$ by $P_{lm}(\mu)$, where l is the yet unspecified parameter, which will be used to specify the eigenstates of the Hamiltonian operator. The solution of the equation (5.44), in the special case when $m = 0$, is then denoted by $P_{l0}(\mu) = P_l(\mu)$. Thus, the two functions $P_{lm}(\mu)$ and $P_l(\mu)$ satisfy following differential equations

$$\frac{d}{d\mu}\left[\left(1 - \mu^2\right)\frac{dP_{lm}}{d\mu}\right] + \left(\lambda - \frac{m^2}{1 - \mu^2}\right)P_{lm} = 0, \tag{5.45}$$

$$\frac{d}{d\mu}\left[\left(1 - \mu^2\right)\frac{dP_l}{d\mu}\right] + \lambda P_l = 0. \tag{5.46}$$

For this type of differential equations, it is possible to derive a fairly simple formula that relates the two functions $P_{lm}(\mu)$ and $P_l(\mu)$, so to solve the equation (5.45), it is sufficient solve the simpler equation (5.46). In order to derive the formula that relates the two functions $P_{lm}(\mu)$ and $P_l(\mu)$, let us first differentiate the equation (5.46) m-times, assuming that $m > 0$ as follows:

$$\frac{d^m}{d\mu^m}\left\{\frac{d}{d\mu}\left[\left(1 - \mu^2\right)\frac{dP_l}{d\mu}\right] + \lambda P_l\right\} = 0, \tag{5.47}$$

or

$$\frac{d^{m+2}P_l}{d\mu^{m+2}} - \frac{d^{m+1}}{d\mu^{m+1}}\left(\mu^2\frac{dP_l}{d\mu}\right) + \lambda\frac{d^m P_l}{d\mu^m} = 0. \tag{5.48}$$

Using now the general result

$$\frac{d^n}{dx^n}\left(x^2 y\right) = x^2\frac{d^n y}{dx^n} + 2nx\frac{d^{n-1}y}{dx^{n-1}} + n(n-1)\frac{d^{n-2}y}{dx^{n-2}}, \tag{5.49}$$

we obtain from (5.48)

$$\left(1 - \mu^2\right)\frac{d^2}{d\mu^2}\frac{d^m P_l}{d\mu^m} - 2(m+1)\mu\frac{d}{d\mu}\frac{d^m P_l}{d\mu^m}$$

$$+ [\lambda - m(m+1)]\frac{d^m P_l}{d\mu^m} = 0. \tag{5.50}$$

Introducing here an auxiliary function Q_{lm} as follows:

$$Q_{lm} = \frac{d^m P_l}{d\mu^m}, \tag{5.51}$$

the equation (5.50) becomes

$$\left(1 - \mu^2\right)\frac{d^2 Q_{lm}}{d\mu^2} - 2(m+1)\mu\frac{dQ_{lm}}{d\mu} + [\lambda - m(m+1)]Q_{lm} = 0. \tag{5.52}$$

On the other hand, if we introduce the auxiliary function Q_{lm} in the equation (5.45) using a different definition

$$P_{lm} = \left(1 - \mu^2\right)^{m/2} Q_{lm}, \tag{5.53}$$

we obtain

$$\left(1 - \mu^2\right) \frac{dP_{lm}}{d\mu} = -m\mu \left(1 - \mu^2\right)^{m/2} Q_{lm} + \left(1 - \mu^2\right)^{m/2+1} \frac{dQ_{lm}}{d\mu} \tag{5.54}$$

and

$$\frac{d}{d\mu}\left[\left(1 - \mu^2\right)\frac{dP_{lm}}{d\mu}\right] = \left(1 - \mu^2\right)^{m/2}\left[\left(1 - \mu^2\right)\frac{d^2Q_{lm}}{d\mu^2}\right.$$

$$\left. - 2(m+1)\mu\frac{dQ_{lm}}{d\mu} + \left(\frac{m^2\mu^2}{1 - \mu^2} - m\right)Q_{lm}\right]. \tag{5.55}$$

Substituting (5.55) into (5.45), we obtain

$$\left(1 - \mu^2\right)\frac{d^2Q_{lm}}{d\mu^2} - 2(m+1)\mu\frac{dQ_{lm}}{d\mu} + [\lambda - m(m+1)]Q_{lm} = 0. \tag{5.56}$$

The two equations (5.56) and (5.52) are identical to each other, which proves that the auxiliary functions Q_{lm}, defined by (5.51) and (5.53) satisfy the same differential equation. Thus, we can assume that they are identical to each other and use the definitions (5.51) and (5.53) to write down the formula, which relates the two solutions $P_{lm}(\mu)$ and $P_l(\mu)$ as follows:

$$P_{lm}(\mu) = \left(1 - \mu^2\right)^{|m|/2} \frac{d^{|m|}P_l(\mu)}{d\mu^{|m|}}, \tag{5.57}$$

where we now allow both positive and negative values of the integer $m = 0, \pm 1, \pm 2, \ldots$ and replace m by $|m|$.

As we have argued above, if we know the solution $P_l(\mu)$ of the equation (5.46), we can calculate the solution $P_{lm}(\mu)$ of the general equation (5.45) using formula (5.57). The equation (5.46) is solved by expanding the function $P_l = P_l(\mu)$ into the power series as follows:

$$P_l(\mu) = \sum_{k=0}^{\infty} a_k \mu^k. \tag{5.58}$$

The equation (5.46) can be written in the following form

$$\left(1 - \mu^2\right)\frac{d^2P_l}{d\mu^2} - 2\mu\frac{dP_l}{d\mu} + \lambda P_l = 0. \tag{5.59}$$

Substituting the series (5.58) into the equation (5.59), we obtain

$$\sum_{k=0}^{\infty}\left[k(k-1)a_k\mu^{k-2} - k(k-1)a_k\mu^k - 2ka_k\mu^k + \lambda a_k\mu^k\right] = 0, \tag{5.60}$$

or

$$\sum_{k=0}^{\infty}\{(k+2)(k+1)a_{k+2} - [k(k+1) - \lambda]a_k\}\mu^k = 0. \tag{5.61}$$

From the equation (5.61), we obtain the recurrence formula for the coefficients of the power series (5.58) as follows:

$$a_{k+2} = \frac{k(k+1) - \lambda}{(k+2)(k+1)}a_k. \tag{5.62}$$

If we know the boundary conditions $a_0 = P_l(0)$ and $a_1 = \dot{P}_l(0)$, the formula (5.62) allows us to calculate the coefficients of the entire infinite series (5.58). As the wave function $\psi = \psi(\theta, \phi)$ is a measure of the probability of finding the rotating body in a certain orientation in Fig. 5.2, an obvious boundary condition for the wave function is that it must be convergent for all values $-\pi \leq \theta \leq +\pi$ or, since $\mu = \cos\theta$, for all values $-1 \leq \mu \leq +1$. However, the power series (5.58) is divergent for $\mu = \pm 1$. The only way to obtain a proper bound-state wave function is to choose the parameter λ in such a way to terminate the power series and reduce it to a polynomial. Let us, therefore, assume that for some integer $k = l$, we have

$$\lambda_l = l(l+1), \quad l = 0, 1, 2, \ldots. \tag{5.63}$$

Substituting (5.63) into (5.62), we obtain

$$a_{k+2} = \frac{k(k+1) - l(l+1)}{(k+2)(k+1)}a_k. \tag{5.64}$$

From (5.64), it is clear that $a_k \equiv 0$ for any $k > l$, and the power series is reduced to a polynomial. The polynomials obtained in this way are known as Legendre polynomials and they satisfy the Legendre differential equation

$$\frac{d}{d\mu}\left[\left(1 - \mu^2\right)\frac{dP_l}{d\mu}\right] + l(l+1)P_l = 0. \tag{5.65}$$

5.3 Quantized Energy of Rotations

Let us now use the definition of the parameter λ in (5.30), where we replace the energy E_R by the quantized energy of rotations E_l. We then obtain from (5.63)

$$\lambda_l = \frac{2IE_l}{\hbar^2} = l(l+1), \quad l = 0, 1, 2, \ldots, \tag{5.66}$$

or

$$E_l = \frac{\hbar^2}{2I} l(l+1), \quad l = 0, 1, 2, \ldots . \tag{5.67}$$

The result (5.67) is the well-known result for the quantized energy eigenstates of the rotating constituent particle of the system. In thermodynamics, we are primarily interested in this result for the energy of rotations of constituent particles in the system (5.67). However, for the sake of completeness, we will derive some results for the wave functions of the rotating constituent particles as well.

5.4   Legendre Polynomials *)

Using the formula (5.64), we can easily calculate the first few Legendre polynomials. They are determined up to an arbitrary multiplicative constant, which is determined by the boundary conditions $a_0 = P_l(0)$ and $a_1 = \dot{P}_l(0)$. In particular, the first two polynomials $P_0(\mu)$ and $P_1(\mu)$ have only one term and are determined by an arbitrary choice of the above boundary conditions. If we choose $a_0 = 1$ for $P_0(\mu)$ and $a_1 = 1$ for $P_1(\mu)$, we have

$$P_0(\mu) = 1, \quad P_1(\mu) = \mu. \tag{5.68}$$

Further, if we choose $a_0 = -1/2$ for $P_2(\mu)$, $a_1 = -12/9$ for $P_3(\mu)$, and $a_0 = 3/8$ for $P_4(\mu)$, respectively, we obtain

$$P_2(\mu) = \frac{1}{2}\left(3\mu^2 - 1\right),$$

$$P_3(\mu) = \frac{4}{9}\left(5\mu^3 - 3\mu\right), \tag{5.69}$$

$$P_4(\mu) = \frac{1}{8}\left(35\mu^4 - 30\mu^2 + 3\right).$$

The particular choices of the above boundary conditions were made to fit a practical formula, which can also be used for calculation of the Legendre polynomials. It is known as Rodrigues' formula, and it reads

$$P_l(\mu) = \frac{1}{2^l\, l!} \frac{d^l}{d\mu^l}\left(\mu^2 - 1\right)^l. \tag{5.70}$$

It is easily shown, by direct calculation, that the Rodrigues' formula reproduces the results (5.68) and (5.69). Using the Legendre differential equation (5.65), it is easily shown that the Legendre polynomials are orthogonal to each other, i.e., that

$$\int_{-1}^{+1} d\mu\, P_l(\mu) P_j(\mu) = 0, \quad l \neq j. \tag{5.71}$$

In order to prove the orthogonality of the Legendre polynomials, let us rewrite the equation (5.65) for $P_l(\mu)$ and $P_j(\mu)$, respectively, as follows:

$$\frac{d}{d\mu}\left[\left(1-\mu^2\right)\dot{P}_l\right]+l(l+1)P_l=0, \tag{5.72}$$

$$\frac{d}{d\mu}\left[\left(1-\mu^2\right)\dot{P}_j\right]+j(j+1)P_j=0. \tag{5.73}$$

Multiplying the equation (5.72) by $P_j(\mu)$ and the equation (5.46) by $P_l(\mu)$, we obtain

$$P_j\frac{d}{d\mu}\left[\left(1-\mu^2\right)\dot{P}_l\right]+l(l+1)P_lP_j=0, \tag{5.74}$$

$$P_l\frac{d}{d\mu}\left[\left(1-\mu^2\right)\dot{P}_j\right]+j(j+1)P_lP_j=0, \tag{5.75}$$

or

$$\frac{d}{d\mu}\left[\left(1-\mu^2\right)P_j\dot{P}_l\right]-\left(1-\mu^2\right)\dot{P}_l\dot{P}_j+l(l+1)P_lP_j=0, \tag{5.76}$$

$$\frac{d}{d\mu}\left[\left(1-\mu^2\right)P_l\dot{P}_j\right]-\left(1-\mu^2\right)\dot{P}_l\dot{P}_j+j(j+1)P_lP_j=0. \tag{5.77}$$

Subtracting the equation (5.77) from the equation (5.76), we obtain

$$-\frac{d}{d\mu}\left[\left(1-\mu^2\right)\left(P_j\dot{P}_l-P_l\dot{P}_j\right)\right]=[l(l+1)-j(j+1)]P_lP_j. \tag{5.78}$$

Integrating the equation (5.78) gives

$$[l(l+1)-j(j+1)]\int_{-1}^{+1}d\mu\,P_l(\mu)P_j(\mu)$$

$$=-\int_{-1}^{+1}d\mu\,\frac{d}{d\mu}\left[\left(1-\mu^2\right)\left(P_j\dot{P}_l-P_l\dot{P}_j\right)\right]$$

$$=-\left[\left(1-\mu^2\right)\left(P_j\dot{P}_l-P_l\dot{P}_j\right)\right]\Big|_{-1}^{+1}=0, \tag{5.79}$$

or

$$[l(l+1)-j(j+1)]\int_{-1}^{+1}d\mu\,P_l(\mu)P_j(\mu)=0. \tag{5.80}$$

In order to satisfy the equation (5.80), we see that for any $j \neq l$, we must have

$$\int\limits_{-1}^{+1} d\mu \, P_l(\mu) P_j(\mu) = 0, \quad l \neq j, \tag{5.81}$$

which is identical with the initial assumption (5.71). Thus, the Legendre polynomials are indeed orthogonal to each other. For $j = l$, the integral in the equation (5.80) is not equal to zero, and its value is needed to determine the normalization constant for the Legendre polynomials. Let us, therefore, as the next step, calculate the integral

$$I_l = \int\limits_{-1}^{+1} d\mu \, [P_l(\mu)]^2 . \tag{5.82}$$

Substituting the Rodrigues' formula (5.70) into (5.82), we obtain

$$I_l = \frac{1}{[2^l \, l!]^2} \int\limits_{-1}^{+1} d\mu \, \frac{d^l}{d\mu^l} \left(\mu^2 - 1\right)^l \frac{d^l}{d\mu^l} \left(\mu^2 - 1\right)^l . \tag{5.83}$$

Using now the general formula

$$\frac{d^l}{d\mu^l} \left(\mu^2 - 1\right)^l \frac{d^l}{d\mu^l} \left(\mu^2 - 1\right)^l$$

$$= \frac{d}{d\mu} \left[\frac{d^{l-1}}{d\mu^{l-1}} \left(\mu^2 - 1\right)^l \frac{d^l}{d\mu^l} \left(\mu^2 - 1\right)^l \right]$$

$$- \frac{d^{l-1}}{d\mu^{l-1}} \left(\mu^2 - 1\right)^l \frac{d^{l+1}}{d\mu^{l+1}} \left(\mu^2 - 1\right)^l , \tag{5.84}$$

and the result

$$\frac{d^{l-1}}{d\mu^{l-1}} \left(\mu^2 - 1\right)^l = 0 \quad \text{for} \quad \mu = \pm 1, \tag{5.85}$$

we can integrate (5.83) by parts to obtain

$$I_l = \frac{(-1)^1}{[2^l \, l!]^2} \int\limits_{-1}^{+1} d\mu \, \frac{d^{l-1}}{d\mu^{l-1}} \left(\mu^2 - 1\right)^l \frac{d^{l+1}}{d\mu^{l+1}} \left(\mu^2 - 1\right)^l . \tag{5.86}$$

Repeating this procedure l-times yields

$$I_l = \frac{(-1)^l}{[2^l \, l!]^2} \int\limits_{-1}^{+1} d\mu \, \left(\mu^2 - 1\right)^l \frac{d^{2l}}{d\mu^{2l}} \left(\mu^2 - 1\right)^l . \tag{5.87}$$

On the other hand, we know that

$$\frac{d^{2l}}{d\mu^{2l}}\left(\mu^2 - 1\right)^l = \frac{d^{2l}}{d\mu^{2l}}\mu^{2l} = (2l)!,$$ (5.88)

and (5.87) reduces to the following form

$$I_l = \frac{(-1)^l(2l)!}{[2^l\,l!]^2}\int\limits_{-1}^{+1} d\mu\left(\mu^2 - 1\right)^l,$$ (5.89)

or

$$I_l = \frac{(-1)^l(2l)!}{[2^l\,l!]^2}\int\limits_{-1}^{+1} d\mu\,(\mu - 1)^l\,(\mu + 1)^l.$$ (5.90)

Let us now drop the constant factor in (5.90) and consider the integral

$$J_l = \int\limits_{-1}^{+1} d\mu\,(\mu - 1)^l(\mu + 1)^l.$$ (5.91)

Integrating by parts, we obtain

$$J_l = (\mu - 1)^l\frac{(\mu + 1)^{l+1}}{l+1}\Bigg|_{-1}^{+1} - \int\limits_{-1}^{+1} d\mu\,l(\mu - 1)^{l-1}\frac{(\mu + 1)^{l+1}}{l+1}.$$ (5.92)

The first term in (5.92) vanishes for $\mu = \pm 1$, and we have

$$J_l = (-1)^1\frac{l}{l+1}\int\limits_{-1}^{+1} d\mu(\mu - 1)^{l-1}(\mu + 1)^{l+1}.$$ (5.93)

Integrating by parts once again gives

$$J_l = (-1)^2\frac{l(l-1)}{(l+1)(l+2)}\int\limits_{-1}^{+1} d\mu(\mu - 1)^{l-2}(\mu + 1)^{l+2}.$$ (5.94)

Repeating this procedure l-times, we finally get

$$J_l = (-1)^l\frac{l!}{(2l)!/l!}\int\limits_{-1}^{+1} d\mu(\mu + 1)^{2l} = \frac{(-1)^l(l!)^2}{(2l)!}\frac{2^{2l+1}}{2l+1}.$$ (5.95)

Substituting (5.95) into (5.90), we obtain

$$I_l = \frac{(-1)^l (2l)!}{[2^l\, l!]^2}\; \frac{(-1)^l (l!)^2\; 2^{2l+1}}{(2l)!\; 2l+1} = \frac{2}{2l+1}. \tag{5.96}$$

Thus, we finally have the result

$$\int\limits_{-1}^{+1} d\mu\, [P_l(\mu)]^2 = \frac{2}{2l+1}. \tag{5.97}$$

5.5 Normalization of the Wave Function *)

Using the results (5.57) and (5.70), we obtain the Rodrigues' formula for the solutions P_{lm} of the equation (5.45), for arbitrary values of the integer m as follows:

$$P_{lm}(\mu) = \frac{1}{2^l\, l!} \left(1 - \mu^2\right)^{|m|/2} \frac{d^{l+|m|}}{d\mu^{l+|m|}} \left(\mu^2 - 1\right)^l. \tag{5.98}$$

From the result (5.98), we see that $P_{lm}(\mu)$ and consequently the wave function $\psi_{lm}(\theta, \phi)$ vanishes when $|m| > l$. Thus, we must have $-l \le m \le +l$. This implies that every energy eigenstate with a given quantum number l has $g(l) = 2l+1$ substates with different values of the quantum number m, i.e., different values of the angular momentum vector component in z-direction L_z.

The wave function for the rotating constituent particle is then given by (5.32), with (5.36) and (5.98). Thus, we have

$$\psi_{lm}(\theta, \phi) = A_{lm} P_{lm}(\cos\theta) \frac{1}{\sqrt{2\pi}} \exp(\pm i |m| \phi), \tag{5.99}$$

where A_{lm} are the normalization constants for the functions P_{lm}, given by (5.98). In order to calculate the normalization constants A_{lm}, we need to calculate the integral

$$I_{lm} = \int\limits_{-1}^{+1} d\mu\, [P_{lm}(\mu)]^2. \tag{5.100}$$

Let us first recall the definition (5.57) with $|m| = m > 0$, i.e.,

$$P_{lm}(\mu) = \left(1 - \mu^2\right)^{m/2} \frac{d^m P_l(\mu)}{d\mu^m}. \tag{5.101}$$

From (5.101), we can calculate

$$\frac{dP_{lm}}{d\mu} = -m\mu \left(1 - \mu^2\right)^{m/2-1} \frac{d^m P_l(\mu)}{d\mu^m} + \left(1 - \mu^2\right)^{m/2} \frac{d^{m+1} P_l}{d\mu^{m+1}}, \tag{5.102}$$

or

$$\left(1-\mu^2\right)^{1/2}\frac{dP_{lm}}{d\mu} = -m\mu\left(1-\mu^2\right)^{(m-1)/2}\frac{d^m P_l(\mu)}{d\mu^m}$$

$$+\left(1-\mu^2\right)^{(m+1)/2}\frac{d^{m+1}P_l}{d\mu^{m+1}}, \tag{5.103}$$

or using the definition (5.101)

$$\left(1-\mu^2\right)^{1/2}\frac{dP_{lm}}{d\mu} = P_{lm+1} - m\mu\left(1-\mu^2\right)^{-1/2}P_{lm}. \tag{5.104}$$

From (5.104), we can write down the following recurrence relation

$$P_{lm+1} = \left(1-\mu^2\right)^{1/2}\frac{dP_{lm}}{d\mu} + m\mu\left(1-\mu^2\right)^{-1/2}P_{lm}, \tag{5.105}$$

or

$$P_{lm+1}^2 = \left(1-\mu^2\right)\left(\frac{dP_{lm}}{d\mu}\right)^2 + 2m\mu P_{lm}\frac{dP_{lm}}{d\mu} + \frac{m^2\mu^2}{1-\mu^2}P_{lm}^2. \tag{5.106}$$

Using (5.106), we can now calculate the integral

$$I_{lm+1} = \int_{-1}^{+1} d\mu\,[P_{lm+1}(\mu)]^2 = \int_{-1}^{+1} d\mu\left(1-\mu^2\right)\frac{dP_{lm}}{d\mu}\frac{dP_{lm}}{d\mu}$$

$$+ \int_{-1}^{+1} d\mu\,2m\mu P_{lm}\frac{dP_{lm}}{d\mu} + \int_{-1}^{+1} d\mu\,\frac{m^2\mu^2}{1-\mu^2}P_{lm}^2. \tag{5.107}$$

Using now the identities

$$\left(1-\mu^2\right)\frac{dP_{lm}}{d\mu}\frac{dP_{lm}}{d\mu} = \frac{d}{d\mu}\left[\left(1-\mu^2\right)P_{lm}\frac{dP_{lm}}{d\mu}\right]$$

$$- P_{lm}\frac{d}{d\mu}\left[\left(1-\mu^2\right)\frac{dP_{lm}}{d\mu}\right], \tag{5.108}$$

and

$$2m\mu P_{lm}\frac{dP_{lm}}{d\mu} = m\mu\frac{d}{d\mu}P_{lm}^2 = \frac{d}{d\mu}\left(m\mu P_{lm}^2\right) - mP_{lm}^2, \tag{5.109}$$

as well as the fact that $(1 - \mu^2) = 0$ and $mP_{lm} = 0$ for $\mu = \pm 1$, we can integrate the first and the second integral on the right-hand side of the equation (5.107) by parts, to obtain

$$I_{lm+1} = \int_{-1}^{+1} d\mu \, P_{lm} \left\{ -\frac{d}{d\mu} \left[\left(1 - \mu^2 \right) \frac{dP_{lm}}{d\mu} \right] \right.$$

$$\left. + \left(-m + \frac{m^2 \mu^2}{1 - \mu^2} \right) P_{lm} \right\}. \tag{5.110}$$

With the identity

$$-m + \frac{m^2 \mu^2}{1 - \mu^2} = -m(m+1) + \frac{m^2}{1 - \mu^2}, \tag{5.111}$$

the equation (5.110) becomes

$$I_{lm+1} = \int_{-1}^{+1} d\mu \, P_{lm} \left\{ -\frac{d}{d\mu} \left[\left(1 - \mu^2 \right) \frac{dP_{lm}}{d\mu} \right] \right.$$

$$\left. + \left[-m(m+1) + \frac{m^2}{1 - \mu^2} \right] P_{lm} \right\}. \tag{5.112}$$

From the equation (5.45) with $\lambda = l(l+1)$, we can write

$$l(l+1)P_{lm} = -\frac{d}{d\mu} \left[\left(1 - \mu^2 \right) \frac{dP_{lm}}{d\mu} \right] + \frac{m^2}{1 - \mu^2} P_{lm}. \tag{5.113}$$

Substituting (5.113) into (5.112), we obtain

$$I_{lm+1} = [l(l+1) - m(m+1)] \int_{-1}^{+1} d\mu \, [P_{lm}]^2$$

$$= (l+m+1)(l-m) \int_{-1}^{+1} d\mu \, [P_{lm}]^2. \tag{5.114}$$

Thus, we obtain the following recurrence formula

$$\int_{-1}^{+1} d\mu [P_{lm+1}(\mu)]^2 = (l+m+1)(l-m) \int_{-1}^{+1} d\mu [P_{lm}]^2. \tag{5.115}$$

From the formula (5.115), we can write the following results

$$\int_{-1}^{+1} d\mu [P_{lm}(\mu)]^2 = (l+m)(l-m+1) \int_{-1}^{+1} d\mu [P_{lm-1}]^2$$

$$\int_{-1}^{+1} d\mu [P_{lm-1}(\mu)]^2 = (l+m-1)(l-m+2) \int_{-1}^{+1} d\mu [P_{lm-2}]^2$$

$$\vdots$$

$$\int_{-1}^{+1} d\mu [P_{l1}(\mu)]^2 = (l+1)l \int_{-1}^{+1} d\mu [P_{l0}]^2. \tag{5.116}$$

Combining all of the results (5.116) and using (5.97), we can write

$$\int_{-1}^{+1} d\mu [P_{lm}(\mu)]^2 = [(l+m)(l+m-1)\cdots(l+1)]$$

$$\times [(l-m+1)(l-m+2)\cdots l] \int_{-1}^{+1} d\mu [P_l]^2$$

$$= \frac{(l+m)!}{l!} \frac{l!}{(l-m)!} \int_{-1}^{+1} d\mu [P_l]^2$$

$$= \frac{(l+m)!}{(l-m)!} \int_{-1}^{+1} d\mu [P_l]^2 = \frac{(l+m)!}{(l-m)!} \frac{2}{2l+1}. \tag{5.117}$$

Finally, if we now allow both positive and negative values of the integer $m = 0$, $\pm 1, \pm 2, \ldots$ and replace m by $|m|$, we obtain the following result

$$\int_{-1}^{+1} d\mu [P_{lm}(\mu)]^2 = \frac{(l+|m|)!}{(l-|m|)!} \frac{2}{2l+1}. \tag{5.118}$$

The normalization constant A_{lm} in (5.99) is now obtained from the condition

$$\int_{-1}^{+1} d\mu [A_{lm} P_{lm}(\mu)]^2 = A_{lm}^2 \frac{(l+|m|)!}{(l-|m|)!} \frac{2}{2l+1} = 1, \tag{5.119}$$

and it is given by

$$A_{lm} = \sqrt{\frac{(l-|m|)!}{(l+|m|)!} \frac{2l+1}{2}}. \tag{5.120}$$

Finally, the complete result for the normalized wave function of the rotating constituent particle is given by

$$\psi_{lm}(\theta,\phi) = \frac{1}{2^l l!} \sqrt{\frac{(l-|m|)!}{(l+|m|)!} \frac{2l+1}{2}} \left(1 - \cos^2\theta\right)^{|m|/2}$$

$$\times \frac{d^{l+|m|}}{d(\cos\theta)^{l+|m|}} \left(\cos^2\theta - 1\right)^l \frac{1}{\sqrt{2\pi}} \exp(\pm i|m|\phi). \tag{5.121}$$

The wave functions (5.121) are called the spherical harmonics and are usually denoted by $Y_{lm}(\theta,\phi)$. These functions are of importance in both quantum mechanics and classical field theories.

5.6 Spin Angular Momentum *)

In the quantum-mechanical analysis of rotations of the constituent composite particles such as diatomic molecules, we have shown that the angular momentum quantum numbers (l and m) classify their states with respect to their rotations about their center of mass in the ordinary three-dimensional space. This angular momentum originates from the ordinary, "orbital" motion of the parts of composite particles about the center of mass of composite particles, so it is usually called *orbital angular momentum*. A composite particle occupying a given rotational energy level (5.67), defined by the quantum number l, can have $2l+1$ different projections of the orbital angular momentum on a selected direction in space (z-axis), i.e., the energy level, defined by the quantum number l, has $2l+1$ sublevels.

On the other hand, the elementary particles, which can be envisioned as point-like particles (e.g., electrons), classically would have zero moments of inertia, and, due to this, zero internal angular momentum. Thus, for true elementary particles, classical physics would predict no internal motion such as the rotational motion encountered in composite particles. However, it turns out that genuinely quantum nature of true elementary particles assigns them a certain "intrinsic" angular momentum called *spin*, which is unrelated to some putative "internal" motion in the ordinary three-dimensional space. The spin is a quantum-mechanical property of all elementary particles that disappears in the limit when $\hbar \to 0$. Because of this, the spin does not have a classical interpretation.

Analogously to the case of orbital angular momentum, for a particle with a spin quantum number σ, we have $-\sigma < \sigma_z < +\sigma$. In other words, there are $2\sigma + 1$ different projections of the spin angular momentum, on a selected direction in space (z-axis), i.e., the quantum state defined by the quantum number σ, has $2\sigma + 1$ spin states. One peculiar property of the spin quantum number σ is that it is not necessarily an integer, i.e., there are particles with both integer and half-integer spins. The electrons have $\sigma = 1/2$ and their states are twofold degenerate, i.e., they have two spin states. The two spin states of an electron, i.e., the two z-projections of the spin angular momentum, are usually called "spin up" ($\sigma_z = +1/2$) and "spin down" ($\sigma_z = -1/2$).

For microscopic particles with spin, the quantum-mechanical description of a state by means of the wave function must include both the probability to find a particle in different positions in the ordinary three-dimensional space and the probability of the possible orientations of the spin. The wave function is then a function of the three continuous space co-ordinates (x, y, z) and a discrete spin variable σ, i.e.,

$$\psi = \psi(x, y, z; \sigma) = \psi(\vec{r}, \sigma) = \psi_r(\vec{r})\, \psi_\sigma(\sigma), \tag{5.122}$$

where $\vec{r} = (x, y, z)$ is the position vector in the three-dimensional space. For example, for the "spin down" state, the spin part $\psi_\sigma(\sigma)$ of the full wave function in (5.122) is zero for $\sigma_z = +1/2$ whereas it is one for $\sigma_z = -1/2$.

5.7   Problems with Solutions *)

Problem

Verify that the function $\Phi(\phi) = \frac{1}{\sqrt{2\pi}} \exp(im\phi)$ is the eigenfunction of the rotation operator $\hat{R}(\alpha)$, defined by

$$\hat{R}(\alpha) = \exp\left(i \frac{\hat{L}_z}{\hbar} \alpha \right), \tag{5.123}$$

i.e., it satisfies the eigenvalue equation

$$\hat{R}(\alpha)\Phi(\phi) = R(\alpha, m)\, \Phi(\phi), \tag{5.124}$$

and determine, the eigenvalue $R(\alpha, m)$ of the operator $\hat{R}(\alpha)$.

Solution

The operator $\hat{R}(\alpha)$ can be written in the form of the Taylor series

$$\hat{R}(\alpha) = \sum_{j=0}^{\infty} \frac{1}{j!} \left(i \frac{\hat{L}_z}{\hbar} \alpha \right)^j = \sum_{j=0}^{\infty} \frac{1}{j!} \left(\alpha \frac{\partial}{\partial \phi} \right)^j. \tag{5.125}$$

Substituting (5.125) and (5.124), we obtain

$$\hat{R}(\alpha)\Phi(\phi) = \sum_{j=0}^{\infty} \frac{1}{j!} \left(\alpha \frac{\partial}{\partial \phi} \right)^j \frac{1}{\sqrt{2\pi}} \exp(im\phi), \tag{5.126}$$

or

$$\hat{R}(\alpha)\Phi(\phi) = \frac{1}{\sqrt{2\pi}} \exp(im\phi) \sum_{j=0}^{\infty} \frac{1}{j!} (im\alpha)^j, \tag{5.127}$$

or

$$\hat{R}(\alpha)\Phi(\phi) = \frac{1}{\sqrt{2\pi}} \exp(im\phi) \exp(im\alpha), \tag{5.128}$$

or finally

$$\hat{R}(\alpha)\Phi(\phi) = \exp(im\alpha)\,\Phi(\phi) = R(\alpha, m)\Phi(\phi). \tag{5.129}$$

Thus, the function $\Phi(\phi)$ indeed satisfies the equation (5.124), and it is an eigenfunction of the operator $\hat{R}(\alpha)$. From the result (5.129), we see that the eigenvalue of the operator $\hat{R}(\alpha)$ is given by

$$R(\alpha, m) = \exp(im\alpha). \tag{5.130}$$

There is an alternative solution to this problem that emphasizes the more general properties of the operator $\hat{R}(\alpha)$. We first note that the standard Taylor expansion of any function $f(\phi + \alpha)$, about the value $\alpha = 0$, can be written as

$$f(\phi + \alpha) = \sum_{j=0}^{\infty} \frac{1}{j!} \frac{\partial^j f(\phi)}{\partial \phi^j} \alpha^j = \sum_{j=0}^{\infty} \frac{1}{j!} \left(\alpha \frac{\partial}{\partial \phi} \right)^j f(\phi). \tag{5.131}$$

Using now the equation (5.125), we can identify the operator $\hat{R}(\alpha)$ in the second sum on the right-hand side of (5.131). Thus, for any function $f(\phi)$, we have

$$f(\phi + \alpha) = \hat{R}(\alpha)f(\phi) \Rightarrow \hat{R}(\alpha)f(\phi) = f(\phi + \alpha). \tag{5.132}$$

In view of the equation (5.132), the operator $\hat{R}(\alpha)$ is known as the *shift operator*. Its effect is to "shift" by α the argument of any function $f(\phi)$. Let us apply the equation (5.132) to our particular function $\Phi(\phi) = \frac{1}{\sqrt{2\pi}} \exp(im\phi)$. By equation (5.132) and simple algebra, we obtain

$$\hat{R}(\alpha)\Phi(\phi) = \Phi(\phi + \alpha) = \frac{1}{\sqrt{2\pi}} \exp(im(\phi + \alpha))$$

$$= \exp(im\alpha) \frac{1}{\sqrt{2\pi}} \exp(im\phi) = R(\alpha, m)\,\Phi(\phi). \tag{5.133}$$

Thus, we find again the relation (5.124), telling us that the the function $\Phi(\phi)$ is the eigenfunction of the shift operator $\hat{R}(\alpha)$, with the eigenvalue equal to

$$R(\alpha, m) = \exp(im\alpha). \tag{5.134}$$

Part Two

Thermodynamics of Systems

6 Number of Accessible States and Entropy

6.1 Introduction and Definitions

Equilibrium macrostate of a thermodynamic system is usually specified by a few easily measurable macroscopic quantities. For example, for gases and liquids, these quantities may include pressure (P), volume (V), and temperature (T). On the other hand, a microstate of a thermodynamic system is defined by giving positions and velocities of all constituent particles of the system. As the number of constituent particles of a system is extremely large ($\sim 10^{23}$), an experimental measurement of all these positions and velocities is clearly an impossible task.

Fortunately, such a detailed information (needed to describe an actual microstate) turns out to be unnecessary for the description of the system's actual macrostate (described just by telling a few thermodynamic quantities such as the temperature, volume, and pressure). The reason for this significant simplification in the description of a macrostate is that any given macrostate can be realized through a large number of microstates. Indeed, an experimentalist interested only in the pressure, volume, and temperature of a gas in thermodynamic equilibrium would measure these three quantities as being constant in time, whereas the actual gas molecules change all the time their positions and velocities (due to collisions). In this way, the equilibrium system visits a huge number of different microstates while remaining in the same macrostate characterized with time-independent thermodynamic variables such as pressure, temperature, and volume.

Because of the large number of constituent particles, the number of the accessed microstates, visited in time by a system in a given macrostate, is huge. Let us denote the number of accessible microstates as W. For realistic physical systems,

$$W \gg 1. \tag{6.1}$$

To elucidate this quantity, let us consider two thermodynamic systems that do not interact with each other in any way. Let us assume that the number of accessible states of the first system is W_1 and the number of accessible states of the second system is W_2. The two systems are independent of each other (due to the absence of mutual interaction), so the first system can be in any of its accessible microstates while the second system is in any of its accessible microstates. Thus, the total number of accessible states of the entire system consisting of both of these systems is simply the product

$$W = W_1 W_2. \tag{6.2}$$

Since the numbers W_1 and W_2 are extremely large numbers (making their product enormous), it is convenient to introduce an additive property of the thermodynamic systems called the *entropy* (S), using the following definition

$$S = k \cdot \ln W, \tag{6.3}$$

where k is some yet unspecified constant. The total entropy of a system consisting of the two noninteracting thermodynamic systems described above is then equal to the sum of the entropies of the two constituent systems

$$S = k \cdot \ln W = k \cdot \ln(W_1 W_2) = k \cdot \ln W_1 + k \cdot \ln W_2, \tag{6.4}$$

or

$$S = S_1 + S_2. \tag{6.5}$$

6.2 Calculation of the Number of Accessible States

In order to be able to calculate the macroscopic properties of a thermodynamic system, we need to calculate the number of accessible states and the entropy of the system. Let us consider an ideal system in which the constituent particles are not interacting with each other, and let n_i denote the number of particles that occupy the quantum states with energies within the interval $(E_i, E_i + \Delta E_i)$. Let the number of states within the interval $(E_i, E_i + \Delta E_i)$ be denoted as g_i. For convenience, we will call these g_i states as "sublevels." Our task now is to arrange n_i particles into the g_i sublevels within the energy interval $(E_i, E_i + \Delta E_i)$. This is illustrated by the Fig. 6.1.

This task depends on the detailed description of the constituent particles. For example, in monatomic ideal gases, the constituent particles have no vibrational and rotational degrees of freedom, and the wave function of a single constituent particle in the cubic box, with co-ordinates $\vec{r}_1 = (x_1, y_1, z_1)$, is given by the result (3.40), i.e.,

$$\psi_{1r}(\vec{r}_1) = \left\{\frac{2}{a}\right\}^{3/2} \sin\left(\frac{n_{1x}\pi}{a}x_1\right) \sin\left(\frac{n_{1y}\pi}{a}y_1\right) \sin\left(\frac{n_{1z}\pi}{a}z_1\right). \tag{6.6}$$

If we incorporate the spin of the constituent particles according to the equation (5.122), the single-particle wave function (6.6) is generalized to

$$\psi_1(\vec{r}_1, \sigma_1) = \psi_1(\rho_1) = \psi_{1r}(\vec{r}_1)\psi_{1\sigma}(\sigma_1), \tag{6.7}$$

1	2	3	$\cdots$	$g_i - 1$	g_i

Figure 6.1 The g_i sublevels in the energy interval $(E_i, E_i + \Delta E_i)$.

where $\rho = (\vec{r}, \sigma)$ is the label for all the variables needed to fully specify the state of the constituent particle. These include both the three-dimensional space position variables $\vec{r}$ and the internal spin variable σ.

Our ideal system consists of N such noninteracting particles in translational motion within the cubic box. Let us first consider the system in the classical limit, in particular, let us ignore the spin. Due to the absence of mutual interactions, the wave function for the entire system might be written as the product of the single-particle wave functions (6.6), for all N particles of the system, i.e.,

$$\Psi(\vec{r}_1, \vec{r}_1 \ldots \vec{r}_N) = \prod_{a=1}^{N} \psi_a(\vec{r}_a). \tag{6.8}$$

By the equation (6.8), it is obvious that in such a state the particles would be distinguishable: the value of the wave function (6.8) changes if we exchange the positions of any two particles. So one can distinguish any particle from other particles by performing measurements on a system that would be described by the wave function (6.8). However, the actual quantum-mechanical description of identical particles requires that they are indistinguishable in any measurement performed on the many particle systems. Two identical particles such as, for example, two out of N electrons in a metal, cannot be distinguished from each other, and the correct many-particle wave functions must reflect this requirement. That is, exchanging any two particles co-ordinates and spin variables must not change the value of the squared magnitude of the wave function. Thus, in the many-body quantum mechanics, the result (6.8) is not applicable to identical particles, and the wave function for the entire system

$$\Psi = \Psi(\rho_1, \rho_2 \ldots \rho_N) \tag{6.9}$$

is a more complex combination of the single-particle wave functions (6.7), which satisfies the above permutational symmetry requirements from the many-body quantum mechanics. These symmetry requirements are related to the spin of the constituent particles. With regard to their spin, there are two kinds of microscopic particles in quantum mechanics:

1. Particles with integral spin (0, 1, 2, ...) called *bosons*.
2. Particles with half-integral spin (1/2, 3/2, ...) called *fermions*.

The symmetry requirements on these two types of microscopic particles are the following:

1. The total wave function (6.9) for bosons (e.g., He^4-atoms or photons) must be symmetric with respect to the exchange of any two particles of the system, i.e., the wave function (6.9) must remain unchanged when we interchange any two particles of the system. In mathematical terms, this means

$$\Psi(\ldots \rho_a \ldots \rho_b \ldots) = \Psi(\ldots \rho_b \ldots \rho_a \ldots). \tag{6.10}$$

In the special case of the exchange of two particles in the same single-particle state
($\rho_a = \rho_b$), the condition (6.10) is an identity, and there are no restrictions on how
many particles can be in the same single-particle state.

2. The total wave function (6.9) for fermions (e.g., He^3-atoms or electrons) must be
antisymmetric with respect to the exchange of any two particles of the system, i.e.,
the wave function (6.9) changes sign when we exchange any two particles of the
system. In mathematical terms, this means

$$\Psi(\ldots \rho_a \ldots \rho_b \ldots) = -\Psi(\ldots \rho_b \ldots \rho_a \ldots). \tag{6.11}$$

In the special case of the exchange of two particles in the same single-particle state
($\rho_a = \rho_b$), we must obviously have

$$\Psi(\ldots \rho_a \ldots \rho_b \ldots) = \Psi(\ldots \rho_b \ldots \rho_a \ldots), \tag{6.12}$$

which contradicts the condition (6.11). For $\rho_a = \rho_b$, the conditions (6.11) and (6.12)
can be both valid only if the total wave function is equal to zero. In the language of
quantum mechanics, the probability of finding the two fermions in the same single-
particle state is equal to zero ($\Psi = 0$). Thus, there can be no more than one fermion
in a single-particle state of the system. This conclusion is known as *Pauli exclusion
principle*.

From the above discussions, we expect that there are three different ways to count
the number of accessible states W of the system, i.e., (1) classical count of accessible states, (2) quantum-mechanical count of accessible states for fermions, and
(3) quantum-mechanical count of accessible states for bosons.

In order to illustrate the differences between these three different approaches, let
us consider a simple example of arranging of two ($n_i = 2$) particles, denoted by the
letters A and B, into three ($g_i = 3$) possible sublevels denoted by the integers 1, 2,
and 3. Based on the analysis of this simple example, we will reach the general results
for the numbers of possible ways to distribute n_i particles into the g_i sublevels. These
results will help us to calculate the number of accessible states W in the above three
distinct cases.

6.2.1 Classical Number of Accessible States

All possible arrangements of $n_i = 2$ distinguishable classical particles into $g_i = 3$ sublevels are shown in Fig. 6.2.

From Fig. 6.2, we see that there are $3^2 = 9$ arrangements of two classical particles
into three sublevels. In mathematical terms, the number of arrangements is equal to the
number of variations of $n_i = 2$ elements of the order $g_i = 3$. In general, for arbitrary n_i
and g_i, the number of possible arrangements is given by

$$v_i = g_i^{n_i}. \tag{6.13}$$

However, all of the variations (6.13) do not necessarily represent different physical
states if the particles (such as A and B in Fig. 6.2) are identical. For the variations

Figure 6.2 Classical arrangements of two particles into three sublevels.

1	2	3
AB		
	AB	
		AB
A	B	
B	A	
A		B
B		A
	A	B
	B	A

counted in (6.13), any permutation of the n_i identical particles will not produce a new state. In other words, the number of variations v_i must be divided by the number of permutations of the n_i particles, which is equal to $n_i!$ Thus, the number of the actual classical arrangements that give rise to the different states is given by

$$w_i = \frac{g_i^{n_i}}{n_i!}. \tag{6.14}$$

The division by $n_i!$ is a correction of the classical count (6.13) to account for the fact that the exchange of the identical constituent particles of the system does not produce a new physical state. However, since this correction factor did not emerge from an actual count of states, the result (6.14) is not necessarily an integer. For example, for the above simple case of two particles distributed into three states, by equation (6.14), we obtain $w_i = 9/2 = 4.5$.

The arrangements counted in (6.14) are for the n_i identical particles with energies within the interval $(E_i, E_i + \Delta E_i)$ in which there are g_i states. If there is only one energy level E_i in this interval, then g_i is simply the degeneracy of this level (recall the Problem in Section 3.6, "Problems with Solutions *)" in Section 3.6). The next question to ask is, how many different arrangements does one have for all identical particles distributed in all different energy intervals of this system? We can easily answer this question for the important case when the numbers n_i, counting particles with energies within the interval $(E_i, E_i + \Delta E_i)$, are given for all the intervals. Indeed, the arrangements of the particles occupying different energy intervals are independent from each other, so the total number of arrangements is obtained simply by multiplying the quantities w_i given by (6.14). Thus, for a macrostate specified by given occupation numbers n_i, we obtain the following result for the number of accessible microstates (of the entire system)

$$W = \prod_i \frac{g_i^{n_i}}{n_i!}. \tag{6.15}$$

6.2.2 Number of Accessible States for Bosons

In quantum mechanics, the two identical particles A and B are genuinely indistinguishable and we have $A = B$. All possible arrangements of $n_i = 2$ indistinguishable

1	2	3
AA		
	AA	
		AA
A	A	
A		A
	A	A

Figure 6.3 All arrangements of two bosons into three sublevels.

bosons into $g_i = 3$ sublevels are shown in Fig. 6.3. Note that more than one boson can be found in the same sublevel, i.e., in the same one-particle state.

From Fig. 6.3, we see that there are

$$\frac{(3+2-1)!}{(3-1)!\,2!} = \frac{4!}{2!\,2!} = 6 \tag{6.16}$$

possible arrangements of two bosons into three sublevels. In general, for arbitrary n_i and g_i, the number of possible arrangements is given by

$$w_i = \frac{(g_i + n_i - 1)!}{(g_i - 1)!\,n_i!}. \tag{6.17}$$

The arrangements counted in (6.17) are for the n_i identical particles occupying the energy level with energy E_i and degeneracy g_i. The next question to ask is, how many different arrangements one has for all identical particles distributed on all different energy levels of this system? We can easily answer this question for the important case when the numbers n_i, counting the particles occupying the energy levels E_i, are given for all the levels. Indeed, the arrangements of the particles occupying different energy levels are independent from each other, so the total number of arrangements is obtained simply by multiplying the quantities w_i given by (6.17). Thus, for a macrostate speci-fied by given occupation numbers n_i, we obtain the following result for the number of accessible microstates (of the entire system)

$$W = \prod_i \frac{(g_i + n_i - 1)!}{(g_i - 1)!\,n_i!}. \tag{6.18}$$

6.2.3 Number of Accessible States for Fermions

All possible arrangements of $n_i = 2$ indistinguishable fermions into $g_i = 3$ sublevels are shown in Fig. 6.4. Note that not more than one fermion can be found in the same sublevel, i.e., in the same one-particle state.

From Fig. 6.3, we see that there are

$$\binom{3}{2} = \frac{3!}{2!\,1!} = 3 \tag{6.19}$$

possible arrangements of two fermions into three sublevels. Mathematically, we can say that the number of arrangements is equal to the number of combinations of $g_i = 3$

Figure 6.4 All arrangement of two fermions into three sublevels.

1	2	3
A	A	
A		A
	A	A

elements, $n_i = 2$ at the time. In general, for arbitrary n_i and g_i, the number of possible arrangements is given by

$$w_i = \binom{g_i}{n_i} = \frac{g_i!}{n_i!(g_i - n_i)!}. \tag{6.20}$$

The arrangements counted in (6.20) are for the n_i identical particles occupying the energy level with energy E_i and degeneracy g_i. The next question to ask is, how many different arrangements does one have for all identical particles distributed on all different energy levels of this system? We can easily answer this question for the important case when the numbers n_i, counting the particles occupying the energy levels E_i, are given for all the levels. Indeed, the arrangements of the particles occupying different energy levels are independent from each other, so the total number of arrangements is obtained simply by multiplying the quantities w_i given by (6.20). Thus, for a macrostate specified by given occupation numbers n_i, we obtain the following result for the number of accessible microstates (of the entire system)

$$W = \prod_i \frac{g_i!}{n_i!(g_i - n_i)!}. \tag{6.21}$$

6.3 Problems with Solutions

Problem

As will be shown in Chapter 9, "Macroscopic Thermodynamics," the heat ΔQ added to a thermodynamic system at a given temperature T is proportional to the entropy increase of the system ΔS, i.e., $\Delta Q = T\Delta S$. Determine the heat that has to be added to a system at the room temperature $T = 298K$, in order to increase the number of accessible microstates by a factor of 10^6, given that the proportionality constant in the definition of entropy is $k = 1.3807 \cdot 10^{-23} \frac{J}{K}$.

Solution

The problem can be solved using the equation (6.3), with the given proportionality constant (Boltzmann constant). The initial entropy of the system is given by

$$S_0 = k \ln W_0, \tag{6.22}$$

where W_0 is the initial number of accessible microstates. After the amount of heat ΔQ is added, the entropy increases to

$$S = k \ln W = k \ln(W_0 \times 10^6) = k \ln W_0 + k \ln 10^6, \tag{6.23}$$

or

$$S = S_0 + 6k \ln 10. \tag{6.24}$$

Thus, the entropy change is

$$\Delta S = S - S_0 = 6k \ln 10. \tag{6.25}$$

The heat that has to be added to achieve this entropy change is then given by

$$\Delta Q = T \Delta S = 6kT \ln 10 \approx 5.684 \times 10^{-20} \, \text{J}, \tag{6.26}$$

or

$$\Delta Q \approx 0.35 \, \text{eV}. \tag{6.27}$$

7 Equilibrium States of Systems

7.1 Equilibrium Conditions

In this chapter, we will discuss the thermal equilibrium state of a thermodynamic system with a fixed number of the constituent particles (N = Constant) and a fixed internal energy of the system (U = Constant). These two conditions can be both enforced by placing the system in an insulating box keeping the system energy constant in time and preventing the particles to escape from the system. To proceed, we will focus our attention on ideal systems discussed in the Chapter 6, "Number of Accessible States and Entropy," in which we calculated the system for given occupation numbers n_i for the energy levels E_i. By the above discussion, for such a system, we have the following two conditions

$$U = \sum_i n_i E_i = \text{Constant}, \tag{7.1}$$

$$N = \sum_i n_i = \text{Constant}, \tag{7.2}$$

Our macroscopic equilibrium state may be thus specified by giving the values of the constants U and N. Notably, by equations (7.1) and (7.2), there are many choices of the occupation number n_i that would satisfy these two conditions. They correspond to various states for which we calculated the entropy $S = k \cdot \ln W$ in Chapter 6 in terms of the occupation numbers. All these states are in principle accessible in the given (by U and N) macrostate. In the course of time, the equilibrium system exhibits transitions between all these states (by maintaining constant U and N). However, the macroscopic system spends most of its time by fluctuating about the state that can be realized with the largest possible number of *microstates*. This special state will determine the average *values* of the occupation numbers n_i. Thus, to find the average values of the occupation numbers n_i, we need to solve the following maximization problem: for *given* values of U and N in (7.1) and (7.2), find the set of the occupation numbers n_i that maximizes the value of the entropy $S = k \cdot \ln W$ calculated in Chapter 6 (for any given set of n_i). We will denote this *entropy maximization* condition briefly as

$$S = S_{\text{max}} = k \cdot \ln W_{\text{max}}. \tag{7.3}$$

Let us denote by dn_i the variations of the occupations numbers n_i around their average values maximizing the entropy. Because U and N are fixed [by equations (7.1) and (7.2)] in this maximization problem and because the entropy is at its maximum (for $dn_i = 0$), we arrive at the following three conditions, in differential form

$$dU = \sum_i d(n_i E_i) = \sum_i E_i dn_i + n_i dE_i = \sum_i E_i dn_i = 0, \tag{7.4}$$

$$dN = \sum_i dn_i = 0, \tag{7.5}$$

$$dS = k \cdot d(\ln W) = 0, \tag{7.6}$$

for small variations dn_i of the occupation numbers. In equation (7.4), we note that the energy levels E_i as well as the number of sublevels g_i remain constant regardless of the changes in their occupation numbers n_i. E_i and g_i have the same values for all microstates, hence they are simply constants in the present maximization problem. We, therefore, have $dE_i = 0$ and $dg_i = 0$. In order to calculate $d(\ln W)$, we need the result for the number of accessible states W obtained in Chapter 6. In this chapter and the Part Three of this book, we are concerned with the systems that are well described by classical physics, e.g., ordinary gases in normal conditions. Thus, we will use the classical result for the number of accessible states (6.15), to obtain

$$\ln W = \ln \prod_i \frac{g_i^{n_i}}{n_i!} = \sum_i \ln \frac{g_i^{n_i}}{n_i!} = \sum_i n_i \ln g_i - \ln (n_i!). \tag{7.7}$$

The occupation number n_i is typically a large number and we can use the so-called Stirling approximation for the expression $\ln(n!)$, i.e., we can calculate

$$\ln(n!) = \ln \prod_{k=1}^{n} k = \sum_{k=1}^{n} \ln k \approx \int_1^n \ln k \, dk =$$

$$= k \ln k \Big|_1^n - \int_1^n dk = n \ln n - n + 1 \approx n \ln n - n. \tag{7.8}$$

Substituting (7.8) into (7.7), we obtain

$$\ln W = \sum_i n_i \ln g_i - n_i \ln n_i + n_i = \sum_i n_i \ln \frac{e \, g_i}{n_i}, \tag{7.9}$$

where the last step follows by recalling that $\ln(e) = 1$, and using the standard properties of the ln-function. The differential of the $\ln(W)$ in (7.9) is then given by

$$d\ln W = \sum_i d[n_i \ln g_i - n_i \ln n_i] + dn_i, \tag{7.10}$$

where, as noted above, the numbers g_i are to be treated as constants. Thus, each of the terms in the sum in (7.10) can be transformed as follows:

$$d[n_i \ln g_i - n_i \ln n_i] = dn_i \ln g_i - dn_i \ln n_i - n_i d[\ln n_i]$$

$$dn_i \ln g_i - dn_i \ln n_i - n_i \frac{dn_i}{n_i} = dn_i [n_i \ln g_i - n_i \ln n_i] - dn_i. \tag{7.11}$$

Substituting (7.11) into (7.10), we obtain

$$d \ln W = \sum_i dn_i [n_i \ln g_i - n_i \ln n_i]. \tag{7.12}$$

Using here the identity $n_i \ln g_i - n_i \ln n_i = n_i \ln (g_i/n_i)$, we obtain our final result for $d(\ln W)$ as

$$d \ln W = \sum_i dn_i \ln \frac{g_i}{n_i}. \tag{7.13}$$

The three equilibrium conditions (7.4–7.6) can be rewritten as follows:

$$\sum_i E_i dn_i = 0 \qquad / \cdot \frac{1}{\theta}, \tag{7.14}$$

$$\sum_i dn_i = 0 \qquad / \cdot \lambda, \tag{7.15}$$

$$\sum_i dn_i \ln \frac{g_i}{n_i} = 0 \qquad / \cdot (-1). \tag{7.16}$$

7.2   Occupation Numbers of Energy Levels

Using now the method of Lagrange multipliers, as indicated in the equations (7.14–7.16), we get

$$\sum_i \left(\frac{E_i}{\theta} + \lambda - \ln \frac{g_i}{n_i} \right) dn_i = 0. \tag{7.17}$$

On the other hand, the constituent particles of a thermodynamic system are free to move between levels. Thus, in general, $dn_i \neq 0$. In order to satisfy the equation (7.17), we must, therefore, have

$$\ln \frac{g_i}{n_i} = \lambda + \frac{E_i}{\theta}, \tag{7.18}$$

or

$$n_i = g_i \exp\left[-\left(\lambda + \frac{E_i}{\theta}\right)\right].$$

(7.19)

The result (7.19) gives the occupation number of levels within the energy interval $(E_i, E_i + \Delta E_i)$, within which there are g_i sublevels. In particular, if there is only one energy level E_i within this interval, then g_i is simply the degeneracy of this level. To apply (7.19) to concrete macrostate, we need to specify the two Lagrange multipliers λ and θ. Let us first observe that the total number of particles in the system is given by

$$N = \sum_i n_i = e^{-\lambda} \sum_i g_i \exp\left(-\frac{E_i}{\theta}\right) = e^{-\lambda} Z,$$

(7.20)

where we define the partition function Z of a thermodynamic system as follows:

$$Z = \sum_i g_i \exp\left(-\frac{E_i}{\theta}\right).$$

(7.21)

The result (7.20) can be rewritten as

$$e^{-\lambda} = \frac{N}{Z}.$$

(7.22)

Substituting (7.22) into (7.19), we obtain

$$n_i = g_i \frac{N}{Z} \exp\left(-\frac{E_i}{\theta}\right).$$

(7.23)

The probability of finding a constituent particle within the energy interval $(E_i, E_i + \Delta E_i)$ is thus given by

$$\omega_i = \frac{n_i}{N} = \frac{g_i}{Z} \exp\left(-\frac{E_i}{\theta}\right).$$

(7.24)

Using the definition (6.3) with (7.2) and (7.9), in combination with the simple fact that $\ln(e) = 1$, we can now calculate the entropy of the system as follows:

$$S = k \ln W = k \sum_i n_i \ln \frac{e g_i}{n_i} = k \sum_i n_i \ln \frac{g_i}{n_i} + kN.$$

(7.25)

Using here (7.23), we obtain

$$S = k \sum_i n_i \ln \left[g_i \frac{Z}{N g_i} \exp\left(\frac{E_i}{\theta}\right) \right] + kN$$

$$= k \sum_i n_i \ln \frac{Z}{N} + \frac{k}{\theta} \sum_i n_i E_i + kN.$$

(7.26)

Using here (7.1) and (7.2), we finally obtain the result for the entropy of the system

$$S = \frac{k}{\theta}U + kN\ln\frac{Z}{N} + kN. \tag{7.27}$$

7.3 Concept of Temperature

Let us now elucidate the nature of the Lagrange multiplier θ. Recall that the entropy of a thermodynamic system is, by equation (6.3), a measure of the number of microstates that correspond to the same macrostate. In this section, we will discuss the following significant property of the entropy: when two systems are brought into contact (as in the Fig. 7.1), then the entropy of the whole system can only increase as the system reaches the final equilibrium state.

In other words, the total entropy of the two systems before they are brought into contact with each other $S_0 = S_1 + S_2$ is less than the final entropy S of the merged system composed of these two systems. The reason for this increase in entropy is the increase of the number of microstates that correspond to the same macrostate of the total system. Before the two systems were brought into contact with each other, we had a higher degree of order, i.e., all the constituent particles of the first system could only occupy the states of the first system, and all the constituent particles of the second system could only occupy the states of the second system. When the two systems are brought into contact, the entire system eventually evolves in a less ordered physical macrostate, which can be realized through a much larger number of microstates.

Consequently, the total entropy as a measure of the number of microstates that correspond to the same macrostate of the system increases too. In mathematical terms, we can say that the time derivative of the total entropy, for the two systems brought into contact with each other, is a positive quantity:

$$\frac{dS}{dt} = \frac{dS_1}{dt} + \frac{dS_2}{dt} = \frac{\partial S_1}{\partial U_1}\frac{dU_1}{dt} + \frac{\partial S_2}{\partial U_2}\frac{dU_2}{dt} > 0. \tag{7.28}$$

The two systems are assumed to exchange energy with each other only, and they do not exchange energy with the surroundings. Therefore, the change in the total energy is assumed to be zero, i.e.,

$$dU = dU_1 + dU_2 = \left(\frac{dU_1}{dt} + \frac{dU_2}{dt}\right)dt = 0, \tag{7.29}$$

Figure 7.1 Two systems brought into contact with each other.

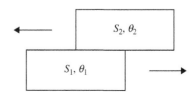

or

$$\frac{dU_2}{dt} = -\frac{dU_1}{dt}. \tag{7.30}$$

Substituting (7.30) into (7.28) gives

$$\frac{dS}{dt} = \frac{dU_1}{dt}\left(\frac{\partial S_1}{\partial U_1} - \frac{\partial S_2}{\partial U_2}\right) > 0. \tag{7.31}$$

Using now the result (7.27), we can write

$$\frac{\partial S}{\partial U} = \frac{k}{\theta}. \tag{7.32}$$

Substituting (7.32) into (7.31) gives

$$\frac{dS}{dt} = k\frac{dU_1}{dt}\left(\frac{1}{\theta_1} - \frac{1}{\theta_2}\right) > 0. \tag{7.33}$$

Using the result (7.33), we can make the analysis of the two possible cases as follows:

1. If we had $\theta_1 > \theta_2$ before the two systems were merged then, in order to satisfy the inequality (7.33), we must have

$$\frac{dU_1}{dt} < 0, \quad \theta_1 > \theta_2. \tag{7.34}$$

2. On the other hand, if we had $\theta_1 < \theta_2$ before the two systems were merged, then, in order to satisfy the inequality (7.33), we must have

$$\frac{dU_1}{dt} > 0, \quad \theta_1 < \theta_2. \tag{7.35}$$

The results (7.34) and (7.35) show that the internal energy of the system with the lower value of θ increases while the internal energy of the system with the higher value of θ decreases. By comparison with common experience, we see that the parameter θ, with the dimension of energy, behaves in the same way as temperature of physical macro-objects. In fact, the parameter θ is proportional to the temperature of a thermodynamic system, and we can write

$$\theta = kT, \quad k = 1.3807 \cdot 10^{-23}\frac{J}{K}, \tag{7.36}$$

where the constant k is called the Boltzmann constant. The relation (7.36) was established by comparing the phenomenological ideal gas law with the corresponding statistical mechanics result, which will be derived in Chapter 11, "Ideal Monoatomic

Gases." The final result in (7.24) for the probability of finding a constituent particle in the state with energy $(E_i, E_i + \Delta E_i)$, is thus given by

$$\omega_i = \frac{g_i}{Z} \exp\left(-\frac{E_i}{kT}\right). \tag{7.37}$$

The probability (7.37) is subject to the normalization condition

$$\sum_i \omega_i = \frac{1}{Z} \sum_i g_i \exp\left(-\frac{E_i}{kT}\right) = \frac{Z}{Z} = 1, \tag{7.38}$$

where the partition function Z can be rewritten as follows:

$$Z = \sum_i g_i \exp\left(-\frac{E_i}{kT}\right). \tag{7.39}$$

Now, we can identify the constant of proportionality in the definition of entropy (6.3) as the Boltzmann constant specified by (7.36). Substituting now (7.36) into (7.27), we obtain the following expression for the entropy of the system

$$S = \frac{U}{T} + kN \ln \frac{Z}{N} + kN. \tag{7.40}$$

7.4   Problems with Solutions

Problem 1

Calculate the partition function Z, the internal energy U, and the entropy S for a system consisting of N indistinguishable constituent particles, which can only occupy two nondegenerate quantum levels with energies $E_1 = 0$ and $E_2 = E$.

Solution

The partition function for this simple system is given by

$$Z = \sum_i g_i \exp\left(-\frac{E_i}{kT}\right) = 1 + \exp\left(-\frac{E}{kT}\right). \tag{7.41}$$

The number of constituent particles with energy $E_2 = E$ ($g_2 = 1$) is

$$n_2 = g_2 \frac{N}{Z} \exp\left(-\frac{E_2}{kT}\right) = \frac{N}{Z} \exp\left(-\frac{E}{kT}\right), \tag{7.42}$$

or

$$n_2 = \frac{N}{\exp\left(\frac{E}{kT}\right) + 1}. \tag{7.43}$$

The internal energy of the system is then given by

$$U = \sum_i n_i E_i = n_1 E_1 + n_2 E_2 = n_2 E_2, \tag{7.44}$$

or

$$U = \frac{NE}{\exp\left(\frac{E}{kT}\right) + 1}. \tag{7.45}$$

The entropy of the system is given by

$$S = \frac{U}{T} + kN \ln \frac{Z}{N} + kN, \tag{7.46}$$

or

$$S = \frac{NE}{T} \frac{1}{\exp\left(\frac{E}{kT}\right) + 1} + kN \ln \frac{1 + \exp\left(-\frac{E}{kT}\right)}{N} + kN. \tag{7.47}$$

Problem 2

Calculate the partition function Z, the internal energy U, and the entropy S for a system consisting of N indistinguishable constituent particles, which can only occupy two nondegenerate quantum levels with energies $E_1 = -E$ and $E_2 = +E$.

Solution

The partition function for this simple system is given by

$$Z = \sum_i g_i \exp\left(-\frac{E_i}{kT}\right) = \exp\left(\frac{E}{kT}\right) + \exp\left(-\frac{E}{kT}\right), \tag{7.48}$$

or

$$Z = 2 \cosh\left(\frac{E}{kT}\right). \tag{7.49}$$

The numbers of constituent particles with energies $E_1 = -E$ ($g_1 = 1$) and $E_2 = +E$ ($g_2 = 1$) are given by

$$n_1 = \frac{N}{Z} \exp\left(\frac{E}{kT}\right), \quad n_2 = \frac{N}{Z} \exp\left(-\frac{E}{kT}\right). \tag{7.50}$$

The internal energy of the system is then given by

$$U = \sum_i n_i E_i = n_1 E_1 + n_2 E_2. \tag{7.51}$$

Using the results for the occupation numbers (7.50), we obtain

$$U = \frac{NE}{Z}\left[-\exp\left(\frac{E}{kT}\right) + \exp\left(-\frac{E}{kT}\right)\right], \tag{7.52}$$

or

$$U = -2\,\frac{NE}{Z}\sinh\left(\frac{E}{kT}\right), \tag{7.53}$$

or

$$U = -NE\tanh\left(\frac{E}{kT}\right). \tag{7.54}$$

The entropy of the system is given by

$$S = \frac{U}{T} + kN\ln\frac{Z}{N} + kN, \tag{7.55}$$

or

$$S = -\frac{NE}{T}\tanh\left(\frac{E}{kT}\right) + kN\ln\frac{2\cosh\left(\frac{E}{kT}\right)}{N} + kN. \tag{7.56}$$

Problem 3

Consider a system with only two accessible energy levels E_1 and E_2 at a constant temperature $T = 150\,\text{K}$. The energy gap between the two levels is $\Delta E = E_2 - E_1 = 3.2 \times 10^{-21}\,\text{J}$. The lower level is nondegenerate, while the upper level is twofold degenerate. Determine the probability that the lower level is occupied.

Solution

The partition function for this system is given by

$$Z = \exp\left(-\frac{E_1}{kT}\right) + 2\exp\left(-\frac{E_2}{kT}\right), \tag{7.57}$$

where $E_2 = E_1 + \Delta E$ and the factor of 2 in the second term is due to the twofold degeneracy of the upper energy level. Thus, we have

$$Z = \exp\left(-\frac{E_1}{kT}\right) + 2\exp\left(-\frac{E_1 + \Delta E}{kT}\right), \tag{7.58}$$

or

$$Z = \exp\left(-\frac{E_1}{kT}\right)\left[1 + 2\exp\left(-\frac{\Delta E}{kT}\right)\right]. \tag{7.59}$$

Thus, by equation (7.24), the probability that the lower level is occupied is

$$P_1 = \frac{\exp\left(-\frac{E_1}{kT}\right)}{Z} = \frac{1}{1 + 2\exp\left(-\frac{\Delta E}{kT}\right)} \approx 0.696. \tag{7.60}$$

Problem 4

The number of accessible states W of a system with internal energy U occupying the volume V is given by

$$W = a\exp\left(b\sqrt{UV}\right), \tag{7.61}$$

where a and b are constants. Determine the entropy as a function of temperature T, volume V, and the constants a and b.

Solution

The entropy, as a function of internal energy U and volume V, is readily obtained from the definition $S = k\ln W$, yielding here

$$S = k\ln a + bk\sqrt{UV}. \tag{7.62}$$

Using now the equation (7.32) with (7.36), i.e.,

$$\frac{\partial S}{\partial U} = \frac{1}{T}, \tag{7.63}$$

we obtain

$$\frac{1}{T} = \frac{\partial S}{\partial U} = \frac{bk}{2}\sqrt{\frac{V}{U}}. \tag{7.64}$$

Using the result (7.64), we can express the internal energy U as a function of temperature T as follows:

$$U = V\left(\frac{bkT}{2}\right)^2. \tag{7.65}$$

Substituting (7.65) into (7.62), we obtain the entropy as a function of temperature T, volume V, and the constants a and b as follows:

$$S = k\ln a + \frac{b^2 k^2}{2}TV. \tag{7.66}$$

Problem 5

The entropy of a system, consisting of N constituent particles within a volume V having the internal energy U, is given by the expression

$$S(N, U, V) = a - b\frac{N^2 V^2}{U^3}, \tag{7.67}$$

where a and b are constants. Derive the expression for internal energy U as a function of temperature T, volume V, number of particles N, and the constants a and b.

Solution

Using the result (7.63), we have

$$\frac{1}{T} = \frac{\partial S}{\partial U} = 3b\frac{N^2 V^2}{U^4}. \tag{7.68}$$

Rearranging the result (7.68), we obtain the following expression for internal energy U as a function of temperature T, volume V, number of particles N, and the constants a and b:

$$U = (3bT)^{1/4} \sqrt{NV}. \tag{7.69}$$

8 Thermodynamic Variables

8.1 Free Energy and the Partition Function

The equation (7.40) in Chapter 7, "Equilibrium States of Systems," specified the entropy S as the function of the internal energy U, temperature T, the partition function Z, and the total number of particles N. Unlike the internal energy U and the temperature T, the partition function Z and the total number of particles N are not best suited quantities for the macroscopic description of a thermodynamic system. In thermodynamics, it is, therefore, customary to introduce a new macroscopic variable called the *free energy* F. For an ideal system of N indistinguishable particles discussed in Chapter 6, "Number of Accessible States and Entropy," and Chapter 7, the free energy is defined as follows:

$$F = -NkT \ln \frac{Z}{N} - NkT. \tag{8.1}$$

Substituting the definition (8.1) into (7.40) and rearranging, we obtain an important macroscopic definition of the free energy of a thermodynamic system as follows:

$$S = \frac{U}{T} - \frac{F}{T} = \frac{U - F}{T} \Rightarrow F = U - TS. \tag{8.2}$$

Free energy F in (8.1) has a significant dependence on the number of particles N. Of special interest for the forthcoming discussions is the change of the free energy upon changing N by $\Delta N = 1$, i.e., upon adding one particle to the system, $\Delta F = F(N+1) - F(N)$. For a large $N \gg 1$, one can approximate this change as follows:

$$\Delta F = \frac{\partial F}{\partial N} \Delta N = \frac{\partial F}{\partial N} = \mu, \tag{8.3}$$

where we define the so-called *chemical potential* μ by $\mu = \partial F / \partial N$. For the case of ideal systems, with the free energy as in (8.1), a simple differentiation gives

$$\mu = \frac{\partial F}{\partial N} = -kT \ln \frac{Z}{N} + NkT \frac{1}{N} - kT = -kT \ln \frac{Z}{N}. \tag{8.4}$$

From (8.4), we readily obtain

$$Z = N \exp\left(-\frac{\mu}{kT}\right). \tag{8.5}$$

Introductory Statistical Thermodynamics

Substituting (8.5) into (7.37), we obtain an alternative result for the probability of finding a constituent particle with energy within the interval $(E_i, E_i + \Delta E_i)$ as follows:

$$\omega_i = \frac{n_i}{N} = \frac{g_i}{N} \exp\left(\frac{\mu - E_i}{kT}\right). \tag{8.6}$$

Substituting (8.5) into (7.23) (with $\theta = kT$), we obtain

$$n_i = g_i \exp\left(\frac{\mu - E_i}{kT}\right). \tag{8.7}$$

So, the mean (average) number of particles occupying a given sublevel is

$$\bar{n}_i = \frac{n_i}{g_i} = \exp\left(\frac{\mu - E_i}{kT}\right). \tag{8.8}$$

The result (8.8) is known as the *Maxwell–Boltzmann formula*. We remark that the chemical potential μ is of particular significance for studying the systems with variable numbers of particles. Such systems will be discussed in more detail in Chapter 10, "Variable Number of Particles."

The free energy F of a thermodynamic system can also be defined in a more general way, which is suitable for the analysis of the systems of interacting particles. In order to obtain such a generalized definition of the free energy, let us rewrite the equation (8.1) as follows:

$$F = -kT \left(\ln Z^N - N \ln N + N\right). \tag{8.9}$$

Using here the Stirling formula (7.8), we obtain

$$F = -kT \left(\ln Z^N - \ln N!\right) = -kT \ln \frac{Z^N}{N!}. \tag{8.10}$$

The partition function $Z = Z_1$ was derived as a sum over the states of a *single* constituent particle of the system. On the other hand, the states of the N noninteracting constituent particles of the system are independent of each other. The quantity Z^N is a potential candidate for the partition function Z_N, which takes into account the entire system of N noninteracting particles. To see this, consider the case $N = 2$, i.e.,

$$Z^2 = \left[\sum_i g_i \exp\left(-\frac{E_i}{kT}\right)\right] \times \left[\sum_j g_j \exp\left(-\frac{E_j}{kT}\right)\right]$$

$$= \sum_i \sum_j g_i g_j \exp\left(-\frac{E_i + E_j}{kT}\right). \tag{8.11}$$

By equation (8.11), we see that Z^2 is essentially a sum over *two-particle* states, with the first particle at the level E_i and the second particle at the level E_j. However, for the indistinguishable particles, the two-particle state with the first particle at the level E_i and the second particle at the level E_j is physically the same as the two-particle state with the first particle at the level E_j and the second particle at the level E_i. Because of this, the equation (8.11) makes an overcount of the two-particle states. To correct for this overcount, the two-particle partition function is defined by

$$Z_2 = \frac{Z^2}{2!}, \tag{8.12}$$

where the factor 2! corrects the overcount of states. Analogously, for the thermodynamic system consisting of N particles, the product Z^N must be divided by the number of permutations $N!$ in order to obtain the corrected N-particle partition function. Thus, the N-particle grand partition function Z_N for the system of noninteracting particles is defined by

$$Z_N = \frac{Z^N}{N!}. \tag{8.13}$$

Substituting the definition (8.13) into (8.10), we obtain the most general definition of the free energy of the system

$$F = -kT \ln Z_N. \tag{8.14}$$

The definition (8.14) is applicable even to the nonideal systems of particles with arbitrary interparticle interactions, where the partition function Z_N is not given by the simple definition (8.13). In the this discussion, we are concerned with the ideal systems of non-interacting particles, and the expression (8.1) for the free energy of the N-particle system, employing the one-particle partition function Z, is sufficient for the complete analysis of the system.

It should also be noted that there are some exceptions from the result (8.13), i.e., the physical situations where $Z_N = Z^N$ and no division by $N!$ is required. This applies particularly to the heavy particles (atoms) that form a solid. The wave functions of such heavy particles are sharply peaked at lattice sites, while they are practically zero at the position of the neighboring lattice sites. Thus, the probability of the exchange permuting such particles between any two sites is negligible. Thereby, the particles of the system, although indistinguishable in principle, become distinguishable by being localized to distinguishable lattice sites. For such situations, the division by $N!$ in (8.13) is not appropriate. Thus, $Z_N = Z^N$, and for such systems, the free energy becomes

$$F = -kT \ln Z_N = -NkT \ln Z. \tag{8.15}$$

The free energy F and consequently the entropy S for such systems is simply proportional to the number of constituent particles N. The physical systems, for which the result (8.15) is directly applicable, include the mutually noninteracting spins localized on different lattice sites (a paramagnetic system). As long as we work with the distinguishable particles bound to N different lattice sites, it is legitimate to calculate the total partition function as a simple product of N single-particle partition functions. Because of this, the free energy in equation (8.15) is just the sum of the N single-particle contributions to the free energy.

The partition function Z contains a complete information about a thermodynamic system. If we know the partition function Z, we can calculate all the other thermodynamic quantities of the system. It is, therefore, of importance to analyze the properties of the partition function of our thermodynamic system. From the definition of the partition function (7.21) with (7.36), we see that the partition function Z is a function of the temperature of the system T. Furthermore, the partition function is a function of the energy levels of a constituent particle E_i and the number of sublevels g_i. As an example, let us now consider an ideal gas, consisting of particles with translational degrees of freedom only. From the result for the translational kinetic energy of a constituent particle (3.43), i.e.,

$$E_i = \frac{h^2}{8mV^{2/3}} \left(n_{ix}^2 + n_{iy}^2 + n_{iz}^2 \right), \tag{8.16}$$

we see that the partition function Z is also a function of the volume V of the system. In other words, the quantities E_i and g_i, and consequently the partition function Z, change if the volume of the system changes. Thus, we can write

$$Z = Z(T, V) \Rightarrow \ln Z = \ln Z(T, V), \tag{8.17}$$

and

$$d(\ln Z) = \frac{\partial}{\partial T}(\ln Z)_V \, dT + \frac{\partial}{\partial V}(\ln Z)_T \, dV. \tag{8.18}$$

The result (8.18) will be useful in the practical calculations of the thermodynamic quantities in the following section.

8.2 Internal Energy: Caloric-State Equation

The objective of this section is to derive an equation which relates the internal energy U and the partition function Z of a thermodynamic system. Such an equation describes the caloric properties of the system at the constant volume, and it is called the *caloric-state equation*. Using this equation, we will be able to eventually calculate experimentally significant quantities such as heat capacities and specific heats of materials. In order to derive the caloric-state equation, let us consider the expression

$$\frac{\partial}{\partial T}(\ln Z)_V = \frac{1}{Z}\left(\frac{\partial Z}{\partial T}\right)_V = \frac{1}{Z}\frac{\partial}{\partial T}\left[\sum_i g_i \exp\left(-\frac{E_i}{kT}\right)\right]_V$$

$$= \frac{1}{Z} \sum_i g_i \frac{\partial}{\partial T} \left[\exp\left(-\frac{E_i}{kT} \right) \right]_V = \frac{1}{Z} \sum_i g_i \frac{E_i}{kT^2} \exp\left(-\frac{E_i}{kT} \right)$$

$$= \frac{1}{NkT^2} \sum_i E_i g_i \frac{N}{Z} \exp\left(-\frac{E_i}{kT} \right). \tag{8.19}$$

If we now recall the result (7.23) with (7.36), i.e.,

$$n_i = g_i \frac{N}{Z} \exp\left(-\frac{E_i}{kT} \right), \tag{8.20}$$

we obtain

$$\frac{\partial}{\partial T} (\ln Z)_V = \frac{1}{NkT^2} \sum_i E_i n_i = \frac{U}{NkT^2}. \tag{8.21}$$

From (8.21), we derive the caloric-state equation in the form

$$U = NkT^2 \frac{\partial}{\partial T} (\ln Z)_V. \tag{8.22}$$

Using the result for the entropy of the system (7.40), i.e.,

$$S = \frac{U}{T} + Nk \ln Z - Nk \ln N + kN, \tag{8.23}$$

and the assumption that in our system the total number of constituent particles is constant ($N = $ Constant), we obtain the following result for the differential of the entropy of the system

$$dS = \frac{dU}{T} - \frac{U}{T^2} dT + Nk \, d(\ln Z). \tag{8.24}$$

Substituting (8.18) into (8.24) yields

$$dS = \frac{dU}{T} - \frac{U}{T^2} dT + Nk \frac{\partial}{\partial T} (\ln Z)_V \, dT + Nk \frac{\partial}{\partial V} (\ln Z)_T \, dV, \tag{8.25}$$

or

$$dS = \frac{dU}{T} - \frac{dT}{T^2} [U - NkT^2 \frac{\partial}{\partial T} (\ln Z)_V] + Nk \frac{\partial}{\partial V} (\ln Z)_T \, dV. \tag{8.26}$$

By the equation (8.22), the second term on the right-hand side of (8.26) vanishes and the differential of the entropy of the system becomes

$$dS = \frac{dU}{T} + Nk\frac{\partial}{\partial V}(\ln Z)_T\, dV. \tag{8.27}$$

The differential of the internal energy dU is now given by

$$dU = T\, dS - NkT\frac{\partial}{\partial V}(\ln Z)_T\, dV. \tag{8.28}$$

On the other hand, we know that the internal energy of the system U can be expressed as a function of entropy S and volume of the system V, i.e., we can write $U = U(S, V)$. Therefore, the differential of the internal energy can also be written in the following form

$$dU = \left(\frac{\partial U}{\partial S}\right)_V dS + \left(\frac{\partial U}{\partial V}\right)_S dV. \tag{8.29}$$

Comparing the expressions (8.28) and (8.29), we can readily derive the following two important relations between the thermodynamic quantities

$$\left(\frac{\partial U}{\partial S}\right)_V = T, \quad \left(\frac{\partial U}{\partial V}\right)_S = -NkT\frac{\partial}{\partial V}(\ln Z)_T. \tag{8.30}$$

8.3 Pressure: Thermal-State Equation

The objective of this section is to introduce the concept of pressure of a thermodynamic system and to derive an important equation, which relates the pressure p and the partition function Z of a thermodynamic system. Such an equation describes the thermal properties of the system at the constant temperature, and it is called the *thermal-state equation*. In order to derive the thermal-state equation, let us first consider the special case of a thermodynamic system enclosed by a cylinder with a movable piston, as shown in Fig. 8.1.

Here, we assume that the area of the piston is equal to A and that the piston can move in x-direction, as indicated in Fig. 8.1. When the piston moves, a certain amount of the mechanical work is performed. A detailed quantitative analysis of the system,

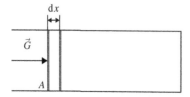

Figure 8.1 A system enclosed by the cylinder with a movable piston.

shown in Fig. 8.1, requires a convention regarding the sign of the mechanical work. A commonly adopted convention is that if the external force $\vec{G}$ performs work on the system (i.e., the system is compressed by an external force), then the mechanical work has a **negative sign**. Conversely, if the system performs work on the surroundings (i.e., the system expands), then the mechanical work has a **positive sign**.

When the piston moves for some distance dx, as indicated in Fig. 8.1, the mechanical work performed by the external force is given by

$$dL = G\,dx. \tag{8.31}$$

The pressure p on the piston, with the total area A, is defined as the force acting on unit area of the piston. As the force $\vec{G}$ acts along the direction of motion and it is evenly distributed over the area of the piston, we have

$$p = \frac{G}{A} \Rightarrow G = pA. \tag{8.32}$$

Substituting (8.32) and (8.31), we obtain

$$dL = pA\,dx = p\,dV. \tag{8.33}$$

Thus, the pressure on the system can be expressed as

$$p = \frac{dL}{dV}. \tag{8.34}$$

To proceed, let us consider a thermally insulated system that cannot exchange any thermal energy (heat) with its environment. By energy conservation, the work done by such system is done at the expense of its internal energy, $dL = -dU$. Thus, if the system expands, then internal energy will decrease and a positive amount of work will be done by the system. Conversely, when the system is compressed, it performs negative work and its internal energy increases. Thus, we have

$$dU = -dL = -p\,dV. \tag{8.35}$$

Here, we are interested in the processes that proceed very slowly so that the system is in a thermodynamic equilibrium at any time. If such a process is done infinitely slowly, quantum-mechanical considerations show that there are no changes of the occupation numbers n_i of the energy levels in an insulated system. On the other hand, the entropy is generally expressible in terms of these occupations numbers; see the equations giving entropy discussed in Chapters 6 and 7, such as the equation (7.25). This leads to the following important theorem: *If a thermodynamic process is done infinitely slowly in a thermally insulated system, then its entropy S must be constant in time.* This result is known as the adiabatic theorem, so the processes going on with constant entropy are called *adiabatic processes*. On the other hand, by equation (8.23) with constant entropy ($dS = 0$), we have

$$dU = \left(\frac{\partial U}{\partial V}\right)_S dV. \tag{8.36}$$

In the function $U = U(S, V)$, the entropy and the volume of the system are independent variables. Thus, by comparison of the equations (8.36) and (8.35), we obtain

$$p = -\left(\frac{\partial U}{\partial V}\right)_S. \tag{8.37}$$

Using now (8.30), we finally obtain the thermal-state equation in the following form

$$p = p(T, V) = -\left(\frac{\partial U}{\partial V}\right)_S = NkT\frac{\partial}{\partial V}(\ln Z)_T. \tag{8.38}$$

Substituting (8.38) into (8.28), we obtain the general result for the differential of the internal energy as follows:

$$dU = T\,dS - p\,dV. \tag{8.39}$$

The equation (8.39) represents one of the fundamental laws of thermodynamics, and its physical significance will be further emphasized in the coming chapters. Using the definition of the free energy (8.1), i.e.,

$$F = -NkT \ln \frac{Z}{N} - NkT. \tag{8.40}$$

we obtain

$$-\left(\frac{\partial F}{\partial V}\right)_T = NkT\frac{\partial}{\partial V}(\ln Z)_T = p, \tag{8.41}$$

or

$$p = -\left(\frac{\partial F}{\partial V}\right)_T. \tag{8.42}$$

The equation (8.42) defines the pressure of the system p in terms of the free energy F. Let us put the above result for the pressure in a broader context that will allow us to derive another important equation expressing the entropy S of the system in terms of the free energy F. Using the equation (8.2), we have

$$dF = d(U - TS) = dU - dTS - T\,dS. \tag{8.43}$$

Expressing here dU from the equation (8.39), we find

$$dF = T\,dS - p\,dV - S\,dT - T\,dS = -p\,dV - S\,dT. \tag{8.44}$$

Here, the differential of $F(V, T)$ can be generally written as

$$dF = \left(\frac{\partial F}{\partial V}\right)_T dV + \left(\frac{\partial F}{\partial T}\right)_V dT. \tag{8.45}$$

By comparing the above two expressions for dF, we find

$$p = -\left(\frac{\partial F}{\partial V}\right)_T, \quad S = -\left(\frac{\partial F}{\partial T}\right)_V. \tag{8.46}$$

Note that the equation (8.46) reproduces the above result for the pressure in equation (8.42), and, in addition, provides a relation that can be used to calculate the entropy S in terms of the free energy F. By using (8.46), the internal energy U can also be expressed purely in terms of the free energy F, via the equation

$$U = F + TS = F - T\left(\frac{\partial F}{\partial T}\right)_V. \tag{8.47}$$

We would like to stress that in writing the above derivatives, we assumed that the number of particles N is fixed. It is, however, easy to generalize the above discussion to the case of a variable number of particles N. Indeed, the differential of $F(V, T, N)$,

$$dF = \left(\frac{\partial F}{\partial V}\right)_{T,N} dV + \left(\frac{\partial F}{\partial T}\right)_{V,N} dT + \left(\frac{\partial F}{\partial N}\right)_{V,T} dN, \tag{8.48}$$

can, by invoking the equations (8.4) and (8.46), be written as

$$dF = -p\,dV - S\,dT + \mu\,dN \tag{8.49}$$

generalizing the above equation for dF to the case of variable number of particles, which is discussed in more detail in Chapter 10.

8.4 Classification of Thermodynamic Variables

All thermodynamic variables can be classified into the following three groups as follows:

1. The first group of thermodynamic variables specifies the amount of matter and includes the number of particles in the system N and the mass of the system m.

2. The second group of thermodynamic variables includes the so-called *extensive state variables*. The extensive state variables (e.g., volume V, internal energy U, and entropy S) are proportional to the mass of the system. Thus, we can write

$$V = v \cdot m, \quad U = u \cdot m, \quad S = s \cdot m, \tag{8.50}$$

where v is the so-called *specific volume* (reciprocal of mass density), u is the so-called *specific internal energy*, and s is the so-called *specific entropy*.

3. The third group of thermodynamic variables are the so-called *intensive state variables*. The intensive state variables (e.g., temperature T and pressure p) are independent on the total mass of the system for given value of system mass density (or specific volume). An intensive variable can always be calculated in terms of other intensive variables. For example, the pressure p of a gas or liquid can always be expressed as a function of its temperature T and specific volume v.

We also note that if the relation between the extensive state variables $U = U(S, V)$ is known, then the intensive state variables are easily calculated as follows:

$$T = \left(\frac{\partial U}{\partial S}\right)_V, \quad p = -\left(\frac{\partial U}{\partial V}\right)_S. \tag{8.51}$$

These two results follow directly from the equation (8.39).

8.5 Problems with Solutions

Problem 1

The partition function of a system is given by $Z = N \exp(bT^3 V)$, where b is a constant. Calculate the pressure p, the entropy S, and the internal energy U of the system.

Solution

The free energy is given by

$$F = -NkT \ln \frac{Z}{N} - NkT = -NbkT^4 V - NkT. \tag{8.52}$$

The pressure of the system is given by

$$p = -\left(\frac{\partial F}{\partial V}\right)_T = NbkT^4. \tag{8.53}$$

The entropy of the system is given by

$$S = -\left(\frac{\partial F}{\partial T}\right)_V = 4NbkT^3 V + kN. \tag{8.54}$$

The internal energy is obtained as $U = F + TS$. Substituting the results for free energy F and entropy S, we obtain

$$U = F + TS = -NbkT^4V - NkT + 4NbkT^4V + kNT,$$ (8.55)

or finally

$$U = 3NbkT^4V.$$ (8.56)

Problem 2

Calculate the free energy F for a system consisting of N constituent particles, which can only occupy two nondegenerate quantum states with energies $E_1 = -E$ and $E_2 = +E$. Using the obtained result for the free energy F, calculate the entropy S and the internal energy $U = F + TS$ of the system.

Solution

The partition function for this simple system is given by

$$Z = \exp\left(\frac{E}{kT}\right) + \exp\left(-\frac{E}{kT}\right) = 2\cosh\left(\frac{E}{kT}\right).$$ (8.57)

The free energy is given by

$$F = -NkT\ln\frac{Z}{N} - NkT,$$ (8.58)

or

$$F = -NkT\ln\frac{2\cosh\left(\frac{E}{kT}\right)}{N} - NkT.$$ (8.59)

The entropy of the system is obtained from the free energy as follows:

$$S = -\left(\frac{\partial F}{\partial T}\right)_V = Nk\ln\frac{2\cosh\left(\frac{E}{kT}\right)}{N} + NkT\frac{\partial}{\partial T}\left[\ln\frac{2\cosh\left(\frac{E}{kT}\right)}{N}\right]_V + kN,$$ (8.60)

or finally

$$S = kN\ln\frac{2\cosh\left(\frac{E}{kT}\right)}{N} - \frac{NE}{T}\tanh\left(\frac{E}{kT}\right) + kN.$$ (8.61)

The internal energy is obtained as $U = F + TS$. Substituting the results for free energy F and entropy S, we obtain

$$U = F + TS = -NE\tanh\left(\frac{E}{kT}\right).$$ (8.62)

The results obtained here, by means of the thermodynamic relations, agree with the results obtained using the generic definitions in the Problem 2 of the Chapter 7.

Problem 3

Consider a system consisting of N constituent particles, which can occupy four spin-$\frac{3}{2}$ quantum states, with energies:

$$E_1 = -\tfrac{3}{2}E$$
$$E_2 = -\tfrac{1}{2}E$$
$$E_3 = +\tfrac{1}{2}E \tag{8.63}$$
$$E_4 = +\tfrac{3}{2}E$$

and degeneracies 1, 3, 3, and 1, respectively. Calculate the partition function Z, the free energy F, the entropy S, and the internal energy $U = F + TS$ of the system.

Solution

The partition function of the system is given by

$$Z = \exp\left(\frac{3E}{2kT}\right) + 3\exp\left(\frac{E}{2kT}\right) + 3\exp\left(-\frac{E}{2kT}\right)$$
$$+ \exp\left(-\frac{3E}{2kT}\right) = \left[\exp\left(\frac{E}{2kT}\right) + \exp\left(-\frac{E}{2kT}\right)\right]^3, \tag{8.64}$$

or finally

$$Z = \left[2\cosh\left(\frac{E}{2kT}\right)\right]^3. \tag{8.65}$$

The free energy is then given by

$$F = -NkT\ln\frac{Z}{N} - NkT, \tag{8.66}$$

or finally

$$F = -3NkT\ln\left[2\cosh\left(\frac{E}{2kT}\right)\right] + NkT\ln N - NkT. \tag{8.67}$$

The entropy of the system is obtained from the free energy as follows:

$$S = -\left(\frac{\partial F}{\partial T}\right)_V = 3Nk\ln\left[2\cosh\left(\frac{E}{2kT}\right)\right]$$

$$+ 3NkT\frac{\partial}{\partial T}\left\{\ln\left[2\cosh\left(\frac{E}{2kT}\right)\right]\right\}_V - Nk\ln N + kN, \tag{8.68}$$

or finally

$$S = 3Nk\ln\left[2\cosh\left(\frac{E}{2kT}\right)\right] - Nk\ln N -$$

$$-\frac{3NE}{2T}\tanh\left(\frac{E}{2kT}\right) + kN. \tag{8.69}$$

The internal energy is obtained as $U = F + TS$. Substituting the results for free energy F and entropy S, we obtain

$$U = F + TS = -\frac{3}{2}NE\tanh\left(\frac{E}{2kT}\right). \tag{8.70}$$

Problem 4

A crystal, consisting of N heavy noninteracting spin-$\frac{1}{2}$ ions, is in a thermal equilibrium at the temperature $T = 1\,K$ and under the influence of a uniform magnetic field with magnetic induction $B = 10\,T$. The presence of the magnetic field gives rise to a two-level quantum system, with energies $E_1 = -\mu B$ and $E_2 = +\mu B$.

(a) Derive the expressions for the partition function Z_N and the entropy S of this system.

(b) If the magnetic induction is adiabatically ($\Delta S = 0$) reduced to $B = 10^{-2}T$, what is the final temperature of the spins?

Solution

The partition function for a single ion in a two-level quantum-mechanical system is given by

$$Z = \exp\left(\frac{\mu B}{kT}\right) + \exp\left(-\frac{\mu B}{kT}\right). \tag{8.71}$$

In this case, we have a crystal consisting of heavy ions at very low temperatures. Thus, the exchange of particles between the lattice sites is practically impossible, and we have a physical situation, where the partition function for N particles is given by $Z_N = Z^N$, and the formula (8.15) for the free energy is applicable. Thus, for the system of N ions, the total partition function is simply

$$Z_N = \left[\exp\left(\frac{\mu B}{kT}\right) + \exp\left(-\frac{\mu B}{kT}\right) \right]^N. \tag{8.72}$$

The free energy $F = -kT \ln Z_N$ is then given by

$$F = -NkT \ln \left[\exp\left(\frac{\mu B}{kT}\right) + \exp\left(-\frac{\mu B}{kT}\right) \right], \tag{8.73}$$

or

$$F = -NkT \ln \left[2\cosh\left(\frac{\mu B}{kT}\right) \right]. \tag{8.74}$$

The entropy S can be calculated as follows:

$$S = -\left(\frac{\partial F}{\partial T}\right)_V = Nk \ln \left[2\cosh\left(\frac{\mu B}{kT}\right) \right]$$
$$+ NkT \frac{\partial}{\partial T} \left\{ \ln \left[2\cosh\left(\frac{\mu B}{kT}\right) \right] \right\}_V, \tag{8.75}$$

or

$$S = Nk \left\{ \ln \left[2\cosh\left(\frac{\mu B}{kT}\right) \right] - \frac{\mu B}{kT} \tanh\left(\frac{\mu B}{kT}\right) \right\}. \tag{8.76}$$

Introducing here a new variable $x = \mu B / kT$, we see that the entropy is a function of the variable x only, i.e.,

$$S(x) = Nk [\ln (2\cosh x) - x \tanh x]. \tag{8.77}$$

Since the entropy is a function of x only, and it is assumed to be constant in the adiabatic process of reducing the magnetic field, it implies that the quantity x must remain constant during the process. Therefore, we must have

$$x_1 = \frac{\mu B_1}{kT_1} = \frac{\mu B_2}{kT_2} = x_2 = \text{Constant}. \tag{8.78}$$

Thus, the final temperature of the spins is

$$T_2 = \frac{B_2}{B_1} T_1 = \frac{10^{-2}\,\text{T}}{10\,\text{T}} \times 1\text{K} = 10^{-3}\text{K}. \tag{8.79}$$

Problem 5

The partition function of a system, as a function of temperature T, volume V, and particle number N, is given by

$$Z = NbV(kT)^{5/2}, \tag{8.80}$$

where b is a constant. Calculate the free energy F, pressure p, entropy S, and the internal energy U of the system.

Solution

The free energy is given by

$$F = -NkT \ln \frac{Z}{N} = -NkT \ln \left[bV(kT)^{5/2} \right] - NkT.$$
(8.81)

The pressure of the system is given by

$$p = -\left(\frac{\partial F}{\partial V} \right)_T = \frac{NkT}{V}.$$
(8.82)

The entropy of the system is given by

$$S = -\left(\frac{\partial F}{\partial T} \right)_V = Nk \ln \left[bV(kT)^{5/2} \right] + \frac{5}{2}Nk + kN.$$
(8.83)

The internal energy is obtained as $U = F + TS$. Substituting the results for free energy F and entropy S, we obtain

$$U = F + TS = \frac{5}{2}NkT.$$
(8.84)

It will be shown later that this partition function describes the classical diatomic ideal gas with five degrees of freedom. Three degrees of freedom are the usual translational degrees of freedom, while the remaining two degrees of freedom correspond to the rotations of the diatomic molecules of the gas.

9 Macroscopic Thermodynamics

9.1 Changes of States. Heat and Work

In practical engineering thermodynamics, one is usually less interested in the equilibrium states of thermodynamic systems. The most interesting phenomena are thermodynamic processes, namely, the changes of the state of a macroscopic, thermodynamic system. If this change is sufficiently slow, it can be considered as a passage through a continuous succession of equilibrium states. Such a change of state is called the *quasi-static* change. A quasi-static change of state is often also a *reversible* change of state, since after such a change, it is possible to return to the initial state passing through the same succession of intermediate states in reversed order. It should, however, be noted that there are thermodynamically irreversible quasi-static processes, e.g., a slow motion of an object on a rough surface in which the work done against friction transforms into irreversibly lost heat energy. In reality, all thermodynamic processes are to some extent irreversible. In practice, however, many quasi-static processes in nature are nearly reversible and the thermodynamics of reversible processes is sufficient for the description of such processes.

Let us now consider an arbitrary thermodynamic process, where the state of the system changes from an initial state (1) to some final state (2), as depicted in the p-V plane shown in Fig. 9.1. The change of the internal energy in this process is given by

$$dU = d\left(\sum_i n_i E_i\right) = \sum_i E_i dn_i + \sum_i n_i dE_i. \tag{9.1}$$

The second term on the right-hand side of the equation (9.1) is zero for processes going on at constant volume. For them, we have $dE_i = 0$.

Indeed, recall that energy levels depend on the volume occupied by the system; see, for example, the equation (3.43), for the case of an ideal gas. When the volume of a system changes, the energy levels E_i also change, i.e., dE_i is nonzero making the second term in the equation (9.1) nonzero. This part of the internal energy change in (9.1) is closely related to the work performed by the system discussed in Chapter 8, "Thermodynamic Variables." In an infinitesimal change of state, the mechanical work performed by the system is given by

$$dL = pdV = -\sum_i n_i dE_i, \tag{9.2}$$

Introductory Statistical Thermodynamics

p

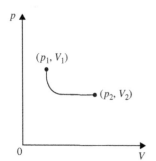

Figure 9.1 A change of state in the p-V plane.

(p_1, V_1)

(p_2, V_2)

0

V

or

$$dL = -\left(\sum_i n_i \frac{\partial E_i}{\partial V}\right)dV = -\left(\frac{\partial U}{\partial V}\right)_S dV. \tag{9.3}$$

Above, we used the fact that n_i = Constant at S = Constant (recall the Adiabatic theorem discussed in Section 8.3, "Pressure: Thermal-state Equation") such that the variation of internal energy U at constant S comes only from the variation of energy levels E_i. The equations (9.2) and (9.3) above are in accordance with the equations (8.34) and (8.35) of Chapter 8. The work performed during a finite change of state is given by

$$L_{12} = \int_1^2 p\,dV. \tag{9.4}$$

In addition to the above discussed internal energy changes at constant S, i.e., n_i = Constant, the equation (9.1) contains also a term reflecting internal energy change due to the change of the occupation numbers n_i, i.e., due to the change of entropy S. This term of the equation (9.1) is the internal energy change due to the heat energy dQ added to the system from its environment, which is of the form

$$dQ = \sum_i E_i dn_i = T dS = \left(\sum_i E_i \frac{\partial n_i}{\partial S}\right)dS = \left(\frac{\partial U}{\partial S}\right)_V dS, \tag{9.5}$$

in accordance with the first equation (8.28). By the equations (9.1), (9.2), and (9.5), we obtain the general relation

$$dU = dQ - dL = T dS - dL \Rightarrow dQ = T dS = dU + dL. \tag{9.6}$$

As it stands, this relation, known as the *first law of thermodynamics*, states that the heat energy dQ added to a system is in part used to increase its internal energy by the amount dU, whereas the remaining part of the added heat energy is used *by the system* to perform the work dL. The heat exchanged during a finite change of state is given by

$$Q_{12} = \int_1^2 T dS. \tag{9.7}$$

Using the equations (9.4) and (9.6), that is, by simply integrating the above displayed relation $dU = dQ - dL$, we obtain the total change of the internal energy of the system during a finite change of state as follows:

$$U_{12} = Q_{12} - L_{12}, \tag{9.8}$$

where we have used the usual convention for the sign of the mechanical work, as in the equation (8.39). The result (9.8) is the mathematical statement of the energy conservation in the First Law of Thermodynamics.

9.2   Laws of Thermodynamics

In the early days of the development of thermodynamics, the four empirical statements were postulated as four *laws of thermodynamics*. These four laws are readily derivable from the microscopic principles discussed here and in the previous three chapters. However, it is appropriate to make the explicit statements of these four laws and discuss their meaning in some detail here.

9.2.1   Zeroth Law of Thermodynamics

If two bodies A and B are separately in equilibrium with a third body C, then the two bodies A and B are in equilibrium with each other. An instance of this is the so-called *thermal equilibrium*, with any two bodies in contact having the same temperature (if in equilibrium). The zeroth law then reduces to the seemingly simple fact that if $T_A = T_C$ and $T_B = T_C$ (that is, if A is in thermal equilibrium with C, and if B is in thermal equilibrium with C), then $T_A = T_B$ (that is, A is in thermal equilibrium with B). Similar statements can be also made about the so-called *mechanical equilibrium* (with two bodies in equilibrium having the same pressure), as well as about the so-called *chemical equilibrium* (with two bodies in equilibrium having the same chemical potential, if allowed to exchange constituent particles).

9.2.2   First Law of Thermodynamics

The change of the internal energy of a system, in the course of some thermodynamic process, is equal to the sum of the heat added to the system during the process and the work performed *on the system* during the process. The first law of thermodynamics is simply the energy conservation law (9.8), i.e., $U_{12} = Q_{12} - L_{12}$. Frequently, this equation is written in the equivalent form, $Q_{12} = U_{12} + L_{12}$, which is read as: during a thermodynamic process, the heat energy added to a system is the sum of its internal energy change and the work performed *by the system*.

9.2.3 Second Law of Thermodynamics

A thermodynamic process, whose only consequence is a decrease of the entropy of the system, is impossible. In other words, in the course of an arbitrary thermodynamic process, the entropy of the system can only increase or remain the same

$$\Delta S = S_{\text{final}} - S_{\text{initial}} \geq 0. \tag{9.9}$$

In the application of the second law to a system composed of several parts (objects), it should be stressed that the S in (9.9) is the entropy of the *entire system*, whence the law states that this entropy must not decrease. In contrast to this, the entropy of individual parts of this system can in principle decrease (or increase). A simple example for this is the thermal equilibration process involving two objects of initially different temperatures (one warmer and one colder) placed in contact with each other. This composite system eventually reaches a final temperature such that the final entropy of the entire system is higher than the initial entropy (in accordance with the second law). However, the entropy of the initially warmer object still decreases during this process simply because its temperature decreases during its cooling toward the final temperature (which is in-between the initial temperatures of the two objects).

9.2.4 Third Law of Thermodynamics

The entropy of a physical system approaches the zero value as the temperature approaches absolute zero ($T = 0\,\text{K}$). Our discussion made so far may suggest that the entropy of a system is a quantity defined up to an arbitrary additive constant (like, say, the electric field potential in electrostatic problems). For example, the equilibrium condition $dS = 0$, discussed in Chapter 6, "Number of Accessible States and Entropy," is satisfied not only by $S = k \cdot \ln W$ but also by the function

$$S = k \cdot \ln W + S_0. \tag{9.10}$$

Interestingly enough, quantum mechanics suggests that the natural choice of the constant S_0 in the equation (9.10) is simply zero. That is, quantum mechanics suggests that the expression for entropy should be written simply as we did before, just by writing

$$S = k \cdot \ln W. \tag{9.11}$$

The reason for this preferred choice stems from a fundamental fact that realistic physical systems can occupy only a single microstate at zero absolute temperature, $T = 0$ (as discussed hereafter). Thus, $W = 1$ at $T = 0$, so the entropy, as defined in (9.11), vanishes at $T = 0$.

In order to explain why $W \to 1$ as $T \to 0$, let us explore the behavior of our system at low temperatures. The number of constituent particles of the system at an arbitrary energy level E_i, is given by (8.20), i.e.,

$$n_i = g_i \frac{N}{Z} \exp\left(-\frac{E_i}{kT}\right). \tag{9.12}$$

The number of constituent particles of the system in the state with the lowest energy $E = E_0$ (ground state) is then equal to

$$n_0 = g_0 \frac{N}{Z} \exp\left(-\frac{E_0}{kT}\right). \tag{9.13}$$

From the results (9.12) and (9.13), we get the ratio

$$\frac{n_i}{n_0} = \frac{g_i}{g_0} \exp\left(-\frac{E_i - E_0}{kT}\right). \tag{9.14}$$

On the other hand, since E_0 is the lowest energy level, one has $E_i - E_0 > 0$ for every other level. Because of this, the exponential on the right-hand side of the equation (9.14) approaches zero as the temperature approaches the absolute zero ($T \to 0$). In other words, for $T = 0$, we have

$$\frac{n_i}{n_0} = 0, \tag{9.15}$$

for any state other than the ground state. The result (9.15) shows that all of the constituent particles of the system are in the ground state, i.e., in a *single state* at $T = 0$. That is, $W = 1$ at $T = 0$, as argued above. Thus, by equation (9.11), we have the conclusion

$$\lim_{T \to 0} S = 0, \tag{9.16}$$

which is just the third law of thermodynamics expressed in mathematical terms.

9.3 Open Systems

Thermodynamic systems in which some working substance enters the system in some initial state and leaves the system in some final state are called *open systems*. In other words, open systems are thermodynamic systems with mass flow. Examples of such systems are turbine systems used in jet engines and power stations. These are open systems, as indicated in Fig. 9.2.

As depicted in Fig. 9.2, the working substance enters the engine in the state (p_1, V_1, T_1) and leaves the engine in the state (p_2, V_2, T_2). The function of the engine is to perform a technologically useful work, which we denote by L_{e12}. In order to make

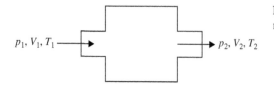

Figure 9.2 An open system with mass flow.

the working substance to enter the engine, the work $p_1 \cdot V_1 < 0$ must be performed. On the other hand to extract the working substance from the engine, it must perform the work $p_2 \cdot V_2 > 0$. The total change of the internal energy of the system is then given by

$$U_2 - U_1 = Q_{12} - L_{e12} - (p_2 V_2 - p_1 V_1), \qquad (9.17)$$

where L_{e12} is the technologically useful work extracted from the engine.

The equation (9.17) is just the energy conservation law. The increase of the internal energy of the system during the change of state from 1 to 2 is equal to the heat added to the engine Q_{12} reduced by the technologically useful work extracted from the engine L_{e12} and the total work performed by the engine to extract and enter the working substance $L_{12} = p_2 V_2 - p_1 V_1$. The equation (9.17) can be rewritten in the following way

$$(U_2 + p_2 V_2) - (U_1 + p_1 V_1) = Q_{12} - L_{e12}. \qquad (9.18)$$

It is now convenient to define a new extensive thermodynamic variable called the *enthalpy* or the *heat function* as follows:

$$H = U + pV. \qquad (9.19)$$

Substituting the definition (9.19) into the equation (9.18), we obtain

$$H_{12} = H_2 - H_1 = Q_{12} - L_{e12}. \qquad (9.20)$$

In the differential form, this equation becomes

$$dH = dQ - dL_e. \qquad (9.21)$$

The differential of the technologically useful work can now be calculated as follows:

$$dL_e = dQ - dH = dQ - d(U + pV)$$
$$= (dQ - pdV) - dU - Vdp = -Vdp, \qquad (9.22)$$

where we have used the result $dU = dQ - pdV$. The total technically useful work extracted from the engine during the change of state from 1 to 2 is then given by

$$L_e = -\int_1^2 Vdp. \qquad (9.23)$$

Figure 9.3 The technologically useful work in the p-V plane.

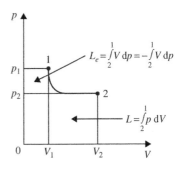

The $p - V$ diagram, which illustrates the meaning of the technically useful work, is shown in the Fig. 9.3.

9.4   Thermal Properties of Systems

The thermal properties of systems are determined by the thermal-state equation $p = p(T, V)$. Solving the thermal-state equation for temperature, we obtain $T = T(p, V)$, and the differential of temperature is given by

$$dT = \left(\frac{\partial T}{\partial p}\right)_V dp + \left(\frac{\partial T}{\partial V}\right)_P dV. \tag{9.24}$$

Using the equation (9.24), we can consider a number of different processes and define the corresponding thermal coefficients suitable for the description of these processes.

9.4.1   Isobaric Expansion

If the temperature of a thermodynamic system changes, while the pressure of the system remains constant ($dp = 0$), the volume of the system changes as well. Specifically, if the system is heated at constant pressure, the volume of the system increases. Conversely, if the system is cooled down at constant pressure, the volume of the system decreases. Using the equation (9.24) with constant pressure ($dp = 0$), we readily obtain the relation

$$dV = \left(\frac{\partial V}{\partial T}\right)_P dT = \alpha V_0 T, \tag{9.25}$$

where we defined the *volume expansion coefficient* as follows:

$$\alpha = \frac{1}{V_0} \left(\frac{\partial V}{\partial T}\right)_P. \tag{9.26}$$

For a finite change of temperature ΔT, we therefore have

$$\Delta V = \left(\frac{\partial V}{\partial T}\right)_P \Delta T = \alpha V_0 \Delta T. \tag{9.27}$$

9.4.2 Isochoric Expansion

If the temperature of a thermodynamic system changes, while the volume of the system remains constant ($dV = 0$), the pressure of the system changes as well. Specifically, if the system is heated at a constant volume, the pressure of the system increases. Conversely, if the system is cooled down within the constant volume, the pressure of the system decreases. Using the equation (9.24) with constant pressure ($dV = 0$), we readily obtain the relation

$$dp = \left(\frac{\partial p}{\partial T}\right)_V dT = \beta p_0 dT, \tag{9.28}$$

where we defined the *isochoric compressibility coefficient* as follows:

$$\beta = \frac{1}{p_0} \left(\frac{\partial p}{\partial T}\right)_V. \tag{9.29}$$

For a finite change of temperature ΔT, we therefore have

$$\Delta p = \left(\frac{\partial p}{\partial T}\right)_V \Delta T = \beta p_0 \Delta T. \tag{9.30}$$

9.4.3 Isothermal Expansion

If the pressure of a thermodynamic system changes, while the temperature of the system remains constant ($dT = 0$), the volume of the system changes as well. Using the equation (9.24) with constant temperature ($dT = 0$), we readily obtain the relation

$$dT = \left(\frac{\partial T}{\partial p}\right)_V dp + \left(\frac{\partial T}{\partial V}\right)_P dV = 0, \tag{9.31}$$

or

$$dV = -\left(\frac{\partial T}{\partial p}\right)_V \left(\frac{\partial V}{\partial T}\right)_P dp. \tag{9.32}$$

In order to calculate the coefficient of proportionality between dV and dp in the relation (9.32), some mathematical prerequisites are needed. If we have two functions $u = u(x, y)$ and $v = v(x, y)$ of two independent variables x and y, we define the Jacobian of these two functions as follows:

$$\frac{\partial(u, v)}{\partial(x, y)} = \begin{vmatrix} \frac{\partial u}{\partial x} & \frac{\partial u}{\partial y} \\ \frac{\partial v}{\partial x} & \frac{\partial v}{\partial y} \end{vmatrix} = \frac{\partial u}{\partial x}\frac{\partial v}{\partial y} - \frac{\partial u}{\partial y}\frac{\partial v}{\partial x}. \tag{9.33}$$

Let us now assume that $u = u(x, y)$ is an arbitrary thermodynamic state function of two independent variables x and y. Using (9.33), we can write

$$\frac{\partial(u, y)}{\partial(x, y)} = \begin{vmatrix} \frac{\partial u}{\partial x} & \frac{\partial u}{\partial y} \\ \frac{\partial y}{\partial x} & \frac{\partial y}{\partial y} \end{vmatrix} = \begin{vmatrix} \frac{\partial u}{\partial x} & \frac{\partial u}{\partial y} \\ 0 & 1 \end{vmatrix} = \left(\frac{\partial u}{\partial x}\right)_y. \tag{9.34}$$

The Jacobian determinant, defined by (9.33), has the following mathematical properties

$$\frac{\partial(u, v)}{\partial(x, y)} = -\frac{\partial(v, u)}{\partial(x, y)}, \quad \frac{\partial(u, v)}{\partial(x, y)} = -\frac{\partial(u, v)}{\partial(y, x)}, \tag{9.35}$$

and

$$\frac{\partial(u, v)}{\partial(x, y)} = \frac{\partial(u, v)}{\partial(t, s)} \frac{\partial(t, s)}{\partial(x, y)}. \tag{9.36}$$

Using (9.36), we can also write

$$\frac{\frac{\partial(u,v)}{\partial(x,y)}}{\frac{\partial(t,s)}{\partial(x,y)}} = \frac{\partial(u, v)}{\partial(t, s)} \tag{9.37}$$

and

$$\frac{\partial(u, v)}{\partial(x, y)} = \left[\frac{\partial(x, y)}{\partial(u, v)}\right]^{-1}. \tag{9.38}$$

Using now the mathematical results (9.34–9.38), we can rewrite the equation (9.32) as follows:

$$dV = -\frac{\partial(T, V)}{\partial(p, V)} \frac{\partial(V, p)}{\partial(T, p)} dp = \frac{\partial(V, T)}{\partial(p, T)} dp, \tag{9.39}$$

or finally

$$dV = \left(\frac{\partial V}{\partial p}\right)_T dp = -\gamma V_0 dp, \tag{9.40}$$

where we defined the *isothermal compressibility coefficient* as follows:

$$\gamma = -\frac{1}{V_0} \left(\frac{\partial V}{\partial p}\right)_T. \tag{9.41}$$

For a finite change of pressure Δp, we therefore have

$$\Delta V = \left(\frac{\partial V}{\partial p}\right)_T \Delta p = -\gamma V_0 \Delta p. \tag{9.42}$$

9.4.4 Relation between Thermal Coefficients

The coefficients α, β, and γ defined by (9.26), (9.29), and (9.41), respectively, are not independent coefficients and a relation between them can be derived. We start with the obvious identity

$$\frac{\partial(T,V)}{\partial(p,V)}\frac{\partial(p,V)}{\partial(p,T)}\frac{\partial(p,T)}{\partial(V,T)} = \frac{\partial(T,V)}{\partial(V,T)} = -1, \tag{9.43}$$

or

$$\left(\frac{\partial T}{\partial p}\right)_V \left(\frac{\partial V}{\partial T}\right)_P \left(\frac{\partial p}{\partial V}\right)_T = -1, \tag{9.44}$$

or

$$\frac{1}{\beta p_0} \cdot \alpha V_0 \cdot \left(-\frac{1}{\gamma V_0}\right) = -1. \tag{9.45}$$

Thus, the relation between the coefficients α, β, and γ reads

$$\alpha = \beta \gamma p_0. \tag{9.46}$$

9.5 Caloric Properties of Systems

The caloric properties of systems are determined by the caloric-state equation $U = U(T,V)$. The differential of internal energy dU, is now given by

$$dU = \left(\frac{\partial U}{\partial T}\right)_V dT + \left(\frac{\partial U}{\partial V}\right)_T dV. \tag{9.47}$$

Since for the system with the mass m, the internal energy $U = u \cdot m$ and the volume $V = v \cdot m$ are the extensive state variables, we also have

$$du = \left(\frac{\partial u}{\partial T}\right)_V dT + \left(\frac{\partial u}{\partial v}\right)_T dv. \tag{9.48}$$

The equation (9.48) defines the differential of the internal energy per unit mass, and it is valid for a system of an arbitrary mass. This is convenient in the present section because we want to define the caloric coefficients per unit mass using a mass independent thermodynamic analysis. Similarly to the thermal coefficients defined in the Section 9.4.4, "Relation between Thermal Coefficients," here we define the caloric coefficients, which are usually called the *specific heats*.

A specific heat with respect to some parameter X, which is to be kept constant, is defined as follows:

$$c_X = \left(\frac{dq}{dT}\right)_X, \tag{9.49}$$

where the subscript X indicates that the differential dq is calculated at constant X. In (9.49), the heat quantity $Q = q \cdot m$ is an extensive state variable, whereas we use the heat quantity per unit mass q to calculate the specific heat per unit mass c_X.

9.5.1 Specific Heat at Constant Volume c_V

The specific heat at constant volume is defined by

$$c_V = \left(\frac{dq}{dT} \right)_V,$$ (9.50)

Using now the equation (8.39) per unit mass, i.e.,

$$du = TdS - pdv = dq - pdv,$$ (9.51)

we can write

$$dq = du + pdv.$$ (9.52)

Using here the result

$$du = \left(\frac{\partial u}{\partial T} \right)_V dT + \left(\frac{\partial u}{\partial v} \right)_T dv,$$ (9.53)

we obtain

$$dq = \left(\frac{\partial u}{\partial T} \right)_V dT + \left[\left(\frac{\partial u}{\partial v} \right)_T + p \right] dv.$$ (9.54)

Thus, by equation (9.54) taken at constant volume ($dv = 0$), we have

$$c_V = \left(\frac{dq}{dT} \right)_V = \left(\frac{\partial u}{\partial T} \right)_V.$$ (9.55)

By the equations (9.53) and (9.55), we then have

$$du = c_V dT + \left(\frac{\partial u}{\partial v} \right)_T dv.$$ (9.56)

For a finite temperature change ΔT at a constant volume of the system ($V = \text{Constant}$) and a temperature-independent specific heat $[c_V \neq c_V(T)]$ by the equation (9.55), we have

$$\Delta u = \Delta q = c_V \Delta T, \quad v = \text{Constant}.$$ (9.57)

For a system with a total mass m, the equation (9.57) gives

$$\Delta U = \Delta Q = mc_V \Delta T, \quad V = \text{Constant}.$$ (9.58)

The result (9.58) is often very useful in practical calculations of caloric changes of state of thermodynamic systems.

9.5.2 Specific Heat at Constant Pressure c_P

The specific heat at constant pressure is defined by

$$c_P = \left(\frac{dq}{dT}\right)_P,$$ (9.59)

Here, it is convenient to use temperature T and pressure p as the two independent variables. From the definition of enthalpy per unit mass (9.19), i.e.,

$$h = u + pv,$$ (9.60)

we have

$$dh = du + pdv + vdp,$$ (9.61)

or using $dq = du + pdv$, we obtain

$$dh = dq + vdp.$$ (9.62)

On the other hand, from the equation $h = h(T, p)$, we can also write

$$dh = \left(\frac{\partial h}{\partial T}\right)_P dT + \left(\frac{\partial h}{\partial p}\right)_T dp.$$ (9.63)

Using (9.62) taken at constant pressure, we have $dq = dh$. Thus, by the equation (9.63), at constant pressure ($dp = 0$), we obtain

$$c_P = \left(\frac{dq}{dT}\right)_P = \left(\frac{\partial h}{\partial T}\right)_P.$$ (9.64)

For a finite temperature change ΔT at a constant pressure of the system ($p = \text{Constant}$) and a temperature-independent specific heat [$c_P \neq c_P(T)$], we therefore have

$$\Delta h = \Delta q = c_P \Delta T, \quad p = \text{Constant}.$$ (9.65)

For a system with a total mass m, the equation (9.65) gives

$$\Delta H = \Delta Q = m c_P \Delta T, \quad p = \text{Constant}.$$ (9.66)

9.5.3 Relation between Specific Heats

It is possible to establish a thermodynamic formula that relates the two specific heats c_V and c_P. By using the fact that the specific entropy s can be represented as a function of (T, p), we can write

$$dq = Tds = T\left(\frac{\partial s}{\partial T}\right)_p dT + T\left(\frac{\partial s}{\partial p}\right)_T dp.$$ (9.67)

By the equation (9.67) and using the definition (9.59), we obtain

$$dq = c_P dT + T\left(\frac{\partial s}{\partial p}\right)_T dp. \tag{9.68}$$

If we now use the thermal-state equation $p = p(T, v)$, we have

$$dp = \left(\frac{\partial p}{\partial T}\right)_V dT + \left(\frac{\partial p}{\partial v}\right)_T dv. \tag{9.69}$$

Substituting (9.69) into (9.68), we obtain

$$dq = c_P dT + T\left(\frac{\partial s}{\partial p}\right)_T \left[\left(\frac{\partial p}{\partial T}\right)_V dT + \left(\frac{\partial p}{\partial v}\right)_T dv\right]. \tag{9.70}$$

From the equation (9.70), we can calculate

$$c_V = \left(\frac{\partial q}{\partial T}\right)_V = c_P + T\left(\frac{\partial s}{\partial p}\right)_T \left(\frac{\partial p}{\partial T}\right)_V, \tag{9.71}$$

or

$$c_P - c_V = -T\left(\frac{\partial s}{\partial p}\right)_T \left(\frac{\partial p}{\partial T}\right)_V. \tag{9.72}$$

Thus, we have obtained the relation between the two specific heats. However, in order to be able to use this relation in practice, we need to express the first partial derivative on the right-hand side of the equation (9.72) in terms of some easily measurable thermodynamic parameters. We can rewrite this partial derivative as follows:

$$\left(\frac{\partial s}{\partial p}\right)_T = \frac{\partial(s, T)}{\partial(p, T)} = \frac{\partial(s, T)}{\partial(p, v)}\frac{\partial(p, v)}{\partial(p, T)} = -\frac{\partial(T, s)}{\partial(p, v)}\left(\frac{\partial v}{\partial T}\right)_P. \tag{9.73}$$

Here, we need to calculate the Jacobian $\partial(T, s)/\partial(p, v)$. It is done by using the equation for the differential of the internal energy per unit mass of a thermodynamic system

$$du = T ds - p dv. \tag{9.74}$$

The left-hand side of the equation (9.74) is the total differential of the function $u = u(s, v)$, whereas the right-hand side of the equation (9.74) is not necessarily a total differential. The general expression

$$A(x, y)dx + B(x, y)dy, \tag{9.75}$$

is a total differential of some function $w = w(x, y)$, if the functions $A(x, y)$ and $B(x, y)$ have the following form

$$A(x, y) = \frac{\partial w(x, y)}{\partial x}, \quad B(x, y) = \frac{\partial w(x, y)}{\partial y}. \tag{9.76}$$

Thus, the condition that the expression (9.75) is a total differential is

$$\frac{\partial A}{\partial y} = \frac{\partial B}{\partial x} = \frac{\partial^2 w}{\partial x \partial y} \Rightarrow \left(\frac{\partial A}{\partial y}\right)_x = \left(\frac{\partial B}{\partial x}\right)_y. \tag{9.77}$$

Consequently, the condition that the expression $du = Tds - pdv$ is a total differential is given by

$$\left(\frac{\partial T}{\partial v}\right)_s = -\left(\frac{\partial p}{\partial s}\right)_v \Rightarrow \frac{\partial(T, s)}{\partial(v, s)} = -\frac{\partial(p, v)}{\partial(s, v)}. \tag{9.78}$$

By the equation (9.78) and the properties of Jacobians discussed in Section 9.4.3, "Isothermal Expansion." we have

$$\frac{\partial(T, s)}{\partial(p, v)} = \frac{\partial(T, s)}{\partial(v, s)} \frac{\partial(v, s)}{\partial(p, v)} = -\frac{\partial(p, v)}{\partial(s, v)} \frac{\partial(v, s)}{\partial(p, v)}$$

$$= \frac{\partial(p, v)}{\partial(s, v)} \frac{\partial(s, v)}{\partial(p, v)} = \frac{\partial(p, v)}{\partial(p, v)} = 1. \tag{9.79}$$

Thus, we have derived the following significant thermodynamics relation

$$\frac{\partial(T, s)}{\partial(p, v)} = 1. \tag{9.80}$$

Substituting (9.80) into (9.73), we obtain

$$\left(\frac{\partial s}{\partial p}\right)_T = -\left(\frac{\partial v}{\partial T}\right)_P. \tag{9.81}$$

Substituting further (9.81) into (9.72), gives

$$c_p - c_V = T \left(\frac{\partial v}{\partial T}\right)_P \left(\frac{\partial p}{\partial T}\right)_V, \tag{9.82}$$

or

$$c_p - c_V = pvT \left[\frac{1}{v}\left(\frac{\partial v}{\partial T}\right)_P\right]\left[\frac{1}{p}\left(\frac{\partial p}{\partial T}\right)_V\right]. \tag{9.83}$$

Using now the definitions (9.26) and (9.29), i.e.,

$$\alpha = \frac{1}{v}\left(\frac{\partial v}{\partial T}\right)_P \quad \text{and} \quad \beta = \frac{1}{p}\left(\frac{\partial p}{\partial T}\right)_V, \tag{9.84}$$

we obtain from (9.83)

$$c_p - c_V = pvT\alpha\beta. \tag{9.85}$$

Using the relation between the thermal coefficients (9.46) in the form $\beta = \alpha/\gamma p$, we finally obtain the relation between the two specific heats, in terms of the easily measurable thermodynamic quantities, as follows:

$$c_p - c_V = vT\frac{\alpha^2}{\gamma}. \tag{9.86}$$

9.6 Relations between Thermodynamic Coefficients

The objective of the present section is to derive a complete set of mathematical relations between the derivatives of various thermodynamic quantities for classical gases with a constant amount of matter ($N = $ Constant). These mathematical relations are used to simplify the quantitative analysis of state changes for such thermodynamic systems. Let us first recall the equation for the differential of the internal energy per unit mass of a thermodynamic system

$$du = Tds - pdv. \tag{9.87}$$

In thermodynamics, it is, therefore, customary to study the changes of state of a system using the $T - S$ and $p - V$ diagrams, as shown in Fig. 9.4.

The two pairs of variables (p, v) and (T, s) are sometimes called the *canonically conjugated variables*. The thermodynamic coefficients are defined as the derivatives of one of the four variables (p, v, T, s) with respect to any other of these four variables. In total, there are 12 such coefficients and they can be arranged in the following

Figure 9.4 Change of state in T-S and p-V planes.

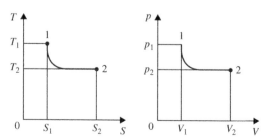

rectangular array

$$
\begin{vmatrix}
\left(\frac{\partial v}{\partial p}\right)_T & \left(\frac{\partial p}{\partial T}\right)_v & \left(\frac{\partial T}{\partial v}\right)_p \\
\left(\frac{\partial s}{\partial T}\right)_v & \left(\frac{\partial T}{\partial v}\right)_s & \left(\frac{\partial v}{\partial s}\right)_T \\
\left(\frac{\partial s}{\partial T}\right)_p & \left(\frac{\partial T}{\partial p}\right)_s & \left(\frac{\partial p}{\partial s}\right)_T \\
\left(\frac{\partial v}{\partial p}\right)_s & \left(\frac{\partial p}{\partial s}\right)_v & \left(\frac{\partial s}{\partial v}\right)_p
\end{vmatrix} .
\tag{9.88}
$$

Here, it can be shown that only 3 of these 12 coefficients are independent. In other words, there are nine equations that relate these 12 coefficients to each other. In order to find these nine equations, we first note that the array (9.88) is formed in such a way that the product of the three coefficients in each row of the array is equal to -1. For example, the product of the three coefficients in the first row of the array (9.88) is equal to

$$
\frac{\partial(v,T)}{\partial(p,T)} \frac{\partial(p,v)}{\partial(T,v)} \frac{\partial(T,p)}{\partial(v,p)}
$$

$$
= \frac{\partial(v,T)}{\partial(T,v)} \frac{\partial(p,v)}{\partial(v,p)} \frac{\partial(T,p)}{\partial(p,T)} = (-1)^3 = -1.
\tag{9.89}
$$

Thus, we immediately obtain the first four relations between the twelve coefficients listed in (9.88)

$$
\left(\frac{\partial v}{\partial p}\right)_T \left(\frac{\partial p}{\partial T}\right)_v \left(\frac{\partial T}{\partial v}\right)_p = -1
\tag{9.90}
$$

$$
\left(\frac{\partial s}{\partial T}\right)_v \left(\frac{\partial T}{\partial v}\right)_s \left(\frac{\partial v}{\partial s}\right)_T = -1
\tag{9.91}
$$

$$
\left(\frac{\partial s}{\partial T}\right)_p \left(\frac{\partial T}{\partial p}\right)_s \left(\frac{\partial p}{\partial s}\right)_T = -1
\tag{9.92}
$$

$$
\left(\frac{\partial v}{\partial p}\right)_s \left(\frac{\partial p}{\partial s}\right)_v \left(\frac{\partial s}{\partial v}\right)_p = -1.
\tag{9.93}
$$

The fifth relation is just the equation (9.80), i.e.,

$$
\begin{vmatrix}
\left(\frac{\partial T}{\partial p}\right)_v & \left(\frac{\partial T}{\partial v}\right)_p \\
\left(\frac{\partial s}{\partial p}\right)_v & \left(\frac{\partial s}{\partial v}\right)_p
\end{vmatrix} = \left(\frac{\partial T}{\partial p}\right)_v \left(\frac{\partial s}{\partial v}\right)_p - \left(\frac{\partial T}{\partial v}\right)_p \left(\frac{\partial s}{\partial p}\right)_v = 1.
\tag{9.94}
$$

In order to find the remaining four relations between the thermodynamic coefficients, we use the result (9.80) in combination with the properties of Jacobians discussed in

Section 9.4.3, "Isothermal Expansion." We thus obtain:

$$1 = \frac{\partial(T,s)}{\partial(p,v)} = \frac{\partial(T,s)}{\partial(T,v)}\frac{\partial(T,v)}{\partial(p,v)} = \left(\frac{\partial s}{\partial v}\right)_T \left(\frac{\partial T}{\partial p}\right)_v, \tag{9.95}$$

$$-1 = \frac{\partial(T,s)}{\partial(v,p)} = \frac{\partial(T,s)}{\partial(T,p)}\frac{\partial(T,p)}{\partial(v,p)} = \left(\frac{\partial s}{\partial p}\right)_T \left(\frac{\partial T}{\partial v}\right)_p, \tag{9.96}$$

$$-1 = \frac{\partial(T,s)}{\partial(v,p)} = \frac{\partial(T,s)}{\partial(v,s)}\frac{\partial(v,s)}{\partial(v,p)} = \left(\frac{\partial T}{\partial v}\right)_s \left(\frac{\partial s}{\partial p}\right)_v, \tag{9.97}$$

$$1 = \frac{\partial(T,s)}{\partial(p,v)} = \frac{\partial(T,s)}{\partial(p,s)}\frac{\partial(p,s)}{\partial(p,v)} = \left(\frac{\partial T}{\partial p}\right)_s \left(\frac{\partial s}{\partial v}\right)_p. \tag{9.98}$$

We have now finally established all nine relations (9.90–9.98) between the thermodynamic coefficients listed in the array (9.88), and thereby proven that there are only three independent thermodynamic coefficients. These three independent thermodynamic coefficients can be determined either experimentally (phenomenological thermodynamics) or using the partition function Z (statistical thermodynamics).

For example, we can choose the three coefficients in the upper left corner of the array (9.88), i.e.,

$$\left(\frac{\partial v}{\partial p}\right)_T, \quad \left(\frac{\partial p}{\partial T}\right)_v \quad \text{and} \quad \left(\frac{\partial s}{\partial T}\right)_v, \tag{9.99}$$

as the three independent thermodynamic coefficients. We can then use the thermal-state equation (8.38) in the form

$$p = p(T,v) = \frac{NkT}{m}\frac{\partial}{\partial v}(\ln Z)_T \tag{9.100}$$

to determine the first two thermodynamic coefficients from (9.99)

$$\left(\frac{\partial p}{\partial T}\right)_v, \quad \left(\frac{\partial v}{\partial p}\right)_T = \left(\frac{\partial p}{\partial v}\right)_T^{-1}. \tag{9.101}$$

Similarly, we can use the caloric-state equation (8.22) in the form

$$u = u(T,v) = \frac{NkT^2}{m}\frac{\partial}{\partial T}(\ln Z)_V \tag{9.102}$$

and the result (9.55) to determine the remaining thermodynamic coefficient in (9.99)

$$\left(\frac{\partial s}{\partial T}\right)_v = \frac{1}{T}\left(\frac{dq}{dT}\right)_v = \frac{1}{T}\left(\frac{\partial u}{\partial T}\right)_v. \tag{9.103}$$

As a final remark, it should be noted that it is not possible to choose the three coefficients from the same row of the array (9.88) as independent quantities, since they are related by the equations (9.90–9.93). In other words, if we know the two coefficients from a given row, the third coefficient from the same row is uniquely determined by the equations (9.90–9.93).

9.7   Problems with Solutions

Problem 1

A quantity of water with mass $m = 1\,\text{kg}$ at the initial temperature $T_0 = 273\,\text{K}$ is brought into contact with a large heat reservoir with constant temperature $T = 373\,\text{K}$. When the water has reached the temperature $T = 373\,\text{K}$, calculate the entropy change for

(a) the water
(b) the heat reservoir
(c) the entire system.

Solution

(a) According to the first law of thermodynamics, we can write

$$dU = dQ - dL = TdS - pdV. \tag{9.104}$$

In the present case, we have $V = \text{Constant}$, such that we have

$$dU = dQ = mc_w dT = TdS, \tag{9.105}$$

where c_w is the specific heat at constant volume for water. Thus, we can write

$$dS_w = mc_w \frac{dT}{T}, \qquad \Delta S_w = mc_w \ln \frac{T}{T_0}. \tag{9.106}$$

Since $c_w = 4.186\,\text{kJ/kg K}$, we obtain the numerical result

$$\Delta S_w = 1\,\text{kg} \times 4186\,\text{J/kg K} \times \ln \frac{373}{273} = 1306.5\,\text{J/K}. \tag{9.107}$$

(b) The change of the internal energy of the water is given by

$$\Delta U_w = \int dU = mc_w \int dT = mc_w (T - T_0). \tag{9.108}$$

Since the entire system is assumed to be thermally insulated from the surroundings, the change of the internal energy of the reservoir is given by

$$\Delta U_r = -\Delta U_w = -mc_w (T - T_0). \tag{9.109}$$

The entropy change of the reservoir at the fixed temperature T is given by

$$\Delta S_r = \int \frac{dU_r}{T} = \frac{\Delta U_r}{T} = -mc_w\left(1 - \frac{T_0}{T}\right). \qquad (9.110)$$

With $c_w = 4.186$ kJ/kg K, we obtain the numerical result

$$\Delta S_r = 1\,\text{kg} \times 4186\,\text{J/kg K} \times \left(1 - \frac{273}{373}\right) = -1122.25\ \text{J/K}. \qquad (9.111)$$

(c) The entropy change of the entire system is given by

$$\Delta S = \Delta S_w + \Delta S_r = mc_w\left(\ln\frac{T}{T_0} + \frac{T_0}{T} - 1\right). \qquad (9.112)$$

With $c_w = 4.186$ kJ/kg K, we obtain the numerical result

$$\Delta S = 1306.5\ \text{J/K} - 1122.25\ \text{J/K} = 184.25\ \text{J/K}. \qquad (9.113)$$

Problem 2

A quantity of water with mass $m = 1$ kg at the initial temperature $T_0 = 273$ K is heated to $T_2 = 373$ K by first bringing it in contact with one large heat reservoir with constant temperature $T_1 = 323$ K and then bringing it in contact with another large heat reservoir with constant temperature $T_2 = 373$ K. What is the change in entropy of the entire system consisting of the water and the two heat reservoirs?

Solution

During the first heating process from $T_0 = 273$ K to $T_1 = 323$ K, the entropy of the second heat reservoir at $T_2 = 373$ K remains unchanged. The entropy change of the entire system is then equal to the entropy change of the water and the first heat reservoir at $T_1 = 323$ K. Using the result (9.112) obtained in the solution of Problem 1, we have

$$\Delta S_1 = mc_w\left(\ln\frac{T_1}{T_0} + \frac{T_0}{T_1} - 1\right). \qquad (9.114)$$

During the second heating process from $T_1 = 323$ K to $T_2 = 373$ K, the entropy of the first heat reservoir at $T_1 = 323$ K remains unchanged. The entropy change of the entire system is then equal to the entropy change of the water and the second heat reservoir at $T_2 = 373$ K. In analogy with the result (9.114), we obtain

$$\Delta S_2 = mc_w\left(\ln\frac{T_2}{T_1} + \frac{T_1}{T_2} - 1\right). \qquad (9.115)$$

The total entropy change during the entire heating process is then equal to

$$\Delta S = \Delta S_1 + \Delta S_2, \tag{9.116}$$

or

$$\Delta S = mc_w \left(\ln \frac{T_1}{T_0} + \frac{T_0}{T_1} - 1 \right) + mc_w \left(\ln \frac{T_2}{T_1} + \frac{T_1}{T_2} - 1 \right). \tag{9.117}$$

The result (9.117) can be rewritten as

$$\Delta S = mc_w \sum_{j=0}^{1} \left(\ln \frac{T_{j+1}}{T_j} + \frac{T_j}{T_{j+1}} - 1 \right). \tag{9.118}$$

The numerical value of the entropy change is obtained from (9.117) as follows:

$$\Delta S = mc_w \left(\ln \frac{T_2}{T_0} + \frac{T_1}{T_2} + \frac{T_0}{T_1} - 2 \right), \tag{9.119}$$

or

$$\Delta S = 1\,\mathrm{kg} \times 4186\,\mathrm{J/kg\,K} \times \left(\ln \frac{373}{273} + \frac{323}{373} + \frac{273}{323} - 2 \right) = 97.4\,\mathrm{J/K}. \tag{9.120}$$

Comparing the numerical result (9.120) with the numerical result (9.113) obtained in Problem 1, the lesson is that the total entropy change decreases when the heating process is divided into two steps.

Problem 3

Using the result of the Problem 2, show that a quantity of water with mass $m = 1\,\mathrm{kg}$ at the initial temperature $T_0 = 273\,\mathrm{K}$ can be heated to $T = 373\,\mathrm{K}$ with no entropy change of the entire system by dividing the heating process into a very large, number of very small heating steps.

Solution

Let us first recall the result for the total entropy change (9.118) generalized to an arbitrary number of n steps, i.e.,

$$\Delta S(n) = mc_w \sum_{j=0}^{n-1} \left(\ln \frac{T_{j+1}}{T_j} + \frac{T_j}{T_{j+1}} - 1 \right). \tag{9.121}$$

As the number of steps becomes very large, the ratio T_{j+1}/T_j approaches unity, and we can use the mathematical approximation

$$\ln(1+x) \to x, \qquad x \to 0, \tag{9.122}$$

to obtain

$$\ln \frac{T_{j+1}}{T_j} = -\ln \frac{T_j}{T_{j+1}} = -\ln\left[1 + \left(\frac{T_j}{T_{j+1}} - 1\right)\right]$$

$$\approx -\left(\frac{T_j}{T_{j+1}} - 1\right) = -\frac{T_j}{T_{j+1}} + 1. \tag{9.123}$$

Substituting (9.123) into (9.121), we obtain

$$\lim_{n \to \infty} \Delta S(n) \equiv 0. \tag{9.124}$$

Thus, the entropy change for the entire system tends to zero, when the heating process is divided into a very large number of very small heating steps.

Problem 4

At normal atmospheric pressure p_0 and the temperature $T_0 = 273$ K, a liquid has the specific heat at constant pressure equal to $c_P = 28$ J/mol K and a molar volume equal to $v_M = 14.72 \times 10^{-6}$ m³/mol. The volume expansion coefficient of the liquid is $\alpha = 1.81 \times 10^{-4}$ 1/K and its isothermal compressibility coefficient is $\gamma = 3.88 \times 10^{-11}$ m²/N. What is the specific heat at constant volume c_V and the ratio $\kappa = c_P/c_V$ for this liquid.

Solution

Let us start with the obvious identity

$$\frac{\partial(T,v)}{\partial(v,T)} \frac{\partial(p,v)}{\partial(p,v)} \frac{\partial(p,T)}{\partial(p,T)} = -1, \tag{9.125}$$

or

$$\frac{\partial(p,v)}{\partial(p,T)} \frac{\partial(p,T)}{\partial(v,T)} \frac{\partial(T,v)}{\partial(p,v)} = -1, \tag{9.126}$$

or

$$\left(\frac{\partial v}{\partial T}\right)_p \left(\frac{\partial p}{\partial v}\right)_T \left(\frac{\partial T}{\partial p}\right)_V = -1, \tag{9.127}$$

or

$$\left(\frac{\partial p}{\partial T}\right)_V = \frac{1}{v}\left(\frac{\partial v}{\partial T}\right)_p \left[-\frac{1}{v}\left(\frac{\partial v}{\partial p}\right)_T\right]^{-1} = \frac{\alpha}{\gamma}. \qquad (9.128)$$

Using now $s = s(T, v)$, we can write

$$ds = \left(\frac{\partial s}{\partial T}\right)_V dT + \left(\frac{\partial s}{\partial v}\right)_T dv, \qquad (9.129)$$

or

$$\left(\frac{\partial s}{\partial T}\right)_p = \left(\frac{\partial s}{\partial T}\right)_V + \left(\frac{\partial s}{\partial v}\right)_T \left(\frac{\partial v}{\partial T}\right)_p, \qquad (9.130)$$

or

$$T\left(\frac{\partial s}{\partial T}\right)_p = T\left(\frac{\partial s}{\partial T}\right)_V + Tv\left(\frac{\partial s}{\partial v}\right)_T \left[\frac{1}{v}\left(\frac{\partial v}{\partial T}\right)_p\right], \qquad (9.131)$$

or

$$c_p = c_V + Tv\alpha \left(\frac{\partial s}{\partial v}\right)_T. \qquad (9.132)$$

Using now the identity

$$\frac{\partial(p, v)}{\partial(T, s)} \equiv 1, \qquad (9.133)$$

we obtain

$$\left(\frac{\partial s}{\partial v}\right)_T = \frac{\partial(T, s)}{\partial(T, v)}\frac{\partial(p, v)}{\partial(T, s)} = \left(\frac{\partial p}{\partial T}\right)_V = \frac{\alpha}{\gamma}. \qquad (9.134)$$

Thus, we obtain

$$c_p = c_V + Tv\frac{\alpha^2}{\gamma}, \qquad (9.135)$$

or

$$c_V = c_p - \frac{Tv\alpha^2}{\gamma}. \qquad (9.136)$$

Substituting the numerical values, we obtain

$$c_V = 28 \, \text{J/mol K} - 273 \, \text{K} \times 14.72 \times 10^{-6} \, \text{m}^3/\text{mol}$$

$$\times \frac{1.81^2 \times 10^{-8} \, 1/\text{K}^2}{3.88 \times 10^{-11} \, \text{m}^2/\text{N}} = 14.83 \, \text{J/mol K}, \tag{9.137}$$

and

$$\kappa = \frac{c_p}{c_V} = \frac{28}{14.83} = 1.888. \tag{9.138}$$

Problem 5

The pressure of a thermally insulated solid sample is slowly increased by a small amount Δp. Calculate the resulting temperature change ΔT of the sample as a function of the pressure change Δp, the initial temperature T_0, the specific heat at constant pressure c_p, the density ρ, and the volume expansion coefficient α.

Solution

Since the sample is thermally insulated, one has $dq = T ds = 0$, i.e., this process goes at constant entropy. Thus, we can write

$$ds = \left(\frac{\partial s}{\partial T}\right)_p dT + \left(\frac{\partial s}{\partial p}\right)_T dp = 0, \tag{9.139}$$

or

$$c_p \frac{dT}{T} = -\left(\frac{\partial s}{\partial p}\right)_T dp. \tag{9.140}$$

Using now the identity

$$\frac{\partial(p,v)}{\partial(T,s)} \equiv 1, \tag{9.141}$$

we obtain

$$\left(\frac{\partial s}{\partial p}\right)_T = \frac{\partial(T,s)}{\partial(T,p)} = \frac{\partial(T,s)}{\partial(T,p)} \frac{\partial(p,v)}{\partial(T,s)} = \frac{\frac{\partial(p,v)}{\partial(T,s)}}{\frac{\partial(T,p)}{\partial(T,s)}} = \frac{\partial(p,v)}{\partial(T,p)} \tag{9.142}$$

or using $\partial(T,p)/\partial(p,T) = -1$

$$\left(\frac{\partial s}{\partial p}\right)_T = -\frac{\partial(p,v)}{\partial(p,T)} = -\left(\frac{\partial v}{\partial T}\right)_p = -v\alpha. \tag{9.143}$$

Thus, we can write

$$\frac{dT}{T} = \frac{\alpha}{c_p} v dp. \tag{9.144}$$

Integrating this equation over the interval $(T_0, T_0 + \Delta T)$, we obtain

$$\ln \frac{T_0 + \Delta T}{T_0} = \int_{p_0}^{p_0 + \Delta p} \frac{\alpha}{c_p} v dp, \tag{9.145}$$

or

$$\Delta T = T_0 \left[\exp\left(\int_{p_0}^{p_0 + \Delta p} \frac{\alpha}{c_p} v dp \right) - 1 \right]. \tag{9.146}$$

Since Δp is very small, we have

$$\int_{p_0}^{p_0 + \Delta p} \frac{\alpha}{c_p} v dp \approx \frac{\alpha_0}{c_{p0}} v_0 \Delta p, \tag{9.147}$$

where the quantity α_0/c_{p0} is taken for the initial state of the solid sample. Thus, we obtain

$$\Delta T = T_0 \left[\exp\left(\frac{\alpha_0}{c_{p0}} v_0 \Delta p \right) - 1 \right]. \tag{9.148}$$

Using now $\exp(x) \to 1 + x$ for small x, we can write

$$\Delta T = \frac{\alpha_0}{c_{p0}} T_0 v_0 \Delta p. \tag{9.149}$$

Problem 6

A substance at pressure p and constant temperature T has the specific heat at constant pressure (per mole) denoted by c_p and a molar volume v. Let us assume that the temperature dependence of the volume expansion coefficient $\alpha = \alpha(T)$ is a known function. Find the derivative of c_p with respect to the pressure p as a function of T and v as well as $\alpha = \alpha(T)$ and its derivatives with respect to T.

Solution

The objective of the problem is to calculate

$$\left(\frac{\partial c_p}{\partial p}\right)_T = \frac{\partial}{\partial p}\left[T\left(\frac{\partial s}{\partial T}\right)_p\right]_T.$$ (9.150)

At constant T, we can change the order of differentiation, to obtain

$$\left(\frac{\partial c_p}{\partial p}\right)_T = T\frac{\partial}{\partial T}\left[\left(\frac{\partial s}{\partial p}\right)_T\right]_p = T\frac{\partial}{\partial T}\left[\frac{\partial(T,s)}{\partial(T,p)}\right]_p,$$ (9.151)

or by equation (9.143),

$$\left(\frac{\partial c_p}{\partial p}\right)_T = T\frac{\partial}{\partial T}\left[\frac{\partial(p,v)}{\partial(T,p)}\right]_p = -T\frac{\partial}{\partial T}\left[\left(\frac{\partial v}{\partial T}\right)_p\right]_p,$$ (9.152)

or

$$\left(\frac{\partial c_p}{\partial p}\right)_T = -T\frac{\partial}{\partial T}[v\alpha(T)]_p,$$ (9.153)

or

$$\left(\frac{\partial c_p}{\partial p}\right)_T = -T\left(\frac{\partial v}{\partial T}\right)_p\alpha - Tv\left(\frac{\partial \alpha}{\partial T}\right)_p,$$ (9.154)

or

$$\left(\frac{\partial c_p}{\partial p}\right)_T = -Tv\alpha^2 - Tv\frac{d\alpha}{dT}.$$ (9.155)

Finally, we obtain

$$\left(\frac{\partial c_p}{\partial p}\right)_T = -Tv\left(\alpha^2 + \frac{d\alpha}{dT}\right).$$ (9.156)

Problem 7

The tension force of an elastic rod is, around the temperature T, related to the length of the rod L by the expression $F = aT^2(x - x_0)$, where a and x_0 are positive constants and F is the *external force* acting on the rod. The heat capacity at constant volume (i.e., at constant length) of an unstretched rod ($x = x_0$), is given by $c_X = bT$, where b is a positive constant.

(a) Write down the first law of thermodynamics for the rod.
(b) Using $S = S(T, x)$, calculate $(\partial S / \partial x)_T$.
(c) If we know $S(T_0, x_0)$, calculate $S(T, x)$ at any other T and x.
(d) If we stretch a thermally insulated rod from $x = x_I$ at $T = T_I$ to $x = x_F$ at $T = T_F$, calculate T_F. Compare T_F with T_I.
(e) Calculate $c_X = c_X(T, x)$ for an arbitrary length x, instead of for the length $x = x_0$.

Solution

(a) The general form of the first law of thermodynamics is

$$dU = dQ - dL. \tag{9.157}$$

In our case, we have $dQ = TdS$ and $dL = -Fdx$, with F as written above in the problem text. The minus sign here is related to the fact that above (in the problem text), F actually is the external force (a force of an external agent) acting on the rod. This force is in equilibrium with the *internal force* F_{int} of the rod molecules, which is thus negative of F, that is, $F_{\text{int}} = F$. Just as for the case of a gas (for which the work $dL = F_{\text{int}} dx$ is done *by the internal force* of pressure of gas molecules), the work done by the internal force of the rod molecules is *also* given by $dL = F_{\text{int}} dx = -Fdx$. Thus, the first law of thermodynamics for the rod is represented by the standard form

$$dU = TdS - F_{\text{int}} dx = TdS + Fdx, \tag{9.158}$$

or

$$dS = \frac{dU - Fdx}{T} = \frac{dU}{T} - aT(x - x_0) dx. \tag{9.159}$$

(b) Using the identity

$$\frac{\partial(T, s)}{\partial(F, x)} = -1, \tag{9.160}$$

we obtain

$$-1 = \frac{\partial(T, s)}{\partial(F, x)} = \frac{\partial(T, s)}{\partial(T, x)} \frac{\partial(T, x)}{\partial(F, x)} = \frac{\frac{\partial(T, s)}{\partial(T, x)}}{\frac{\partial(F, x)}{\partial(T, x)}} = \frac{\left(\frac{\partial S}{\partial x}\right)_T}{\left(\frac{\partial F}{\partial T}\right)_x}, \tag{9.161}$$

or

$$\left(\frac{\partial S}{\partial x}\right)_T = -\left(\frac{\partial F}{\partial T}\right)_x = -2aT(x - x_0). \tag{9.162}$$

(c) In order to calculate $S(T, x)$ at some T and x, it is suitable to calculate the change of entropy with temperature at constant length $x = x_0$, where we know the heat capacity $c_X = bT$. Thus, we have

$$c_X = bT = T \left(\frac{\partial S}{\partial T} \right)_{x=x_0} \Rightarrow b = \left(\frac{\partial S}{\partial T} \right)_{x=x_0}, \tag{9.163}$$

or

$$\left(\frac{\partial S}{\partial T} \right)_{x=x_0} = \frac{S(T, x_0) - S(T_0, x_0)}{T - T_0} = b. \tag{9.164}$$

Thus, we obtain

$$S(T, x_0) = S(T_0, x_0) + b(T - T_0). \tag{9.165}$$

On the other hand, using the result (9.162), we have

$$\left(\frac{\partial S}{\partial x} \right)_T = -2aT(x - x_0), \tag{9.166}$$

or

$$\int_{S(T,x_0)}^{S(T,x)} dS = -2aT \int_{x_0}^{x} (x - x_0) \, dx, \tag{9.167}$$

or

$$S(T, x) = S(T, x_0) - aT(x - x_0)^2. \tag{9.168}$$

Combining the results (9.165) and (9.168), we finally obtain

$$S(T, x_0) = S(T_0, x_0) + b(T - T_0) - aT(x - x_0)^2. \tag{9.169}$$

(d) Since the thermally insulated rod undergoes a constant-entropy change, we have

$$S(T_F, x_F) = S(T_I, x_I) \tag{9.170}$$

or

$$b(T_I - T_0) - aT_I(x_I - x_0)^2 = b(T_F - T_0) - aT_F(x_F - x_0)^2. \tag{9.171}$$

Thus, we obtain

$$T_F = \frac{b - a(x_I - x_0)^2}{b - a(x_F - x_0)^2} T_I. \tag{9.172}$$

Thus, we see that for $L_F > L_I$, we have $T_F < T_I$, whereas for $L_F < L_I$, we have $T_F > T_I$.

(e) Let us first calculate

$$\left(\frac{\partial c_X}{\partial x}\right)_T = T\frac{\partial}{\partial x}\left[\left(\frac{\partial S}{\partial T}\right)_x\right]_T = T\frac{\partial}{\partial T}\left[\left(\frac{\partial S}{\partial x}\right)_T\right]_x. \tag{9.173}$$

Using now (9.162), i.e.,

$$\left(\frac{\partial S}{\partial x}\right)_T = -2aT(x - x_0), \tag{9.174}$$

we obtain

$$\left(\frac{\partial c_X}{\partial x}\right)_T = T\frac{\partial}{\partial T}[-2aT(x - x_0)]_x = -2aT(x - x_0). \tag{9.175}$$

Thus, we have

$$\int_{c_X(T,x_0)}^{c_X(T,x)} dc_X = -2aT\int_{x_0}^{x}(x - x_0)\,dx, \tag{9.176}$$

or

$$c_X(T, x) = c_X(T, x_0) - aT(x - x_0)^2. \tag{9.177}$$

Thus, we finally obtain

$$c_X(T, x) = bT - aT(x - x_0)^2. \tag{9.178}$$

Problem 8

A substance has the volume expansion coefficient α and the isothermal compressibility coefficient γ given by

$$\alpha = \frac{1}{V}\left(\frac{\partial V}{\partial T}\right)_P = \frac{2bT}{V} \quad \text{and} \quad \gamma = -\frac{1}{V}\left(\frac{\partial V}{\partial p}\right)_T = \frac{a}{V}, \tag{9.179}$$

respectively, where a and b are constants. Determine the change of pressure of the substance when its temperature is changed from T_1 to T_2 within a constant volume. Express the result in terms of the constants a and b, as well as the temperatures T_1 and T_2.

Solution

From the definition of γ in (9.179), we readily obtain

$$\left(\frac{\partial V}{\partial p}\right)_T = -a \Rightarrow V(p, T) = -ap + f(T). \tag{9.180}$$

Substituting the integrated result in (9.180) into the definition of α in (9.179), we further obtain

$$\left(\frac{\partial V}{\partial T}\right)_P = f'(T) = 2bT \Rightarrow f(T) = bT^2 + C, \tag{9.181}$$

where C is an unspecified constant. Thus, when the volume of the substance is kept constant $[V(p, T) = \text{Constant}]$, we have

$$V(p, T) = -ap + bT^2 + C \Rightarrow -ap_1 + bT_1^2 = -ap_2 + bT_2^2. \tag{9.182}$$

From (9.182), we readily obtain the pressure change in the process as follows:

$$\Delta p = p_2 - p_1 = \frac{b}{a}\left(T_2^2 - T_1^2\right). \tag{9.183}$$

Problem 9

A substance has the volume expansion coefficient α and the isothermal compressibility coefficient γ given by

$$\alpha = \frac{1}{V}\left(\frac{\partial V}{\partial T}\right)_P = A(1 - ap) \tag{9.184}$$

and

$$\gamma = -\frac{1}{V}\left(\frac{\partial V}{\partial p}\right)_T = K[1 + b(T - T_0)], \tag{9.185}$$

respectively, where we introduce $A = 4.2 \times 10^{-4}\,1/\text{K}$, $a = 1.2 \times 10^{-9}\,1/\text{Pa}$, $K = 2.52 \times 10^{-10}\,1/\text{Pa}$, and $b = 2 \times 10^{-3}\,1/\text{K}$ as given constants.

(a) Derive the thermal equation of the state of the substance and express the constant a in terms of the constants A, K, and b. Check the above given numerical values for consistency.

(b) Calculate the required temperature change to increase the pressure in the substance to $p = 10^7\,\text{Pa}$, if at the initial temperature of $T_0 = 273\,\text{K}$, the pressure is $p_0 = 10^5\,\text{Pa}$.

Solution

(a) As both the volume expansion coefficient α and the isothermal compressibility coefficient γ are defined as partial derivatives of V divided by V, let us consider the total derivative of $\ln V$, i.e.,

$$d[\ln V] = \frac{1}{V}\left(\frac{\partial V}{\partial p}\right)_T dp + \frac{1}{V}\left(\frac{\partial V}{\partial T}\right)_P dT. \tag{9.186}$$

Using the definitions (9.184) and (9.185), we can write

$$d[\ln V] = -\gamma dp + \alpha dT, \tag{9.187}$$

or

$$d[\ln V] = -K[1 + b(T - T_0)]dp + A(1 - ap)dT. \tag{9.188}$$

Thus, we have

$$\frac{\partial(\ln V)}{\partial p} = -K[1 + b(T - T_0)], \qquad \frac{\partial(\ln V)}{\partial T} = A(1 - ap), \tag{9.189}$$

or

$$\frac{\partial^2(\ln V)}{\partial T \partial p} = -Kb, \qquad \frac{\partial^2(\ln V)}{\partial p \partial T} = -Aa. \tag{9.190}$$

Since the two second partial derivatives in (9.190) are equal to each other, we must have $Aa = Kb$, or

$$a = \frac{Kb}{A} = \frac{2.52 \cdot 10^{-10} \cdot 2 \cdot 10^{-3}}{4.2 \cdot 10^{-4}} = 1.2 \cdot 10^{-9} \frac{1}{\text{Pa}}, \tag{9.191}$$

which confirms the consistency of the numerical values given above. Now, we can integrate the partial derivatives (9.189) to obtain the thermal equation of state. Let us first integrate the equation

$$\frac{\partial(\ln V)}{\partial p} = -K[1 + b(T - T_0)], \tag{9.192}$$

to obtain

$$\ln V = -K[1 + b(T - T_0)]p + f(T). \tag{9.193}$$

Substituting the result (9.193) into the second equation in (9.189), we get

$$\frac{\partial(\ln V)}{\partial T} = -Kbp + f'(T) = A(1 - ap). \tag{9.194}$$

Using $Kbp = Aap$, gives

$$f'(T) = A \Rightarrow f(T) = AT + C, \tag{9.195}$$

where C is a yet unspecified constant. Thus, the thermal equation of state has the form

$$\ln V = -K[1 + b(T - T_0)]p + AT + C. \tag{9.196}$$

On the other hand, we know that at $T = T_0$ and $p = p_0$, we have $V = V_0$. Thus, we can write

$$\ln V_0 = -Kp_0 + AT_0 + C, \tag{9.197}$$

and we obtain the following result for constant C

$$C = \ln V_0 + Kp_0 - AT_0. \tag{9.198}$$

The final result for the thermal equation of state is then

$$\ln \frac{V}{V_0} = -K[p - p_0 + b(T - T_0)p] + A(T - T_0). \tag{9.199}$$

(b) Using the equation of state (9.199) for $p = 100p_0$ and $V = V_0$, we obtain

$$0 = -K[99p_0 + b(T - T_0)100p_0] + A(T - T_0), \tag{9.200}$$

or

$$99Kp_0 = (A - 100Kbp_0)(T - T_0). \tag{9.201}$$

Thus, we obtain the required temperature change

$$T - T_0 = \frac{99Kp_0}{A - 100Kbp_0} = \frac{99p_0}{b(a^{-1} - 100p_0)} \approx 6\,\mathrm{K}. \tag{9.202}$$

10 Variable Number of Particles

10.1 Chemical Potential

In our discussion so far, we have assumed that the number of the constituent particles of a given thermodynamic system is constant ($N = $ Constant). Let us now consider a thermodynamic system with a variable number of particles. In other words, let us assume that the number of constituent particles of a system is not constant, for example, due to creation and annihilation of the constituent particles occurring in the system. The creation and annihilation of the constituent particles of the system can be realized in various ways. One common mechanism for creation and annihilation of the constituent particles of the system is chemical reactions of the matter in the system. Creation and annihilation of particles occur also in all common semiconductors, where quasi-particles, namely, electrons and holes, are created and annihilated in pairs. Another situation with a variable number of particles occurs in considering thermodynamic equilibrium between an open system and its environment. If the system is separated by a permeable membrane from its environment that can let particles in and out of the system, then the number of particles of the system N would fluctuate in time around some average equilibrium value. In all of the above examples, the number of particles is not constant in time.

For the systems with variable number of particles, it is convenient to introduce a new thermodynamic variable μ, which determines the energy increase in the system due to the creation of a single particle in the system. The variable μ is usually called the *chemical potential* of the system. In the course of a thermodynamic process, in which creations and annihilations of constituent particles occur, the infinitesimal change of the internal energy is defined by a generalization of the equation (8.39) of the form

$$dU = T\,dS - p\,dV + \mu\,dN. \tag{10.1}$$

Now, we need to relate the chemical potential μ to the other thermodynamic quantities (e.g., $p, V, T, S, \dots$). The equation (10.1) suggests that in the case of a thermodynamic system with a variable number of particles, we have three pairs of canonically conjugated variables (p, V), (T, S), and (μ, N). Obviously, the left-hand side of the equation (10.1) is a total differential of the internal energy. The internal energy U here is a three-variable function. The first variable can be either V or p, the second variable can be either T or S, and the third variable can be either μ or N. Thus, there are $2^3 = 8$ possible choices for the three variables that can be used to express U or other state functions. Whatever choice is made, the three pairs of variables must be related to each other in such a way that the right-hand side of the equation (10.1) is also a total differential. The general expression

$$A(x, y, z)\, dx + B(x, y, z)\, dy + C(x, y, z)\, dz \tag{10.2}$$

is a total differential of some function $w = w(x, y, z)$, if the functions $A(x, y, z)$, $B(x, y, z)$, and $C(x, y, z)$ have the following form:

$$A(x, y, z) = \frac{\partial w(x, y, z)}{\partial x}, \quad B(x, y, z) = \frac{\partial w(x, y, z)}{\partial y},$$

$$C(x, y, z) = \frac{\partial w(x, y, z)}{\partial z}. \tag{10.3}$$

Thus, the three conditions that the expression (10.2) is a total differential are

$$\frac{\partial A}{\partial y} = \frac{\partial B}{\partial x} = \frac{\partial^2 w}{\partial x \partial y} \quad \Rightarrow \quad \left(\frac{\partial A}{\partial y}\right)_{x,z} = \left(\frac{\partial B}{\partial x}\right)_{y,z}, \tag{10.4}$$

$$\frac{\partial A}{\partial z} = \frac{\partial C}{\partial x} = \frac{\partial^2 w}{\partial x \partial z} \quad \Rightarrow \quad \left(\frac{\partial A}{\partial z}\right)_{x,y} = \left(\frac{\partial C}{\partial x}\right)_{y,z}, \tag{10.5}$$

$$\frac{\partial B}{\partial z} = \frac{\partial C}{\partial y} = \frac{\partial^2 w}{\partial z \partial y} \quad \Rightarrow \quad \left(\frac{\partial B}{\partial z}\right)_{x,y} = \left(\frac{\partial C}{\partial y}\right)_{x,z}. \tag{10.6}$$

In terms of Jacobians, these three conditions become

$$\frac{\partial(A, x)}{\partial(y, x)} = \frac{\partial(B, y)}{\partial(x, y)}, \quad z = \text{Constant}, \tag{10.7}$$

$$\frac{\partial(A, x)}{\partial(z, x)} = \frac{\partial(C, z)}{\partial(x, z)}, \quad y = \text{Constant}, \tag{10.8}$$

$$\frac{\partial(B, y)}{\partial(z, y)} = \frac{\partial(C, z)}{\partial(y, z)}, \quad x = \text{Constant}. \tag{10.9}$$

or, by using the properties of Jacobians discussed in Section 9.4.3, "Isothermal Expansion":

$$\frac{\partial(A, x)}{\partial(B, y)} = -1, \quad z = \text{Constant}, \tag{10.10}$$

$$\frac{\partial(A, x)}{\partial(C, z)} = -1, \quad y = \text{Constant}, \tag{10.11}$$

$$\frac{\partial(B, y)}{\partial(C, z)} = -1, \quad x = \text{Constant}. \tag{10.12}$$

Using the general results (10.10–10.12), the conditions that the right-hand side of the equation (10.1) is a total differential are

$$\frac{\partial(T,S)}{\partial(p,V)} = 1, \qquad N = \text{Constant}. \tag{10.13}$$

$$\frac{\partial(T,S)}{\partial(\mu,N)} = -1, \quad V = \text{Constant}, \tag{10.14}$$

$$\frac{\partial(p,V)}{\partial(\mu,N)} = 1, \qquad S = \text{Constant}. \tag{10.15}$$

The results (10.13–10.15) are some of the relations between the chemical potential μ and the other thermodynamic variables in a system with the variable number of particles. Clearly, there are many more relations between these six thermodynamic variables (p, V, T, S, μ, N), analogous to the nine relations derived in the Chapter 9, "Macroscopic Thermodynamics." In fact, in the case of systems with the variable number of particles, we have 60 thermodynamic coefficients and a systematic mathematical approach to the relations between all of these coefficients is very complex and not necessary for the present analysis.

10.2 Thermodynamic Potential

Since the chemical potential μ is the energy increment due to the creation of a single particle of the system, it is an intensive thermodynamic quantity. In macroscopic thermodynamics, it is, therefore, convenient to introduce a new macroscopic, extensive quantity called the *thermodynamic potential* or *Gibbs potential* as follows:

$$\Phi = N \cdot \mu. \tag{10.16}$$

From the equation (10.16), we obtain

$$d\Phi = \mu \, dN + N \, d\mu. \tag{10.17}$$

On the other hand, from the equation (10.1), it is clear that when all the other thermodynamic quantities are kept constant, we have $dU = \mu \, dN$. At this point, we require that the change of internal energy of the system dU, caused by the change of the number of particles in the system dN only, is equal to the change of the thermodynamic potential of the system. In other words, when all the other thermodynamic quantities are kept constant, we require that

$$dU = d\Phi = \mu \, dN \tag{10.18}$$

or

$$\Phi = \int \frac{d\Phi}{dN} \, dN = \frac{d\Phi}{dN} \int dN = \frac{d\Phi}{dN} N = \mu N. \tag{10.19}$$

In order to satisfy the equation (10.19), the function $\mu = d\Phi/dN$ must be an intensive quantity, independent of the number of particles (i.e., the mass of the system), and consequently it cannot be a function of any extensive thermodynamic quantity. The only way this can be realized is if μ is a function of two other intensive quantities such as the pressure p and temperature T, that is, $\mu = \mu(T,p)$. We can now calculate the differential of the chemical potential $\mu = \mu(T,p)$ as follows:

$$d\mu = \left(\frac{\partial \mu}{\partial T}\right)_p dT + \left(\frac{\partial \mu}{\partial p}\right)_T dp. \tag{10.20}$$

Using the result $\mu = \mu(T,p)$ and Jacobian (10.14), we have

$$\frac{\partial(\mu, N)}{\partial(T, S)} = -1, \quad V = \text{Constant}, \tag{10.21}$$

or

$$\begin{vmatrix} \dfrac{\partial \mu}{\partial T} & \dfrac{\partial \mu}{\partial S} \\[2ex] \dfrac{\partial N}{\partial T} & \dfrac{\partial N}{\partial S} \end{vmatrix} = \begin{vmatrix} \dfrac{\partial \mu}{\partial T} & 0 \\[2ex] \dfrac{\partial N}{\partial T} & \dfrac{\partial N}{\partial S} \end{vmatrix} = \left(\frac{\partial \mu}{\partial T}\right)_{P,V}\left(\frac{\partial N}{\partial S}\right)_V = -1, \tag{10.22}$$

or

$$\left(\frac{\partial \mu}{\partial T}\right)_{P,V} = -\left(\frac{\partial S}{\partial N}\right)_V = -s_0, \tag{10.23}$$

where we introduced the entropy reduced to a single particle of the system s_0 as follows:

$$s_0 = \left(\frac{\partial S}{\partial N}\right)_V. \tag{10.24}$$

Using the result $\mu = \mu(T,p)$ and Jacobian (10.15), we also have

$$\frac{\partial(\mu, N)}{\partial(p, V)} = 1, \quad S = \text{Constant}, \tag{10.25}$$

or

$$\begin{vmatrix} \dfrac{\partial \mu}{\partial p} & \dfrac{\partial \mu}{\partial V} \\[2ex] \dfrac{\partial N}{\partial p} & \dfrac{\partial N}{\partial V} \end{vmatrix} = \begin{vmatrix} \dfrac{\partial \mu}{\partial p} & 0 \\[2ex] \dfrac{\partial N}{\partial p} & \dfrac{\partial N}{\partial V} \end{vmatrix} = \left(\frac{\partial \mu}{\partial p}\right)_{T,S}\left(\frac{\partial N}{\partial V}\right)_S = 1, \tag{10.26}$$

or

$$\left(\frac{\partial \mu}{\partial p}\right)_{T,S} = \left(\frac{\partial V}{\partial N}\right)_S = v_0, \tag{10.27}$$

where we introduced the volume reduced to a single particle of the system v_0 as follows:

$$v_0 = \left(\frac{\partial V}{\partial N}\right)_S . \tag{10.28}$$

Substituting the results (10.23) and (10.27) into (10.20), we obtain

$$d\mu = -s_0 \, dT + v_0 \, dp. \tag{10.29}$$

Substituting (10.29) into (10.17), we further obtain

$$d\Phi = \mu \, dN - N \, s_0 \, dT + N \, v_0 \, dp, \tag{10.30}$$

or

$$d\Phi = \mu \, dN - S \, dT + V \, dp. \tag{10.31}$$

The result (10.31) can be rearranged as follows:

$$d\Phi = - T \, dS - S \, dT + p \, dV + V \, dp + T \, dS - p \, dV + \mu \, dN, \tag{10.32}$$

or using the equation (10.1)

$$d\Phi = -d(TS) + d(pV) + dU = d(-TS + pV + U). \tag{10.33}$$

Thus, we finally obtain the relation between the thermodynamic potential Φ and the other thermodynamic functions as

$$\Phi = U - TS + pV, \tag{10.34}$$

or using the definition of free energy of a thermodynamic system $F = U - TS$, we can also write

$$\Phi = F + pV. \tag{10.35}$$

From the result (10.35) with (10.31), we can calculate the differential of the free energy for a system with a variable number of particles, as follows:

$$dF = d\Phi - d(pV) = \mu \, dN - S \, dT + V \, dp - V \, dp - p \, dV \tag{10.36}$$

or finally

$$dF = -S \, dT - p \, dV + \mu \, dN. \tag{10.37}$$

Recall that we already met this relation in Section 8.3, "Pressure: Thermal-State Equation." In earlier chapters we have considered the thermodynamic systems with fixed number of particles inside a variable volume. In chemical systems, we usually have an entirely different situation with a variable number of particles within the fixed volume. In chemical thermodynamics, it is, therefore, useful to introduce a new thermodynamic variable called the *Helmholtz thermodynamic potential* as follows:

$$\Omega = F - \Phi = F - \mu N = -pV, \tag{10.38}$$

where we made use of the result (10.35). Using the result for the free energy $F = U - TS$, we obtain an alternative expression for the Helmholtz thermodynamic potential

$$\Omega = U - TS - \mu N. \tag{10.39}$$

Using the definition (10.38) and the result (10.37), we can calculate the differential of the Helmholtz thermodynamic potential as follows:

$$d\Omega = dF - \mu\, dN - N\, d\mu = -S\, dT - p\, dV + \mu\, dN - \mu\, dN - N\, d\mu, \tag{10.40}$$

or finally

$$d\Omega = -S\, dT - N\, d\mu - p\, dV. \tag{10.41}$$

The number of particles in a chemical system considered in the above discussion can then be calculated from the Helmholtz thermodynamic potential as follows:

$$N = -\left(\frac{\partial \Omega}{\partial \mu}\right)_{T,V}. \tag{10.42}$$

10.3   Phases and Phase Equilibrium

10.3.1   Latent Heat

The states of matter, which can exist simultaneously in equilibrium and in contact with each other, are described as phases. An example is the coexistence between any two phases of water, such as liquid water and water vapor. If the two phases of the matter are labeled by (1) and (2), the conditions for the equilibrium between these two phases are following

$$T_1 = T_2 = T$$

$$p_1 = p_2 = p \tag{10.43}$$

$$\mu_1(T, p) = \mu_2(T, p)$$

as discussed in Section 9.2.1, "Zeroth Law of Thermodynamics." This equilibrium situation is sometimes described as $T - p - \mu$ distribution (or $T - p - \mu$ ensemble). The change from one phase to another occurs at constant pressure ($dp = 0$) and constant temperature ($dT = 0$), and it is accompanied by the production or absorption of some quantity of heat, which is called the *latent heat*. From the equation (9.22), we see that when a thermodynamic process occurs at constant pressure ($dp = 0$), the quantity of heat that the system exchanges with the surroundings is equal to the change in the enthalpy of the system ($dQ - dH = 0$). Thus, if we denote the latent heat per unit mass by q_L, we can write

$$q_L = h_2 - h_1. \tag{10.44}$$

By convention, the latent heat q_L is positive if the heat is absorbed by the matter that undergoes the phase transition. Conversely, the latent heat q_L is negative if the heat is produced within matter that undergoes the phase transition and absorbed by the surroundings. By definition of enthalpy (9.19), we obtain from (10.44)

$$q_L = (u_2 + p_2 v_2) - (u_1 + p_1 v_1) = (u_2 - u_1) + p(v_2 - v_1). \tag{10.45}$$

Using here the equation $dU = T\,dS - p\,dV$, we have

$$q_L = T(s_2 - s_1). \tag{10.46}$$

10.3.2   Clausius–Clapeyron Formula

Let us now consider the equilibrium condition for the chemical potentials in (10.43) and differentiate both sides with respect to temperature as follows:

$$\left(\frac{\partial \mu_1}{\partial T}\right)_P + \left(\frac{\partial \mu_1}{\partial p}\right)_T \frac{dp}{dT} = \left(\frac{\partial \mu_2}{\partial T}\right)_P + \left(\frac{\partial \mu_2}{\partial p}\right)_T \frac{dp}{dT}. \tag{10.47}$$

Using here the result (10.29), we have

$$\left(\frac{\partial \mu}{\partial T}\right)_P = -s_0, \quad \left(\frac{\partial \mu}{\partial p}\right)_T = v_0. \tag{10.48}$$

Substituting (10.48) into (10.47), we obtain

$$-s_{01} + v_{01}\frac{dp}{dT} = -s_{02} + v_{02}\frac{dp}{dT}, \tag{10.49}$$

or

$$\frac{dp}{dT} = \frac{s_{02} - s_{01}}{v_{02} - v_{01}}. \tag{10.50}$$

Multiplying the right-hand side of the equation (10.50) by unity expressed as N/N, and using $S = N s_0$ and $V = N v_0$, we obtain

$$\frac{dp}{dT} = \frac{s_2 - s_1}{v_2 - v_1}. \tag{10.51}$$

Using now (10.46), we obtain the *Clausius–Clapeyron formula* for phase transitions in the following form

$$\frac{dp}{dT} = \frac{q_L}{T(v_2 - v_1)}. \tag{10.52}$$

The phase transitions are, of course, possible at various pressures and corresponding temperatures. However, the two phases cannot exist in equilibrium with each other at arbitrary pressures and temperatures, since T and p are related to each other along the phase transition (coexistence) line in the (T, p) plane. We have already used this observation while differentiating the equilibrium condition for chemical potentials in the equation (10.47).

The latent heat q_L varies from one pressure (and corresponding temperature) to another, and this variation can be analyzed by calculating the rate of change of the latent heat with temperature. Using (10.44), we obtain

$$\frac{dq_L}{dT} = \frac{dh_2}{dT} - \frac{dh_1}{dT}, \tag{10.53}$$

or

$$\frac{dq_L}{dT} = \left(\frac{\partial h_2}{\partial T}\right)_P + \left(\frac{\partial h_2}{\partial p}\right)_T \frac{dp}{dT} - \left(\frac{\partial h_1}{\partial T}\right)_P - \left(\frac{\partial h_1}{\partial p}\right)_T \frac{dp}{dT}. \tag{10.54}$$

Using here the definition of the specific heat at constant pressure given by (9.64), we further obtain

$$\frac{dq_L}{dT} = c_{P2} - c_{P1} + \left[\left(\frac{\partial h_2}{\partial p}\right)_T - \left(\frac{\partial h_1}{\partial p}\right)_T\right] \frac{dp}{dT}. \tag{10.55}$$

Using here (9.61) in the form $dh = T\,ds + v\,dp$, we obtain

$$\left(\frac{\partial h}{\partial p}\right)_T = T\left(\frac{\partial s}{\partial p}\right)_T + v. \tag{10.56}$$

If we recall the relation (9.96), i.e.,

$$\left(\frac{\partial s}{\partial p}\right)_T \left(\frac{\partial T}{\partial v}\right)_p = -1, \tag{10.57}$$

we obtain

$$\left(\frac{\partial s}{\partial p}\right)_T = -\left(\frac{\partial v}{\partial T}\right)_p.$$ (10.58)

Substituting (10.58) into (10.56), we obtain

$$\left(\frac{\partial h}{\partial p}\right)_T = -T\left(\frac{\partial v}{\partial T}\right)_p + v.$$ (10.59)

Substituting further (10.59) into (10.55), we obtain

$$\frac{dq_L}{dT} = c_{P2} - c_{P1} + (v_2 - v_1)\frac{dp}{dT} - T\left[\left(\frac{\partial v_2}{\partial T}\right)_p - \left(\frac{\partial v_1}{\partial T}\right)_p\right]\frac{dp}{dT}.$$ (10.60)

Using here the Clausius-Clapeyron Formula (10.52), yields

$$\frac{dq_L}{dT} = \frac{q_L}{T} + c_{P2} - c_{P1} - \frac{q_L}{v_2 - v_1}\left[\left(\frac{\partial v_2}{\partial T}\right)_p - \left(\frac{\partial v_1}{\partial T}\right)_p\right].$$ (10.61)

The rate of change of the latent heat with temperature (10.61) is constant for many substances, and, for example, in case of the phase transition between water and ice, it can be calculated to give

$$\frac{dq_L}{dT} \approx 2730\frac{J}{kg\,K} \approx 0.65\frac{cal}{g\,K}.$$ (10.62)

10.4 Problems with Solutions

Problem 1

An electrochemical cell consists of a positive copper (Cu) electrode and a negative zinc (Zn) electrode. When an electromotive force ε is connected between the two electrodes, the cell is in equilibrium and no net current flows through the external circuit. The chemical reaction which occurs in the cell is the following,

$$Zn + Cu\,SO_4 \leftrightarrow Cu + Zn\,SO_4.$$

During this chemical reaction, the cell performs the amount of work

$$dL = dq\varepsilon = Ze\,dN\varepsilon,$$

where Z is the number of valence electrons of copper (Cu). The first law of thermodynamics in this case reads

$$dU = T\,dS - \mu\,dN, \quad \mu = Ze\varepsilon.$$

If we work with the number of moles $dn = dN/N_{AV}$, we introduce the molar Gibbs thermodynamic potential

$$\Phi = N_{AV}\mu = ZN_{AV}e\varepsilon = Zf\varepsilon,$$

where $f = N_{AV}e = 96488.5\,C/mol$ is the Faraday constant. Thus, the first law of thermodynamics, becomes

$$dU = T\,dS - \Phi\,dn, \quad \Phi = Zf\varepsilon.$$

Use the first law of thermodynamics to calculate the change of the internal energy of the cell at the fixed temperature T and pressure p, when 1 mole of copper is produced. Prove that the change of the internal energy (i.e., the heat of chemical reaction) can be determined by measuring the electromotive force ε only.

Solution

From the first law of thermodynamics, we obtain

$$\left(\frac{\partial U}{\partial n}\right)_T = T\left(\frac{\partial S}{\partial n}\right)_T - \Phi. \tag{10.63}$$

Now using the relation

$$\frac{\partial(T,S)}{\partial(\Phi,n)} = 1, \tag{10.64}$$

we can write

$$\frac{\partial(T,S)}{\partial(T,n)} = \frac{\partial(\Phi,n)}{\partial(T,n)} \quad \Rightarrow \quad \left(\frac{\partial S}{\partial n}\right)_T = \left(\frac{\partial \Phi}{\partial T}\right)_n. \tag{10.65}$$

Thus, we have

$$\left(\frac{\partial U}{\partial n}\right)_T = T\left(\frac{\partial \Phi}{\partial T}\right)_n - \Phi, \tag{10.66}$$

or

$$\Delta U = \int_{U(T,0)}^{U(T,1)} dU = \int_0^1 \left[T\left(\frac{\partial \Phi}{\partial T}\right)_n - \Phi \right] dn, \tag{10.67}$$

or

$$\Delta U = Zf \int_0^1 \left[T \left(\frac{\partial \varepsilon}{\partial T} \right)_n - \varepsilon \right] dn. \tag{10.68}$$

The change of ε during the production of 1 mole of copper is very slow, and we can write

$$\Delta U = Zf \left[T \left(\frac{\partial \varepsilon}{\partial T} \right)_n - \varepsilon \right]. \tag{10.69}$$

Furthermore, the temperature dependence of the electromotive force at fixed n is insignificant, and we finally obtain

$$|\Delta U| = Zf\varepsilon. \tag{10.70}$$

Thus, by measuring ε, we directly measure the heat of the chemical reaction ΔU.

Problem 2

A thermally insulated copper can of mass $M = 0.75\,\text{kg}$ with specific heat $c_M = 418\,\text{J/K}$ contains $m_w = 0.2\,\text{kg}$ of water in equilibrium at the temperature $T_w = 293\,\text{K}$. Let us now place $m_I = 0.03\,\text{kg}$ of ice at the temperature $T_0 = 273\,\text{K}$ into the can. The latent heat of melting the ice is $q_{LI} = 3.33 \times 10^5\,\text{J/kg}$ and the specific heat of water is $c_w = 4.186\,\text{kJ/kg K}$. When all the ice has melted, a new equilibrium state is reached.

(a) Calculate the temperature of the water in the new equilibrium.
(b) Calculate the total entropy change in this process.
(c) Calculate the work that has to be performed on the system in the new equilibrium in order to restore the temperature of the water in the copper can back to $T = 293\,\text{K}$.

Solution

(a) Let us denote the final temperature of the system after this process by T_F. The change of the internal energy of the copper can, when it is cooled from $T_w = 293\,\text{K}$ to T_F, is given by

$$\Delta U_M = \int_{T_w}^{T_F} dU_M = Mc_M (T_F - T_w). \tag{10.71}$$

The change of the internal energy of the water, when it is cooled from $T_w = 293\,\text{K}$ to T_F, is given by

$$\Delta U_w = \int_{T_w}^{T_F} dU_w = m_w c_w (T_F - T_w). \tag{10.72}$$

The change of the internal energy of the ice cube, when it is melted and then warmed from $T_0 = 273\,\mathrm{K}$ to T_F, is given by

$$\Delta U_I = m_I q_{LI} + m_I c_w (T_F - T_0). \tag{10.73}$$

Since the system is thermally insulated $\Delta Q = 0$ and there is no work performed $\Delta L = 0$, the total change of the internal energy of the system is equal to zero

$$\Delta U = M c_M (T_F - T_w) + m_w c_w (T_F - T_w)$$
$$+ m_I q_{LI} + m_I c_w (T_F - T_0) = 0. \tag{10.74}$$

Solving this equation for T_F, we obtain

$$T_F = \frac{M c_M T_w + m_w c_w T_w - m_I q_{LI} + m_I c_w T_0}{M c_M + m_w c_w + m_I c_w}. \tag{10.75}$$

Inserting the numerical values, we obtain $T_F = 285\,\mathrm{K}$.

(b) Using now the first law of thermodynamics for a system of mass m, where no work is performed ($V = \mathrm{Constant}$), we have a following general result

$$dU = dQ = T\,dS = m c_V\,dT, \tag{10.76}$$

where c_V is the specific heat of the system. Thus, we can write

$$dS = m c_V \frac{dT}{T}. \tag{10.77}$$

The change of the entropy of the copper can, when it is cooled from $T_w = 293\,\mathrm{K}$ to T_F, is then given by

$$\Delta S_M = \int_{T_w}^{T_F} dS_M = M c_M \ln \frac{T_F}{T_w}. \tag{10.78}$$

The change of the entropy of the water in the can, when it is cooled from $T_w = 293\,\mathrm{K}$ to T_F, is given by

$$\Delta S_w = \int_{T_w}^{T_F} dS_w = m_w c_w \ln \frac{T_F}{T_w}. \tag{10.79}$$

The entropy change of the ice cube, when it is melted at constant temperature T_0 and then warmed from $T_0 = 273\,\text{K}$ to T_F, is given by

$$\Delta S_I = m_I \frac{q_{LI}}{T_0} + m_I c_w \ln \frac{T_F}{T_0}. \tag{10.80}$$

The total entropy change of the system is given by

$$\Delta S = M c_M \ln \frac{T_F}{T_w} + m_w c_w \ln \frac{T_F}{T_w} + m_I \frac{q_{LI}}{T_0} + m_I c_w \ln \frac{T_F}{T_0}. \tag{10.81}$$

Inserting the numerical values, we obtain $\Delta S = 1193\,\text{J/K}$.

(c) The work necessary to restore the temperature of the water in the copper can back to $T = 293\,\text{K}$ is given by

$$\Delta L = \Delta U = M c_M \int_{T_F}^{T_w} dT + (m_w + m_I) c_w \int_{T_F}^{T_w} dT, \tag{10.82}$$

or

$$\Delta L = [M c_M + (m_w + m_I) c_w](T_w - T_F). \tag{10.83}$$

Inserting the numerical values, we obtain $\Delta L = 10210.25\,\text{J}$.

Problem 3

A mixture of water and ice at the freezing temperature $T_0 = 273\,\text{K}$ is put into a refrigerator in order to freeze the remaining $m = 0.1\,\text{kg}$ of water. The heat from the refrigerator is used to warm up a body of mass $M = 1\,\text{kg}$ and constant heat capacity $c = 2.5\,\text{kJ/kg\,K}$, initially at temperature equal to $T_0 = 273\,\text{K}$. The latent heat of freezing the ice is equal to $q_L = 3.33 \times 10^5\,\text{J/kg}$. Use the second law of thermodynamics to calculate the minimum amount of heat that must be removed by the refrigerator in the process.

Solution

The change in entropy of the mixture of water and ice during the freezing process at constant temperature is

$$\Delta S_{WI} = -\frac{m q_L}{T_0}. \tag{10.84}$$

The change in entropy of the body is given by

$$\Delta S_B = \int_{T_0}^{T_F} \frac{M c\, dT}{T} = M c \ln \frac{T_F}{T_0}. \tag{10.85}$$

The total entropy change of the entire system is

$$\Delta S = \Delta S_{WI} + \Delta S_B = -\frac{mq_L}{T_0} + Mc \ln \frac{T_F}{T_0} \geq 0. \tag{10.86}$$

Thus, we have

$$\ln \frac{T_F}{T_0} \geq \frac{mq_L}{McT_0}, \tag{10.87}$$

or

$$T_F \geq T_0 \exp\left(\frac{mq_L}{McT_0}\right). \tag{10.88}$$

On the other hand, the quantity of heat transferred to the body is given by

$$\Delta Q_B = Mc\,(T_F - T_0), \tag{10.89}$$

and it is minimum, when T_F is minimum, i.e., when

$$T_F = T_0 \exp\left(\frac{mq_L}{McT_0}\right). \tag{10.90}$$

Thus, we obtain

$$\Delta Q_{Bmin} = McT_0 \left[\exp\left(\frac{mq_L}{McT_0}\right) - 1\right]. \tag{10.91}$$

Introducing the numerical values, we obtain

$$\Delta Q_{Bmin} \approx 33.3\,\text{kJ}. \tag{10.92}$$

Problem 4

Use the definition of the Helmholtz thermodynamic potential Ω for a system with the variable number of particles to prove the following relation

$$V\,dp - S\,dT - N\,d\mu = 0. \tag{10.93}$$

Solution

From the definition of the Helmholtz thermodynamic potential, we have

$$\Omega = U - TS - \mu N = -pV, \tag{10.94}$$

or

$$U - TS - \mu N + pV = 0. \tag{10.95}$$

Differentiating the equation (10.95) gives

$$dU - T dS - S dT - N d\mu - \mu dN + p dV + V dp = 0. \tag{10.96}$$

Using the equation (10.1), $dU = T dS - p dV + \mu dN$, we readily obtain the required result

$$-S dT - N d\mu + V dp = 0. \tag{10.97}$$

Problem 5

Consider a model of the Earth's atmosphere, where the air pressure in the atmosphere $P(z)$ is the following function of the altitude z

$$p(z) = p(0) \exp\left(-\frac{mgz}{kT}\right). \tag{10.98}$$

(a) Show that the rate of change of the boiling point of the water, due to the variation of the air pressure, can be expressed by the following equation

$$\frac{dT_B}{dz} = -\frac{mgp(z)}{kT}\frac{dT_B}{dp} = -\frac{mgp(z)}{kT}\frac{T_B(\rho_L - \rho_G)}{\rho_L \rho_G q_{LG}}, \tag{10.99}$$

where $\rho_L = 10^3\,\text{kg/m}^3$ is the mass density of the liquid water, $\rho_G = 0.6\,\text{kg/m}^3$ is the mass density of water vapor at the boiling temperature of $T = 373\,\text{K}$, and $q_{LG} = 2.4 \times 10^6\,\text{J/kg}$ is the latent heat of vaporization of water per unit mass.

(b) Given the mass density of air at the Earth's surface $\rho_A = 1.29\,\text{kg/m}^3$, calculate the rate of change of the boiling point of the water with altitude expressed in degrees Kelvin per kilometer.

Solution

(a) The rate of change of the boiling point can be written in the form

$$\frac{dT_B}{dz} = \frac{dp}{dz}\frac{dT_B}{dp}. \tag{10.100}$$

From the given function (10.98), we obtain

$$\frac{dp}{dz} = -\frac{mg}{kT} p(0) \exp\left(-\frac{mgz}{kT}\right) = -\frac{mgp(z)}{kT}. \tag{10.101}$$

Using the Clausius–Clapeyron formula (10.52) for vaporization of water in the following form

$$\frac{dp}{dT_B} = \frac{q_{LG}}{T_B(v_G - v_L)} = \frac{Mq_{LG}}{T_B(V_G - V_L)}, \qquad v = \frac{V}{M}, \tag{10.102}$$

we can write

$$\frac{dT_B}{dp} = \frac{T_B}{Mq_{LG}}(V_G - V_L) = \frac{T_B}{Mq_{LG}}\left(\frac{M}{\rho_G} - \frac{M}{\rho_L}\right), \tag{10.103}$$

or

$$\frac{dT_B}{dp} = \frac{T_B(\rho_L - \rho_G)}{\rho_L\rho_G q_{LG}}. \tag{10.104}$$

Substituting the results (10.101) and (10.104) into (10.100) gives the required result

$$\frac{dT_B}{dz} = -\frac{mgp(z)}{kT}\frac{dT_B}{dp} = -\frac{mgp(z)}{kT}\frac{T_B(\rho_L - \rho_G)}{\rho_L\rho_G q_{LG}}. \tag{10.105}$$

(b) Using the given numerical values of the parameters, we obtain

$$\frac{dT_B}{dz} \approx 0.0033\frac{K}{m} = 3.3\frac{K}{km}. \tag{10.106}$$

Problem 6

The partition function for a system with a variable number of particles, as a function of temperature T, volume V, and the number of particles N, is given by the expression

$$Z(N, V, T) = N\exp\left(\frac{bT^3V^2}{N^2}\right), \tag{10.107}$$

where b is a constant. Derive the expression for the chemical potential μ of the system.

Solution

The free energy of this system, by definition, is given by

$$F = -NkT\ln\frac{Z}{N} - NkT = -\frac{bkT^4V^2}{N} - NkT. \tag{10.108}$$

If we now refer to the equation (10.37), i.e.,

$$dF = -S\,dT - p\,dV + \mu\,dN, \tag{10.109}$$

we obtain the following result for the chemical potential μ of the system

$$\mu = \left(\frac{\partial F}{\partial N}\right)_{T,V} = bkT^4 \left(\frac{V}{N}\right)^2 - kT. \tag{10.110}$$

Problem 7

The internal energy for a system with variable number of particles, as a function of entropy S, volume V, and the number of particles N, is given by the expression

$$U(S,V,N) = aNk \left(\frac{N}{V}\right)^{2/7} \exp\left(\frac{2S}{7Nk}\right), \tag{10.111}$$

where a is a constant. Derive the thermal and caloric-state equations for the system and calculate the specific heat of the system at constant volume c_V.

Solution

Let us first assume that N and V are kept constant. Then, the derivation of the expression for U, given by (10.111), with respect to U itself involves only the derivative of entropy with respect to U on the right-hand side of the equation (10.111). Thus, we obtain

$$1 = \left(\frac{\partial U}{\partial U}\right)_{N,V} = \frac{2}{7Nk}\left(\frac{\partial S}{\partial U}\right)_{N,V} \cdot U. \tag{10.112}$$

Using now the equation (10.1), i.e.,

$$dU = T\,dS - p\,dV + \mu\,dN, \tag{10.113}$$

we see that

$$\left(\frac{\partial S}{\partial U}\right)_{N,V} = \frac{1}{T}. \tag{10.114}$$

Substituting (10.114) into (10.112), we obtain the caloric equation of state of the system in the form

$$1 = \frac{2U}{7NkT} \quad \Rightarrow \quad U = \frac{7}{2}NkT. \tag{10.115}$$

Using equations (10.111) and (10.113), we obtain the pressure of the system as follows:

$$p = -\left(\frac{\partial U}{\partial V}\right)_{N,S} = \frac{2U}{7V}. \tag{10.116}$$

Substituting U from (10.115) into (10.116), we obtain the thermal equation of state of the system in the form

$$p = \frac{NkT}{V}. \tag{10.117}$$

As it will be discussed in a later chapter, this system is a diatomic ideal gas with translational, rotational, and vibrational degrees of freedom. The specific heat at constant volume c_V is then given by

$$c_V = \left(\frac{\partial U}{\partial T}\right)_{V,N} = \frac{7}{2}Nk. \tag{10.118}$$

Part Three

Ideal and Nonideal Gases

11 Ideal Monatomic Gases

11.1 Continuous Energy Spectrum

The objective of the present chapter is to apply the general thermodynamic concepts developed in previous chapters to the special case of an ideal monatomic gas. The ideal monatomic gas considered here satisfies the following requirements:

1. The gas consists of a large number of identical monatomic constituent particles (monatomic molecules);
2. The separation between the constituent particles is large enough, so that there are no substantial interactions, nor statistical and quantum correlations between the particles. The ideal gas concept thus represents an approximation that is valid in dilute limit of realistic gases. In this limit, interparticle interactions are substantial only during infrequent collisions between the constituent particles (assumed here to be elastic in nature);
3. The particles are of very small sizes and can be treated as point-like objects without an internal structure. Such objects have zero moments of inertia, and thus they carry zero internal rotational energy. Their "zero sizes" cannot vibrate, and thus the particles carry no internal vibrational energy. In reality, monatomic molecules of noble, inert gases (He, Ne, Ar, etc.) do behave approximately in this way, in the sense that the effects of their internal structure can be, to a good approximation, ignored under normal conditions ($p \sim 1$ atm, $T \sim$ room temperature).

In view of the above assumptions, the constituent molecules can be treated as mutually noninteracting particles confined in a box with volume V, each having only the translational kinetic energy, given by

$$E_i = \frac{h^2}{8mV^{2/3}}\left(n_x^2 + n_y^2 + n_z^2\right), \tag{11.1}$$

as we discussed in Chapter 3, "Kinetic Energy of Translational Motion." Let us depict this situation in a fictitious three-dimensional n-space, as shown in Fig. 11.1.

Since the numbers n_x, n_y, and n_z are all positive, we study the energy states in the first octant only. As indicated in Fig. 11.1, we introduce the spherical co-ordinates (n, θ, ϕ) in the n-space as follows:

$$n_x = n\sin\theta\cos\phi, \; n_y = n\sin\theta\sin\phi, \; n_z = n\cos\theta. \tag{11.2}$$

Introductory Statistical Thermodynamics

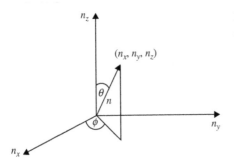

Figure 11.1 Spherical co-ordinates in the n-space.

The translational kinetic energy (11.1) is spherically symmetric and we have

$$E_i = \frac{h^2}{8mV^{2/3}} n^2. \tag{11.3}$$

From the estimate (3.45), one can see that the spacing between the adjacent energy levels is extremely small. One can therefore assume that the energy levels are continuously distributed in the n-space, much like in classical physics. The discrete energy levels (11.3) can then be replaced by classical continuum representation, that is,

$$E_1(n) = \frac{h^2}{8mV^{2/3}} n^2 = \frac{p^2}{2m}, \tag{11.4}$$

where p is the magnitude of the linear momentum vector of a constituent particle, not to be confused with the pressure p of a thermodynamic system. Also, the speed v of a constituent particle, defined by $p = mv$, should not be confused with the volume per unit mass v of a thermodynamic system. The two conflicting notations will never be used in the same context in this book, so there is no risk for confusion. Furthermore, it should be noted that the subscript 1 in $E_1(n)$ is not a label of the first energy state, but merely an indicator that the energy (11.4) is the translational energy of a single constituent particle. The significance of this notation will become clear from the below discussion.

Because the positive numbers n_x, n_y, and n_z in (11.1) are integers, each energy state occupies the volume equal to $1 \times 1 \times 1 = 1$ in the n-space in Fig. 11.1. Thus, the number of states (sublevels), $dg(n)$ that are within the spherical shell with radii n and $n + dn$, is simply equal to the volume of this shell in the first octant of the n-space,

$$dg(n) = \frac{1}{8}(4\pi n^2 dn) = \frac{\pi n^2}{2} dn. \tag{11.5}$$

Using the equation (11.4), we obtain the following relation between the number n and the momentum p of the constituent particle

$$n = \frac{2V^{1/3}}{h} p \Rightarrow dn = \frac{2V^{1/3}}{h} dp. \tag{11.6}$$

Substituting (11.6) into (11.5), we obtain

$$dg(p) = \left(\frac{2V^{1/3}}{h}\right)^3 \frac{\pi p^2}{2} dp, \tag{11.7}$$

or

$$dg(p) = \frac{V}{h^3} 4\pi p^2 dp. \tag{11.8}$$

11.2 Continuous Partition Function

From the above discussions, the partition function sum in (7.39) becomes an integral in the continuous n-space,

$$Z = \sum_i g_i \exp\left(-\frac{E_i}{kT}\right) \approx \int dg(n) \exp\left[-\frac{E_1(n)}{kT}\right]. \tag{11.9}$$

In terms of linear momenta p, the partition function (11.9) reads as

$$Z = \int_0^\infty dg(p) \exp\left[-\frac{E_1(p)}{kT}\right] = \frac{V}{h^3} \int_0^\infty \exp\left[-\frac{E_1(p)}{kT}\right] 4\pi p^2 dp. \tag{11.10}$$

As noted in (11.4), $E_1(p) = p^2/2m$. Using here the obvious integral

$$V = \int_V dV_1 = \int \int \int_V dx \, dy \, dz, \tag{11.11}$$

and rewriting the integration measure in the p-space as

$$\int_0^\infty 4\pi p^2 \, dp = \int_{-\infty}^\infty \int_{-\infty}^\infty \int_{-\infty}^\infty dp_x \, dp_y \, dp_z, \tag{11.12}$$

we can rewrite the partition function (11.10) as follows:

$$Z = \frac{1}{h^3} \int_{\Gamma_1} \exp\left(-\frac{E_1}{kT}\right) dx \, dy \, dz \, dp_x \, dp_y \, dp_z. \tag{11.13}$$

The integral on the right-hand side of the equation (11.13) is over the entire six-dimensional phase space denoted by Γ_1, covering all three spatial co-ordinates of the ordinary space (x, y, z), within the confining box, and the three co-ordinates in the fictitious p-space, i.e., the three components of the linear momentum vector (p_x, p_y, p_z) of a single constituent particle. Introducing here the dimensionless infinitesimal element of the six-dimensional phase space for a single particle as follows:

$$d\Gamma_1 = \frac{1}{h^3} \, dx \, dy \, dz \, dp_x \, dp_y \, dp_z, \tag{11.14}$$

we obtain a compact formal notation for the partition function Z in the following form

$$Z = \int_{\Gamma_1} \exp\left(-\frac{E_1}{kT}\right) d\Gamma_1. \tag{11.15}$$

The partition function Z defined by (11.15) can be defined in the same way for each of the N constituent particles of the system by integrating over the six-dimensional phase space of each of these N particles.

Using the definition (8.13), we can now define the grand partition function for all N particles of the system as follows:

$$Z_N = \frac{Z^N}{N!} = \frac{1}{N!} \int_{\Gamma} \exp\left(-\frac{1}{kT} \sum_{a=1}^{N} E_a\right) d\Gamma_1 d\Gamma_2 \ldots d\Gamma_N. \tag{11.16}$$

Thus, the grand partition function for the entire system can be written in the following compact form

$$Z_N = \frac{1}{N!} \int_{\Gamma} \exp\left(-\frac{E_N}{kT}\right) d\Gamma, \tag{11.17}$$

where

$$E_N = \sum_{a=1}^{N} E_a = \sum_{a=1}^{N} \frac{p_a^2}{2m} \tag{11.18}$$

is the sum of the energies of all the constituent particles of the system, i.e., the total energy of the system. The quantity,

$$d\Gamma = \prod_{a=1}^{N} d\Gamma_a = d\Gamma_1 d\Gamma_2 \ldots d\Gamma_N, \tag{11.19}$$

is the infinitesimal element of the $6N$-dimensional phase space of the entire thermodynamic system with N constituent particles.

11.3   Partition Function of Ideal Monatomic Gases

In the simple case of an ideal monatomic gas, it is sufficient to calculate the partition function (11.10) in order to obtain the thermodynamic equations of state and calculate all the required thermodynamic variables. In order to calculate the integral

$$Z = \frac{V}{h^3} \int_0^\infty \exp\left(-\frac{p^2}{2mkT}\right) 4\pi\, p^2\, dp, \tag{11.20}$$

we introduce a new variable w instead of the momentum p as follows:

$$p = \sqrt{2mkT}\, w \Rightarrow dp = \sqrt{2mkT}\, dw. \tag{11.21}$$

Substituting (11.21) into (11.20), we obtain

$$Z = \frac{V}{h^3}(2mkT)^{3/2} 4\pi \int_0^\infty w^2 e^{-w^2}\, dw, \tag{11.22}$$

or

$$Z = V\left(\frac{2\pi\, mkT}{h^2}\right)^{3/2} \frac{4}{\sqrt{\pi}} I_2, \tag{11.23}$$

where we define the integral I_2 as follows:

$$I_2 = \int_0^\infty w^2 e^{-w^2}\, dw = \frac{1}{2}\int_0^\infty w e^{-w^2}\, d(w^2) = \frac{1}{2}\int_0^\infty w\, d(-e^{-w^2}). \tag{11.24}$$

Integrating by parts, we obtain

$$I_2 = \frac{1}{2}\left(-w e^{-w^2}\Big|_0^\infty + \int_0^\infty e^{-w^2}\, dw\right) = \frac{1}{2}\frac{\sqrt{\pi}}{2} = \frac{\sqrt{\pi}}{4}. \tag{11.25}$$

Substituting (11.25) into (11.23), we obtain the final result for the partition function of an ideal monatomic gas

$$Z = V\left(\frac{2\pi\, mkT}{h^2}\right)^{3/2}. \tag{11.26}$$

11.4   Kinetic Theory of Ideal Monatomic Gases

11.4.1   Maxwell–Boltzmann Speed Distribution

In the present section, we study the kinetic properties of an ideal monatomic gas, e.g., the probability distribution for the velocities of the constituent particles as well as mean values of velocities and kinetic energies of the constituent particles.

Let us first recall the result for the occupation number of the state with the energy in the interval $(E_i, E_i + \Delta E_i)$, given by (8.20), i.e.,

$$n_i = g_i \frac{N}{Z} \exp\left(-\frac{E_i}{kT}\right). \tag{11.27}$$

In the limit of a continuous distribution of the energy levels in the n-space, the number of particles with linear momenta p within an infinitesimal range $(p, p+dp)$ is given by

$$dn(p) = dg(p) \frac{N}{Z} \exp\left[-\frac{E(p)}{kT}\right]. \tag{11.28}$$

The probability for a particle to have the momentum p in the interval $(p, p+dp)$ is thus given by

$$d\omega(p) = \frac{dn(p)}{N} = \frac{dq(p)}{Z} \exp\left(-\frac{p^2}{2mkT}\right). \tag{11.29}$$

Substituting the results (11.8) for $dg(p)$ and (11.26) for Z into the equation (11.28), we obtain

$$d\omega(p) = \frac{1}{V} \frac{h^3}{(2\pi\, mkT)^{3/2}} \frac{V}{h^3} 4\pi\, p^2\, dp \exp\left(-\frac{p^2}{2mkT}\right) \tag{11.30}$$

or finally

$$d\omega(p) = \frac{4\pi\, p^2\, dp}{(2\pi\, mkT)^{3/2}} \exp\left(-\frac{p^2}{2mkT}\right) = v(p)dp. \tag{11.31}$$

If we want to find the probability distribution in terms of particle speeds, we import the relation $p = mv$ into the result (11.31) to obtain the probability for a particle to have a speed v in the interval $(v, v+dv)$, which is thus given by

$$d\omega(v) = \frac{m^3\, 4\pi\, v^2\, dv}{(2\pi\, mkT)^{3/2}} \exp\left(-\frac{mv^2}{2kT}\right), \tag{11.32}$$

or finally

$$d\omega(v) = \left(\frac{m}{2\pi kT}\right)^{3/2} 4\pi v^2 dv \exp\left(-\frac{mv^2}{2kT}\right) = v(v)dv, \tag{11.33}$$

where

$$v(v) = 4\pi \left(\frac{m}{2\pi kT}\right)^{3/2} v^2 \exp\left(-\frac{mv^2}{2kT}\right),$$ (11.34)

is the probability density that a particle in an ideal monatomic gas has the speed v. The distribution function in (11.34) is called the *Maxwell–Boltzmann speed distribution*.

11.4.2 Most Probable Speed of Gas Particles

The Maxwell–Boltzmann speed distribution goes to zero both at zero v and at infinite v. It thus has a maximum at a certain nonzero v representing the most probable speed. It is obtained from the condition

$$\frac{dv(v)}{dv} = 0,$$ (11.35)

where

$$v(v) = \text{Constant} \times v^2 \exp\left(-\frac{mv^2}{2kT}\right).$$ (11.36)

If we omit the multiplication constant in (11.36), we obtain

$$\frac{dv(v)}{dv} = \left(2v - v^2 \frac{2mv}{2kT}\right) \exp\left(-\frac{mv^2}{2kT}\right) = 0,$$ (11.37)

or

$$2v_P - \frac{mv_P^3}{kT} = 0 \Rightarrow \frac{2kT}{m} = v_P^2.$$ (11.38)

The final result for the most probable speed of particles in an ideal monatomic gas is then given by

$$v_P = \sqrt{\frac{2kT}{m}}.$$ (11.39)

11.4.3 Average Speed of Gas Particles

The average speed of particles in an ideal monatomic gas is obtained from

$$<v> = \int_0^\infty v\,v(v)\,dv,$$ (11.40)

or

$$< v > = \left(\frac{m}{2\pi kT} \right)^{3/2} 4\pi \int_0^\infty v^3 dv \exp\left(-\frac{mv^2}{2kT} \right). \tag{11.41}$$

Introducing here a new variable w as follows:

$$v = \sqrt{\frac{2kT}{m}} w \Rightarrow dv = \sqrt{\frac{2kT}{m}} dw, \tag{11.42}$$

we obtain

$$< v > = \left(\frac{m}{2\pi kT} \right)^{3/2} 4\pi \left(\frac{2kT}{m} \right)^{4/2} \frac{1}{2} \int_0^\infty w^2 e^{-w^2} d(w^2), \tag{11.43}$$

Introducing here another new variable $t = w^2$ and rearranging, we further obtain

$$< v > = \frac{2}{\pi} \left(\frac{m}{2\pi kT} \right)^{3/2} \left(\frac{2\pi kT}{m} \right)^2 \int_0^\infty t e^{-t} dt$$

$$= \frac{2}{\pi} \left(\frac{2\pi kT}{m} \right)^{1/2} \int_0^\infty t d(-e^{-t}). \tag{11.44}$$

Integration by parts gives

$$< v > = \sqrt{\frac{8kT}{\pi m}} \left(-t e^{-t} \Big|_0^\infty + \int_0^\infty e^{-t} dt \right). \tag{11.45}$$

The final result for the average speed of particles in an ideal monatomic gas is then given by

$$< v > = \sqrt{\frac{8kT}{\pi m}}. \tag{11.46}$$

11.4.4   Root-Mean-Square Speed of Gas Particles

The average of the square of the particle speed in an ideal monatomic gas is obtained from

$$< v^2 > = \int_0^\infty v^2 v(v) dv, \tag{11.47}$$

or

$$< v^2 > = \left(\frac{m}{2\pi kT} \right)^{3/2} 4\pi \int_0^\infty v^4 dv \exp\left(-\frac{mv^2}{2kT} \right).$$

(11.48)

Introducing here a new variable w as follows:

$$v = \sqrt{\frac{2kT}{m}} w \Rightarrow dv = \sqrt{\frac{2kT}{m}} dw,$$

(11.49)

we obtain

$$< v^2 > = \left(\frac{m}{2\pi kT} \right)^{3/2} 4\pi \left(\frac{2kT}{m} \right)^{5/2} \frac{1}{2} \int_0^\infty w^3 e^{-w^2} d\left(w^2 \right)$$

(11.50)

or

$$< v^2 > = \frac{2}{\sqrt{\pi}} \left(\frac{m}{2kT} \right)^{3/2} \left(\frac{2kT}{m} \right)^{5/2} \int_0^\infty w^3 d\left(-e^{-w^2} \right).$$

(11.51)

Integration by parts gives

$$< v^2 > = \frac{kT}{m} \frac{4}{\sqrt{\pi}} \left(-w^3 e^{-w^2} \Big|_0^\infty + \int_0^\infty 3w^2 e^{-w^2} dw \right),$$

(11.52)

or

$$< v^2 > = \frac{3kT}{m} \frac{4}{\sqrt{\pi}} \int_0^\infty w^2 e^{-w^2} dw = \frac{3kT}{m} \frac{4}{\sqrt{\pi}} I_2,$$

(11.53)

where the integral I_2 was calculated in (11.25) as $I_2 = \sqrt{\pi}/4$.

Thus, the final result for the average of the square of the particle speed in an ideal monatomic gas is given by

$$< v^2 > = \frac{3kT}{m}.$$

(11.54)

The root-mean-square of the particle speed is then obtained as

$$v_{RMS} = \sqrt{< v^2 >} = \sqrt{\frac{3kT}{m}}.$$

(11.55)

11.4.5 Average Kinetic Energy and Internal Energy

The average kinetic energy of a single constituent particle in an ideal monatomic gas is, by the equation (11.54), given by

$$< E > = \frac{1}{2}m < v^2 > = \frac{3}{2}kT. \tag{11.56}$$

The internal energy of the system can then be calculated as follows:

$$U = N < E > = \frac{3}{2}NkT. \tag{11.57}$$

In the next section, we will show that this result from the kinetic theory agrees with the thermodynamic result obtained from the caloric-state equation (8.22) with the partition function (11.26).

11.4.6 Equipartition Theorem

The probability for a particle to have its velocity vector within the velocity space element $dv_x dv_y dv_z$ is simply obtained from the equation (11.33) by replacing the expression $4\pi v^2 dv$ therein by $dv_x dv_y dv_z$, i.e.,

$$4\pi v^2 dv \rightarrow dv_x dv_y dv_z, \tag{11.58}$$

yielding,

$$d\omega\left(v_x, v_y, v_z\right) = \left(\frac{m}{2\pi kT}\right)^{3/2} dv_x dv_y dv_z \exp\left(-\frac{mv^2}{2kT}\right), \tag{11.59}$$

or

$$d\omega\left(v_x, v_y, v_z\right) = v(v_x) v\left(v_y\right) v(v_z) dv_x dv_y dv_z, \tag{11.60}$$

where

$$v\left(v_j\right) = \left(\frac{m}{2\pi kT}\right)^{1/2} \exp\left(-\frac{mv_j^2}{2kT}\right), \quad j = x, y, z. \tag{11.61}$$

The one-dimensional probability distribution functions $v(v_j)$ satisfy the normalization condition

$$\int_{-\infty}^{\infty} v\left(v_j\right) dv_j = 1, \quad j = x, y, z. \tag{11.62}$$

The normalization condition (11.62) is easily verified as

$$
\int\limits_{-\infty}^{\infty} v(v_j)\,dv_j = \left(\frac{m}{2\pi kT}\right)^{1/2} \int\limits_{-\infty}^{\infty} \exp\left(-\frac{mv_j^2}{2kT}\right) dv_j
$$

$$
= \left(\frac{m}{2\pi kT}\right)^{1/2} \left(\frac{2kT}{m}\right)^{1/2} \int\limits_{-\infty}^{\infty} e^{-w^2}\,dw
$$

$$
= \frac{1}{\sqrt{\pi}} \int\limits_{-\infty}^{\infty} e^{-w^2}\,dw = 1, \tag{11.63}
$$

where we introduced a new variable w as follows:

$$
v_j = \sqrt{\frac{2kT}{m}}\,w \Rightarrow dv_j = \sqrt{\frac{2kT}{m}}\,dw. \tag{11.64}
$$

Using now the probability distribution (11.61), we can calculate

$$
<v_j^2> = \left(\frac{m}{2\pi kT}\right)^{1/2} \int\limits_{-\infty}^{\infty} v_j^2 \exp\left(-\frac{mv_j^2}{2kT}\right) dv_j. \tag{11.65}
$$

Introducing a new variable w defined by (11.64), we get

$$
<v_j^2> = \left(\frac{m}{2\pi kT}\right)^{1/2} \left(\frac{2kT}{m}\right)^{3/2} \int\limits_{-\infty}^{\infty} w^2 e^{-w^2}\,dw, \tag{11.66}
$$

or

$$
<v_j^2> = \frac{1}{\sqrt{\pi}} \left(\frac{m}{2kT}\right)^{1/2} \left(\frac{2kT}{m}\right)^{3/2} 2I_2, \tag{11.67}
$$

where the integral I_2 was calculated in (11.25) as

$$
I_2 = \int\limits_{0}^{\infty} w^2 e^{-w^2}\,dw = \frac{1}{2} \int\limits_{-\infty}^{\infty} w^2 e^{-w^2}\,dw = \frac{\sqrt{\pi}}{4}. \tag{11.68}
$$

Substituting (11.68) into (11.67), we obtain

$$
<v_j^2> = \frac{2kT}{m}\,\frac{1}{\sqrt{\pi}}\,2\,\frac{\sqrt{\pi}}{4} = \frac{kT}{m}, \quad j = x, y, z. \tag{11.69}
$$

Using the result (11.69), we can calculate the average kinetic energy of a single particle (11.56) using the following definition

$$< E > = \frac{1}{2} m \left(< v_x^2 > + < v_y^2 > + < v_z^2 > \right). \tag{11.70}$$

Substituting (11.69) into (11.70), we obtain

$$< E > = \frac{1}{2} m \left(\frac{kT}{m} + \frac{kT}{m} + \frac{kT}{m} \right) = \frac{3}{2} kT. \tag{11.71}$$

The internal energy of the system can then be calculated as follows:

$$U = N < E > = N \left(\frac{kT}{2} + \frac{kT}{2} + \frac{kT}{2} \right) = \frac{3}{2} NkT. \tag{11.72}$$

From the result (11.72), we see that each of the three translational degrees of freedom makes the same contribution, equal to $\frac{1}{2}NkT$, to the internal energy of the system. This observation is a special case of the so-called *equipartition theorem* in classical thermodynamics.

According to the equipartition theorem, each term in a nonrelativistic expression $E = E(x, y, z, p_x, p_y, p_z)$ for the energy of a constituent particle that is quadratic in a dynamical variable contributes the amount of $\frac{1}{2}NkT$ to the internal energy of the system.

For example, for N classical simple harmonic oscillators with classical energy

$$E_V (x, p_x) = \frac{p_x^2}{2m} + \frac{1}{2} m\omega^2 x^2, \tag{11.73}$$

there are, by (11.73), two such quadratic terms. By the equipartition theorem, we thus expect that the internal energy of any system with N classical harmonic oscillators is given by $U = 2 \times \frac{1}{2}NkT = NkT$. This result for the vibrational degrees of freedom in the classical limit, will be derived from quantum statistical physics in Chapter 12, "Ideal Diatomic Gases." In Section 11.6.2, "Harmonic and Anharmonic Oscillators," we will derive this result from the classical statistical physics of oscillators in thermal equilibrium.

11.5   Thermodynamics of Ideal Monatomic Gases

11.5.1   Caloric-State Equation

Now, we turn to the study of the thermodynamic properties of an ideal monatomic gas and calculate its main thermodynamic variables. Let us start with the general caloric-state equation (8.22). Substituting the partition function for an ideal monatomic gas

(11.26) into the caloric-state equation (8.22), we obtain the internal energy of the system as follows:

$$U = NkT^2 \frac{\partial}{\partial T} (\ln Z)_V,$$ (11.74)

with

$$\ln Z = \ln V + \ln T^{3/2} + \ln \left(\frac{2\pi m k}{h^2} \right)^{3/2}.$$ (11.75)

Using (11.75), we can calculate

$$\frac{\partial}{\partial T} (\ln Z)_V = \frac{d}{dT} \left(\frac{3}{2} \ln T \right) = \frac{3}{2T}.$$ (11.76)

Substituting (11.76) into (11.74), we obtain the result for the internal energy of an ideal monatomic gas

$$U = NkT^2 \frac{3}{2T} = \frac{3}{2} NkT.$$ (11.77)

The result (11.77) is in agreement with the results (11.57) and (11.72) obtained from the kinetic theory of ideal monatomic gases.

11.5.2 Thermal-State Equation

The free energy of an ideal monatomic gas is then obtained from the general result (8.1) with the partition function for an ideal monatomic gas (11.26). Thus, we have

$$F = -NkT \ln \frac{Z}{N} - NkT.$$ (11.78)

Substituting (11.26) into (11.78), we obtain

$$F = -NkT \ln \left[\frac{V}{N} \left(\frac{2\pi m k T}{h^2} \right)^{3/2} \right] - NkT.$$ (11.79)

or

$$F = -\frac{3}{2} NkT \left[\frac{2}{3} \ln \frac{V}{N} + \ln \left(\frac{2\pi m k T}{h^2} \right) \right] - NkT.$$ (11.80)

From the result (8.2), we can calculate the entropy of an ideal monatomic gas

$$S = \frac{U - F}{T} = \frac{3}{2} Nk \left[1 + \frac{2}{3} \ln \frac{V}{N} + \ln \left(\frac{2\pi m k T}{h^2} \right) \right] + kN.$$ (11.81)

Using now the result (8.42), we obtain the thermal-state equation for an ideal monatomic gas

$$p = -\left(\frac{\partial F}{\partial V}\right)_T = NkT \frac{\partial}{\partial V}[\ln V + f(T, N)], \tag{11.82}$$

or finally

$$p = \frac{NkT}{V} \Rightarrow pV = NkT. \tag{11.83}$$

This important result for the pressure of a classical ideal gas is known as the *ideal gas law*. Here, we derived it for an ideal monatomic gas; however, it holds also for ideal gases of polyatomic molecules, provided that N in (11.83) is understood as the total number of molecules (see Chapter 12). The equation (11.83) applies also to a mixture of several different ideal gases, provided that N is understood as the total number of molecules (of whatever sort) in the mixture. It should be stressed that the equation (11.83) is exact for classical ideal gases only. However, the classical ideal gas concept does correspond to the physically significant dilute limit of realistic gases, in which one can ignore the interparticle interactions (discussed in Chapter 13, "Nonideal Gases") and quantum correlations (discussed in the Part Four of this book). The enthalpy for an ideal monatomic gas is then given by (9.19), i.e.,

$$H = U + pV = \frac{3}{2}NkT + NkT = \frac{5}{2}NkT. \tag{11.84}$$

11.5.3   Universal and Particular Gas Constants

In the expressions for the thermodynamic variables of an ideal monatomic gas, the product Nk is often encountered. The number of particles N is a very large number, not suitable for the practical analysis. On the other hand, the Boltzmann constant k, given in (7.36), is a very small number, which is not very convenient for practical calculations either. It is, therefore, of interest to express the product Nk in terms of some more convenient macroscopic quantities.

In order to achieve that goal, we first note that 1 mole of a gas has a mass M in grams equal to the atomic/molecular number of that gas. For example, the chemical element helium (He^4) has the atomic weight 4. Thus, the mass of 1 mole of an ideal monatomic gas, consisting of helium atoms as constituent particles, is $M = 4g$. On the other hand, 1 mole of any substance consists of N_{AV} particles, where

$$N_{AV} = 6.023 \cdot 10^{23} \frac{\text{Molecules}}{\text{Mole}}, \tag{11.85}$$

is the so-called *Avogadro number*. The number of moles in an ideal monatomic gas, with an arbitrary mass m_{gas} and an arbitrary number of constituent particles N, is equal to

$$\frac{N}{N_{AV}} = \frac{m_{gas}}{M}. \tag{11.86}$$

From the equation (11.86), we can calculate the product Nk as

$$Nk = \frac{N}{N_{AV}} N_{AV} k = \frac{m_{gas}}{M} N_{AV} k. \tag{11.87}$$

Let us now introduce the *universal gas constant,* denoted by R, using the following definition

$$R = N_{AV} k, \tag{11.88}$$

or

$$R = 6.023 \cdot 10^{23} \times 1.3807 \cdot 10^{-23} \frac{J}{K} = 8.314 \frac{J}{K \, Mole}. \tag{11.89}$$

We can also introduce the *particular gas constant,* denoted by R_g, as

$$R_g = \frac{R}{M}, \tag{11.90}$$

which is different for different gases. Using (11.89) and (11.90), we finally obtain

$$Nk = \frac{m_{gas}}{M} R = m_{gas} R_g. \tag{11.91}$$

Substituting (11.91) into (11.77), the caloric-state equation becomes

$$U = \frac{3}{2} \frac{m_{gas}}{M} RT \Rightarrow u = \frac{U}{m_{gas}} = \frac{3}{2} R_g T, \tag{11.92}$$

where $u = U/m_{gas}$ is the internal energy per unit mass. Similarly the thermal-state equation (11.83) becomes

$$pV = \frac{m}{M} RT \Rightarrow pV = R_g T, \tag{11.93}$$

where $v = V/m_{gas}$ is the volume per unit mass. The enthalpy (11.84) is given by

$$H = \frac{5}{2} \frac{m_{gas}}{M} RT \Rightarrow h = \frac{5}{2} R_g T, \tag{11.94}$$

where $h = H/m_{gas}$ is the enthalpy per unit mass.

11.5.4 Caloric and Thermal Coefficients

The specific heat at constant volume (9.55) for ideal monatomic gas becomes

$$c_V = \left(\frac{\partial u}{\partial T}\right)_V = \frac{3}{2}R_g. \tag{11.95}$$

The specific heat at constant pressure (9.64) for ideal monatomic gas becomes

$$c_P = \left(\frac{\partial h}{\partial T}\right)_P = \frac{5}{2}R_g. \tag{11.96}$$

The relation between the specific heats for ideal monatomic gas is now given by

$$c_V - c_P = R_g. \tag{11.97}$$

The volume expansion coefficient (9.26) for ideal monatomic gas is calculated as follows:

$$\alpha = \frac{1}{v_0}\left(\frac{\partial v}{\partial T}\right)_P = \frac{1}{v_0}\frac{\partial}{\partial T}\left(\frac{R_g T}{p_0}\right) = \frac{R_g}{p_0 v_0} = \frac{1}{T_0}. \tag{11.98}$$

The volume expansion itself is described by a linear equation

$$v = v_0\left(1 + \frac{T}{T_0}\right) = v_0\left(1 + \frac{T}{273.16}\right), \tag{11.99}$$

where the temperature T is in Centigrade. The isochoric compressibility coefficient (9.29) for ideal monatomic gas is calculated as follows:

$$\beta = \frac{1}{p_0}\left(\frac{\partial p}{\partial T}\right)_V = \frac{1}{p_0}\frac{\partial}{\partial T}\left(\frac{R_g T}{v_0}\right) = \frac{R_g}{p_0 v_0} = \frac{1}{T_0}. \tag{11.100}$$

The isochoric compression itself is described by a linear equation

$$p = p_0\left(1 + \frac{T}{T_0}\right) = p_0\left(1 + \frac{T}{273.16}\right), \tag{11.101}$$

where the temperature T is in Centigrade. The isothermal compressibility coefficient (9.41) for ideal monatomic gas is calculated using the relation (9.46) as follows:

$$\gamma = \frac{1}{p_0}\frac{\alpha}{\beta} = \frac{1}{p_0}. \tag{11.102}$$

The isothermal compression itself is described by a linear equation

$$v = v_0\left(1 - \frac{p}{p_0}\right). \tag{11.103}$$

Thus, we have calculated a number of important thermodynamic variables for an ideal monatomic gas. It is of course not difficult to calculate a number of other thermodynamic variables, but such an extensive analysis goes beyond the scope of an introductory book like this one.

11.6 Ideal Gases in External Potentials

In this section, we will consider a classical ideal gas of identical particles in the presence of an external potential acting on all particles. A familiar realization is the Earth's atmosphere in which all the molecules are under the influence of the gravitational field potential. It is well known that, because of gravity, atmospheric gases become nonuniform, with the gas density and pressure decreasing with altitude. Here, we will discuss these phenomena by assuming the gas in thermal equilibrium. In this section, we will also discuss closely related classical statistical mechanics of harmonic and anharmonic oscillators in thermal equilibrium.

11.6.1 General Maxwell–Boltzmann Distribution

Energy of N mutually noninteracting, identical particles placed in an external potential $V(\vec{r}) = V(x, y, z)$ acting on *all* particles is a simple generalization of the equation (11.18), being a sum of all particles, energies of the form

$$E_N = \sum_{a=1}^{N} E_a = \sum_{a=1}^{N} \left[\frac{p_a^2}{2m} + V(\vec{r}_a) \right], \tag{11.104}$$

where $(\vec{r}_a, \vec{p}_a) = (x_a, y_a, z_a, p_{xa}, p_{ya}, p_{za})$ specifies the position and momentum of the a-the particle. For example, in a uniform gravitational field along the z-direction, one has the familiar result $V(\vec{r}_a) = mg z_a$. Let us rewrite the equation (11.104) as

$$E_N = \sum_{a=1}^{N} H_1(\vec{r}_a, \vec{p}_a) = \sum_{a=1}^{N} H_1(x_a, y_a, z_a, p_{xa}, p_{ya}, p_{za}), \tag{11.105}$$

where we introduced the one-particle classical Hamiltonian

$$H_1(\vec{r}, \vec{p}) = \frac{(\vec{p})^2}{2m} + V(\vec{r}) = \frac{p_x^2 + p_y^2 + p_z^2}{2m} + V(x, y, z). \tag{11.106}$$

Note that $H_1(\vec{r}_a, \vec{p}_a) = E_a$ is the energy of the a-th particle. Because there are no interactions between particles, the N-particle partition function still has the form as in the equation (11.16),

$$Z_N = \frac{Z^N}{N!}, \tag{11.107}$$

with Z being the one-particle partition function that, by the equation (11.13), can be written as

$$Z = \frac{1}{h^3} Z_{\text{classical}}, \tag{11.108}$$

with

$$Z_{\text{classical}} = \int \int \int \int \int \int dx\,dy\,dz\,dp_x\,dp_y\,dp_z$$
$$\times \exp\left[-\frac{H_1(x,y,z,p_x,p_y,p_z)}{kT}\right]. \tag{11.109}$$

Introducing here the infinitesimal element of the classical six-dimensional phase space for a single particle as follows:

$$d\Gamma_{C1} = dx\,dy\,dz\,dp_x\,dp_y\,dp_z = d^3r\,d^3p, \tag{11.110}$$

we obtain a compact formal notation for the partition function $Z_{\text{classical}}$ in the following form

$$Z_{\text{classical}} = \int_{\Gamma_{C1}} \exp\left(-\frac{H_1(\vec{r},\vec{p})}{kT}\right) d\Gamma_{C1}. \tag{11.111}$$

The internal energy of the gas is, by the equation (8.22), given by $U = N U_1$, with U_1 being the internal energy per particle,

$$U_1 = kT^2 \frac{\partial}{\partial T}(\ln Z) = kT^2 \frac{\partial}{\partial T}(\ln Z_{\text{classical}}). \tag{11.112}$$

Note that, by the equation (11.112), our result for U_1 does not depend on the Planck constant, hence it represents a prediction based purely on classical physics. By the equation (11.112),

$$U_1 = kT^2 \frac{1}{Z_{\text{classical}}} \frac{\partial Z_{\text{classical}}}{\partial T}. \tag{11.113}$$

Using here

$$\frac{\partial Z_{\text{classical}}}{\partial T} = \int_{\Gamma_{C1}} d\Gamma_{C1} \frac{\partial}{\partial T} \exp\left(-\frac{H_1(\vec{r},\vec{p})}{kT}\right)$$
$$= \int_{\Gamma_{C1}} d\Gamma_{C1} \exp\left(-\frac{H_1(\vec{r},\vec{p})}{kT}\right) \frac{H_1(\vec{r},\vec{p})}{kT^2}, \tag{11.114}$$

the single-particle internal energy (11.113) can be written as

$$U_1 = \int_{\Gamma_{C1}} d\Gamma_{C1} \frac{1}{Z_{classical}} \exp\left(-\frac{H_1(\vec{r},\vec{p})}{kT}\right) H_1(\vec{r},\vec{p}), \qquad (11.115)$$

or in a brief form, as

$$U_1 = \int_{\Gamma_{C1}} d\Gamma_{C1} F_1(\vec{r},\vec{p}) H_1(\vec{r},\vec{p}), \qquad (11.116)$$

with

$$F_1(\vec{r},\vec{p}) = \frac{1}{Z_{classical}} \exp\left(-\frac{H_1(\vec{r},\vec{p})}{kT}\right). \qquad (11.117)$$

By the equation (11.117), the function $F_1(\vec{r},\vec{p})$ is positive and satisfies the condition

$$\int_{\Gamma_{C1}} d\Gamma_{C1} F_1(\vec{r},\vec{p}) = \int_{\Gamma_{C1}} d\Gamma_{C1} \frac{1}{Z_{classical}} \exp\left(-\frac{H_1(\vec{r},\vec{p})}{kT}\right)$$

$$= \frac{1}{Z_{classical}} \int_{\Gamma_{C1}} d\Gamma_{C1} \exp\left(-\frac{H_1(\vec{r},\vec{p})}{kT}\right) = \frac{Z_{classical}}{Z_{classical}} = 1, \qquad (11.118)$$

or

$$\int_{\Gamma_{C1}} d\Gamma_{C1} F_1(\vec{r},\vec{p}) = 1. \qquad (11.119)$$

In view of these facts and the form of the equation (11.116), the $F_1(\vec{r},\vec{p})$ has the meaning of a single-particle probability density such that the quantity $d^3r d^3p F_1(\vec{r},\vec{p})$ represents the probability to find the particle within the six-dimensional phase-space volume element $d\Gamma_{C1} = d^3r d^3p$ around the phase-space point $(\vec{r},\vec{p})$. In view of this, it is natural to write the equation (11.116), as

$$U_1 = <H_1(\vec{r},\vec{p})> = \int_{\Gamma_{C1}} d\Gamma_{C1} F_1(\vec{r},\vec{p}) H_1(\vec{r},\vec{p}), \qquad (11.120)$$

with the notation

$$<b_1(\vec{r},\vec{p})> = \int_{\Gamma_{C1}} d\Gamma_{C1} F_1(\vec{r},\vec{p}) b_1(\vec{r},\vec{p}), \qquad (11.121)$$

for the average value of any single particle quantity $b_1(\vec{r}, \vec{p})$. For example, the average kinetic energy of the particle is given by

$$< \frac{(\vec{p})^2}{2m} > = \int_{\Gamma_{C1}} d\Gamma_{C1} F_1(\vec{r}, \vec{p}) \frac{(\vec{p})^2}{2m}, \tag{11.122}$$

whereas the average value of the potential energy of the particle is given by

$$< V(\vec{r}) > = \int_{\Gamma_{C1}} d\Gamma_{C1} F_1(\vec{r}, \vec{p}) V(\vec{r}). \tag{11.123}$$

Since $d^3r\, d^3p\, F_1(\vec{r}, \vec{p})$ represents the probability of finding any of the N particles within the six-dimensional phase-space volume element $d\Gamma_{C1} = d^3r\, d^3p$ around the phase-space point $(\vec{r}, \vec{p})$, the quantity,

$$dN = N \cdot d^3r\, d^3p\, F_1(\vec{r}, \vec{p}) = N \cdot d\Gamma_{C1} F_1(\vec{r}, \vec{p}), \tag{11.124}$$

must be the average number of particles present within the six-dimensional phase-space volume element $d\Gamma_{C1} = d^3r\, d^3p$ around the phase-space point $(\vec{r}, \vec{p})$. By adding all these dN's over the entire phase-space volume, we must get N, i.e.,

$$\int_{\Gamma_{C1}} dN = \int_{\Gamma_{C1}} N\, d\Gamma_{C1} F_1(\vec{r}, \vec{p}) = N. \tag{11.125}$$

Indeed, note that the equation (11.125) is equivalent to the equation (11.119). By $d\Gamma_{C1} = d^3r\, d^3p$, let us rewrite (11.125) in the following form

$$\int_{\vec{r}} d^3r \int_{\vec{p}} d^3p\, N F_1(\vec{r}, \vec{p}) = N, \tag{11.126}$$

or as

$$\int_{\vec{r}} d^3r\, n(\vec{r}) = N, \tag{11.127}$$

where the symbols $\int_{\vec{r}}$ and $\int_{\vec{p}}$ denote triple integrals over the volume of the ordinary $\vec{r}$-space and the $\vec{p}$-space, respectively. In the equation (11.127), we introduced a field like variable $n(\vec{r})$ via

$$n(\vec{r}) = \int_{\vec{p}} d^3p\, N F_1(\vec{r}, \vec{p}). \tag{11.128}$$

Note that, since the integration in the equation (11.128) is carried over $\vec{p}$ only, the result is a function of $\vec{r}$ only. As suggested by the equation (11.127), the quantity $n(\vec{r})$ is the average number of all particles that are within the ordinary three-dimensional volume element $d^3 r$ around the three-dimensional space point $\vec{r}$. In view of this, the field $n(\vec{r})$ is the *local* particle number density at the space point $\vec{r}$.

In the applications to N-particle systems, the quantity,

$$f_1(\vec{r}, \vec{p}) = N \cdot F_1(\vec{r}, \vec{p}), \tag{11.129}$$

the so-called *one-particle reduced distribution*, plays a significant role, as shown by the above equations that could have been written in terms of $f_1(\vec{r}, \vec{p})$ only. For example,

$$n(\vec{r}) = \int_{\vec{p}} d^3 p f_1(\vec{r}, \vec{p}). \tag{11.130}$$

Likewise, the internal energy of the ideal gas in an external field can also be written in terms of $f_1(\vec{r}, \vec{p})$. Indeed, by $U = N U_1$ and equations (11.116) and (11.129), we have

$$U = N \cdot U_1 = N \cdot \int_{\Gamma_{C1}} d\Gamma_{C1} F_1(\vec{r}, \vec{p}) H_1(\vec{r}, \vec{p}), \tag{11.131}$$

or, by $d\Gamma_{C1} = d^3 r d^3 p$, we have

$$U = \int_{\vec{r}} d^3 r \int_{\vec{p}} d^3 p f_1(\vec{r}, \vec{p}) H_1(\vec{r}, \vec{p}). \tag{11.132}$$

This result can be rewritten in a physically more appealing form, as

$$U = \int_{\vec{r}} d^3 r \tilde{u}(\vec{r}), \tag{11.133}$$

where we introduced a field like variable,

$$\tilde{u}(\vec{r}) = \int_{\vec{p}} d^3 p f_1(\vec{r}, \vec{p}) H_1(\vec{r}, \vec{p}). \tag{11.134}$$

Note that, since the integration in the equation (11.134) is carried over $\vec{p}$ only, the result is a function of $\vec{r}$ only. As suggested by the equation (11.133), the quantity $d^3 r \tilde{u}(\vec{r})$ is the internal energy of all particles that are within the ordinary three-dimensional volume element $d^3 r$ around the three-dimensional space point $\vec{r}$. In view of this, the field $\tilde{u}(\vec{r})$ is the *local* energy density (per unit volume) at the point $\vec{r}$. Likewise, for the total internal kinetic energy of the gas, we have

$$U_K = < \sum_{a=1}^{N} \frac{p_a^2}{2m} > = N \cdot < \frac{(\vec{p})^2}{2m} >$$

$$= N \cdot \int_{\Gamma_{C1}} d\Gamma_{C1} F_1(\vec{r}, \vec{p}) \frac{(\vec{p})^2}{2m} = \int_{\Gamma_{C1}} d\Gamma_{C1} N F_1(\vec{r}, \vec{p}) \frac{(\vec{p})^2}{2m}$$

$$= \int_{\Gamma_{C1}} d\Gamma_{C1} f_1(\vec{r}, \vec{p}) \frac{(\vec{p})^2}{2m}. \tag{11.135}$$

This result can be rewritten in a physically more appealing form

$$U_K = \int_{\vec{r}} d^3 r \, \tilde{u}_K(\vec{r}), \tag{11.136}$$

where we introduced a field like variable,

$$\tilde{u}_K(\vec{r}) = \int_{\vec{p}} d^3 p f_1(\vec{r}, \vec{p}) \frac{(\vec{p})^2}{2m}. \tag{11.137}$$

Much like in the previous examples, it is easy to see that $\tilde{u}_K(\vec{r})$ is the *local* density of the kinetic energy.

The above concept of the one-particle distribution function is applicable even to systems that are out of equilibrium, in which case $f_1(\vec{r}, \vec{p})$ is also a function of time t. Interactions between particles also do not modify any of the ideas expressed in the equations (11.125) through (11.135), as they all stem from the idea behind the equation (11.124), i.e., the equation

$$dN = d^3 r \, d^3 p f_1(\vec{r}, \vec{p}), \tag{11.138}$$

giving the number of particles within the six-dimensional phase-space volume element $d\Gamma_{C1} = d^3 r \, d^3 p$ around the phase-space point $(\vec{r}, \vec{p})$. Studying the time evolution of $f_1(\vec{r}, \vec{p}, t)$ was one of the first tasks of the nonequilibrium statistical mechanics undertaken by Ludwig Boltzmann, who proposed the first transport equation for the one-particle reduced distribution function for dilute gases. The Boltzmann transport equation and its successors generally predict that the system evolves into the thermal equilibrium state. In this state, $f_1(\vec{r}, \vec{p})$ is, for very dilute (ideal) gases, given by equations (11.117) and (11.129) yielding

$$f_1(\vec{r}, \vec{p}) = N \cdot F_1(\vec{r}, \vec{p}) = \frac{N}{Z_{\text{classical}}} \exp\left(-\frac{H_1(\vec{r}, \vec{p})}{kT} \right), \tag{11.139}$$

with $Z_{\text{classical}}$ as given in the equation (11.109) or (11.111). With the one-particle Hamiltonian in (11.106), being the sum of a kinetic and a potential energy, one has

$$Z_{\text{classical}} = \int_{\vec{r}} d^3r \int_{\vec{p}} d^3p \, \exp\left(-\frac{H_1(\vec{r},\vec{p})}{kT}\right), \tag{11.140}$$

or

$$Z_{\text{classical}} = \int_{\vec{p}} d^3p \, \exp\left(-\frac{p^2}{2mkT}\right) \int_{\vec{r}} d^3r \, \exp\left(-\frac{V(\vec{r})}{kT}\right). \tag{11.141}$$

Thus,

$$Z_{\text{classical}} = Z_{\text{KIN}} \times Z_{\text{POT}}, \tag{11.142}$$

with the "kinetic" partition function,

$$Z_{\text{KIN}} = \int_{\vec{p}} d^3p \, \exp\left(-\frac{p^2}{2mkT}\right)$$

$$= \left[\int_{-\infty}^{+\infty} dp_x \, \exp\left(-\frac{p_x^2}{2mkT}\right)\right]^3 = (2\pi\, mkT)^{3/2}, \tag{11.143}$$

and the "potential" partition function,

$$Z_{\text{POT}} = \int_{\vec{r}} d^3r \, \exp\left(-\frac{V(\vec{r})}{kT}\right). \tag{11.144}$$

Thus,

$$F_1(\vec{r},\vec{p}) = F_{\text{KIN}}(\vec{p}) \cdot F_{\text{POT}}(\vec{r}) \Rightarrow f_1(\vec{r},\vec{p}) = N \cdot F_{\text{KIN}}(\vec{p}) \cdot F_{\text{POT}}(\vec{r}), \tag{11.145}$$

with

$$F_{\text{KIN}} = \frac{1}{Z_{\text{KIN}}} \exp\left(-\frac{p^2}{2mkT}\right) = \frac{1}{(2\pi\, mkT)^{3/2}} \exp\left(-\frac{p^2}{2mkT}\right), \tag{11.146}$$

and

$$F_{\text{POT}} = \frac{1}{Z_{\text{POT}}} \exp\left(-\frac{V(\vec{r})}{kT}\right).$$

(11.147)

By virtue of the equations (11.143–11.147), we have that

$$\int_{\vec{p}} d^3p\, F_{\text{KIN}}(\vec{p}) = 1, \quad \int_{\vec{r}} d^3r\, F_{\text{POT}}(\vec{r}) = 1.$$

(11.148)

The one-particle reduced distribution function $f_1(\vec{r},\vec{p})$ in the equation (11.145) is known as the *Maxwell–Boltzmann equilibrium distribution*. It does not contain the Planck constant at any place, so it is a classical distribution. The major feature of the probability density $F_1(\vec{r},\vec{p})$ for the position $\vec{r}$ and the momentum $\vec{p}$ of a single particle is that $\vec{r}$ and $\vec{p}$ are statistically independent because of the product form of $F_1(\vec{r},\vec{p})$ in the equation (11.145). The main consequence of that is that if, for example, one considers the functions $b_1(\vec{p})$ of one-particle momentum only, then, by equations (11.121) and (11.148), one has

$$< b_1(\vec{p}) > = \int_{\vec{r}} d^3r \int_{\vec{p}} d^3p\, F_1(\vec{r},\vec{p})\, b_1(\vec{p})$$

$$= \int_{\vec{r}} d^3r \int_{\vec{p}} d^3p\, F_{\text{POT}}(\vec{r})\, F_{\text{KIN}}(\vec{p})\, b_1(\vec{p})$$

$$= \int_{\vec{r}} d^3r\, F_{\text{POT}}(\vec{r}) \int_{\vec{p}} d^3p\, F_{\text{KIN}}(\vec{p})\, b_1(\vec{p})$$

$$= \int_{\vec{p}} d^3p\, F_{\text{KIN}}(\vec{p})\, b_1(\vec{p})$$

$$= \int_{\vec{p}} d^3p\, \frac{1}{Z_{\text{KIN}}} \exp\left(-\frac{p^2}{2mkT}\right) b_1(\vec{p}).$$

(11.149)

From the equation (11.149), we see that this result is completely insensitive to the form of the external potential $V(\vec{r})$. So, the final result for the average calculated in the equation (11.149) is the same as for the particles confined in a box. Thus, for example, for the average kinetic energy of a single particle, for whatever $V(\vec{r})$, we have the same result as the one in the equation (11.71),

$$< \frac{p^2}{2m} > = \int_{\vec{p}} \frac{d^3p}{Z_{\text{KIN}}} \exp\left(-\frac{p^2}{2mkT}\right) \frac{p^2}{2m} = \frac{3}{2}kT.$$

(11.150)

Thus, within classical physics, the equilibrium probability density distribution of particle momenta is universal, and it is, by the equation (11.146), given by the Maxwell formula

$$F_{KIN}(\vec{p}) = \frac{1}{(2\pi m k T)^{3/2}} \exp\left(-\frac{p^2}{2m k T}\right). \tag{11.151}$$

It should be stressed that the content of the equation (11.151) is the same as that of the velocity distribution (11.59), as can be seen by using $\vec{p} = m\vec{v}$, i.e.,

$$F_{KIN}(\vec{p})\, d^3p = F_{KIN}(m\vec{v})\, m^3\, d^3v$$

$$= \left(\frac{m}{2\pi k T}\right)^{3/2} \exp\left(-\frac{m v^2}{2 k T}\right) d^3v, \tag{11.152}$$

coinciding with (11.59).

Though not affecting the momentum distribution, the external potential does have a significant effect on the ideal gases. It makes them nonuniform. Indeed, by (11.130) with the Maxwell–Boltzmann function $f_1(\vec{r}, \vec{p})$ in the equation (11.145), we find for the local gas density,

$$n(\vec{r}) = \int_{\vec{p}} d^3 p f_1(\vec{r}, \vec{p}) = \int_{\vec{p}} d^3 p N F_{KIN}(\vec{p}) F_{POT}(\vec{r})$$

$$= N F_{POT}(\vec{r}) \int_{\vec{p}} d^3 p F_{KIN}(\vec{p}) = N F_{POT}(\vec{r}), \tag{11.153}$$

where for the last step we employed the equation (11.148). From the equation (11.153) with (11.147), we obtain

$$n(\vec{r}) = N F_{POT}(\vec{r}) = \frac{N}{Z_{POT}} \exp\left(-\frac{V(\vec{r})}{k T}\right). \tag{11.154}$$

The result (11.154) is well known as the *Boltzmann formula*. It shows that the local gas density crucially depends on the external potential through the second, exponential term in the equation (11.154), the so-called *Boltzmann factor*. By writing the equation (11.154) at two different points, $\vec{r}$ and $\vec{r}_0$, we obtain

$$\frac{n(\vec{r})}{n(\vec{r}_0)} = \exp\left(-\frac{V(\vec{r}) - V(\vec{r}_0)}{k T}\right). \tag{11.155}$$

For example, in a uniform gravitational field along the z-direction, one has the familiar result $V(\vec{r}) = V(x, y, z) = mgz$, so by the equation (11.154), $n(\vec{r})$ will depend on z only. Thus, by (11.155), we easily find

$$\frac{n(z)}{n(z_0)} = \exp\left(-\frac{mgz - mgz_0}{k T}\right), \tag{11.156}$$

or with $z_0 = 0$,

$$n(z) = n(0) \exp\left(-\frac{mgz}{kT}\right),$$ (11.157)

showing that the local density $n(\vec{r})$ exponentially decreases with increasing height. As another example, consider a gas in a container with volume V, of whatever shape. If the walls of the container are impenetrable, the container can be modeled as a particle in a box potential, with

$$V(\vec{r}) = 0 \Rightarrow \exp\left(-\frac{V(\vec{r})}{kT}\right) = 1,$$ (11.158)

inside the box, and

$$V(\vec{r}) = +\infty \Rightarrow \exp\left(-\frac{V(\vec{r})}{kT}\right) = 0,$$ (11.159)

outside the box. From the equations (11.158) and (11.159), one can easily see that the potential partition function (11.144) is, for this case,

$$Z_{\mathrm{POT}} = \int_{\vec{r}} d^3r \exp\left(-\frac{V(\vec{r})}{kT}\right) = \int_{\vec{r} \in \mathrm{inside}} d^3r = V,$$ (11.160)

i.e., it is equal to the volume V of the space inside the container. The Boltzmann formula (11.154) with (11.158) and (11.160), thus, gives

$$n(\vec{r}) = \frac{N}{V},$$ (11.161)

for an $\vec{r}$ inside the container, and, by equations (11.159) and (11.160),

$$n(\vec{r}) = 0,$$ (11.162)

for an $\vec{r}$ outside the container.

Let us now calculate the local density of the kinetic energy. For a uniform gas, this density is expected to be uniform and given by

$$\tilde{u}_K = \frac{U}{V} = \frac{\frac{3}{2}NkT}{V} = \frac{3}{2}\frac{N}{V}kT = \frac{3}{2}NkT.$$ (11.163)

For the general case of a classical gas in an external potential, from the equation (11.137), we have

$$\tilde{u}_K(\vec{r}) = \int_{\vec{p}} d^3p f_1(\vec{r}, \vec{p}) \frac{(\vec{p})^2}{2m}$$ (11.164)

yielding, with the Maxwell–Boltzmann distribution (11.145), the result,

$$
\tilde{u}_K(\vec{r}) = \int_{\vec{p}} d^3 p \, N \, F_{KIN}(\vec{p}) \, F_{POT}(\vec{r}) \frac{(\vec{p})^2}{2m}
$$

$$
= N \, F_{POT}(\vec{r}) \int_{\vec{p}} d^3 p \, F_{KIN}(\vec{p}) \frac{(\vec{p})^2}{2m}
$$

$$
= N \, F_{POT}(\vec{r}) \frac{3}{2} k T = \frac{3}{2} n(\vec{r}) k T, \tag{11.165}
$$

where we made use of equations (11.145), (11.146), (11.150), and (11.154). Thus, the local kinetic energy density is given by

$$
\tilde{u}_K(\vec{r}) = \frac{3}{2} n(\vec{r}) k T, \tag{11.166}
$$

with $n(\vec{r})$ given by the Boltzmann formula. Note that the equation (11.166) is of the same form as the equation (11.163) for a uniform gas with, however, the uniform density replaced by the local particle density. Though not so obvious, the result in the equation (11.166) is not surprising, in view of the previous observation that the presence of the potential does not affect the average kinetic energy of a particle (it is still $= \frac{3}{2} k T$), and, as there are $n(\vec{r})$ molecules per unit volume, the energy density must be given by the equation (11.166). Finally, in view of the result (11.166), one may expect that the nonuniform gas may be assigned a nonuniform pressure field, following the local density in the same way as for uniform gases, in the form

$$
p(\vec{r}) = n(\vec{r}) k T = \frac{2}{3} \tilde{u}_K(\vec{r}). \tag{11.167}
$$

We will prove the validity of this guess in the Problem 8 of this chapter. In combination with the Boltzmann formula (11.154) applied to the Earth's atmosphere gases in the equation (11.157), the equation (11.167) yields the well-known *Barometric formula*

$$
p(z) = k T n(z) = k T n(0) \exp\left(-\frac{mgz}{kT}\right)
$$

$$
= p(0) \exp\left(-\frac{mgz}{kT}\right), \tag{11.168}
$$

showing that the local pressure $p(z)$ exponentially decreases with increasing height.

11.6.2 Harmonic and Anharmonic Oscillators

Here, we discuss a single particle moving in a one-dimensional potential $V(x)$, with the energy

$$
H_1(x, p) = \frac{p^2}{2m} + V(x). \tag{11.169}
$$

The discussion of the classical equilibrium statistics of this system is in large part similar to that of the previous section for the particle moving in three dimensions. Of special interest is the harmonic oscillator case, corresponding to the potential

$$V(x) = \frac{m\omega^2}{2} x^2.$$ (11.170)

Here, ω is the angular frequency of the harmonic oscillator. For the one-dimensional system, the one-particle partition function can be written as follows:

$$Z = \frac{1}{h} Z_{classical},$$ (11.171)

with

$$Z_{classical} = \int \int dx\, dp\, \exp\left(-\frac{H_1(x,p)}{kT}\right).$$ (11.172)

The internal energy of this single-particle system is, by the equation (8.22), given by

$$U_1 = kT^2 \frac{\partial}{\partial T}(\ln Z) = kT^2 \frac{\partial}{\partial T}(\ln Z_{classical}).$$ (11.173)

Note that, by the equation (11.173), our result for U_1 does not depend on the Planck constant, hence it represents a prediction based purely on classical physics. Similarly to the case of a particle moving in three dimensions, here we find that

$$U_1 = <H_1(x,p)> = \int \int dx\, dp\, F_1(x,p) H_1(x,p),$$ (11.174)

with the probability density given by

$$F_1(x,p) = \frac{1}{Z_{classical}} \exp\left(-\frac{H_1(x,p)}{kT}\right).$$ (11.175)

Again, we have

$$<b_1(x,p)> = \int \int dx\, dp\, F_1(x,p) b_1(x,p),$$ (11.176)

for the average value of any single particle quantity $b_1(x,p)$. For example, the average kinetic energy of the particle is given by

$$<\frac{p^2}{2m}> = \int \int dx\, dp\, F_1(x,p) \frac{p^2}{2m},$$ (11.177)

whereas the average value of the potential energy of the particle is given by

$$< V(x) > = \int \int dx\, dp\, F_1(x, p)\, V(x). \tag{11.178}$$

By (11.169) and (11.172),

$$Z_{\text{classical}} = Z_{\text{KIN}} \times Z_{\text{POT}}, \tag{11.179}$$

with the "kinetic" partition function

$$Z_{\text{KIN}} = \int\limits_{-\infty}^{+\infty} dp \exp\left(-\frac{p^2}{2mkT}\right) = (2\pi\, mkT)^{1/2}, \tag{11.180}$$

and the "potential" partition function

$$Z_{\text{POT}} = \int\limits_{-\infty}^{+\infty} dx \exp\left(-\frac{V(x)}{kT}\right). \tag{11.181}$$

Thus,

$$F_1(x, p) = F_{\text{KIN}}(p) \cdot F_{\text{POT}}(x), \tag{11.182}$$

with

$$F_{\text{KIN}} = \frac{1}{Z_{\text{KIN}}} \exp\left(-\frac{p^2}{2mkT}\right)$$

$$= \frac{1}{(2\pi\, mkT)^{1/2}} \exp\left(-\frac{p^2}{2mkT}\right), \tag{11.183}$$

and

$$F_{\text{POT}} = \frac{1}{Z_{\text{POT}}} \exp\left(-\frac{V(x)}{kT}\right). \tag{11.184}$$

By virtue of the equations (11.180–11.181) and (11.183–11.184), we have that

$$\int\limits_{-\infty}^{+\infty} dp\, Z_{\text{KIN}}(p) = 1, \quad \int\limits_{-\infty}^{+\infty} dx\, Z_{\text{POT}}(x) = 1. \tag{11.185}$$

As in the case of a particle in three dimensions, the probability density in the equation (11.182) is of the Maxwell–Boltzmann type, with the position and momentum being independent random variables, with probability densities given in equations (11.183) and (11.184). In relation to this, inserting (11.182) into the equations (11.177) and (11.178), gives the results

$$
< \frac{p^2}{2m} > = \int\limits_{-\infty}^{+\infty} \frac{dp}{Z_{\text{KIN}}} \exp\left(-\frac{p^2}{2mkT}\right) \frac{p^2}{2m} = \frac{kT}{2} \tag{11.186}
$$

and

$$
< V(x) > = \int\limits_{-\infty}^{+\infty} \frac{dx}{Z_{\text{POT}}} \exp\left(-\frac{V(x)}{kT}\right) V(x). \tag{11.187}
$$

In evaluating the integral in the equation (11.180), we used the result

$$
\int\limits_{-\infty}^{+\infty} dz \exp\left(-\frac{\alpha z^2}{2}\right) = \left(\frac{2\pi}{\alpha}\right)^{1/2}, \tag{11.188}
$$

whereas, in evaluating the integral in the equation (11.186), we used the result

$$
\int\limits_{-\infty}^{+\infty} z^2 dz \exp\left(-\frac{\alpha z^2}{2}\right) = \left(\frac{2\pi}{\alpha}\right)^{1/2} \frac{1}{\alpha}. \tag{11.189}
$$

For the case of the harmonic oscillator potential in the equation (11.170), we have

$$
Z_{\text{POT}} = \int\limits_{-\infty}^{+\infty} dx \exp\left(-\frac{m\omega^2 x^2}{2kT}\right) = \left(\frac{2\pi kT}{m\omega^2}\right)^{1/2}. \tag{11.190}
$$

So, by (11.187),

$$
< V(x) > = < \frac{m\omega^2 x^2}{2} > = \int\limits_{-\infty}^{+\infty} dx \left(\frac{m\omega^2}{2\pi kT}\right)^{1/2}
$$

$$
\times \exp\left(-\frac{m\omega^2 x^2}{2kT}\right) \frac{m\omega^2 x^2}{2} = \frac{kT}{2}. \tag{11.191}
$$

In evaluating the integrals in equations (11.190) and (11.191), we again used the results (11.188) and (11.189). In summary, by equations (11.186) and (11.191), for the classical harmonic oscillator, there is an equipartition of energy, with

$$
< \frac{p^2}{2m} > = < \frac{m\omega^2 x^2}{2} > = \frac{kT}{2}. \tag{11.192}
$$

So the net harmonic oscillator internal energy in the equation (11.174) is

$$U_1 = <H_1(x,p)> = <\frac{p^2}{2m}> + <\frac{m\omega^2 x^2}{2}> = kT. \tag{11.193}$$

By equations (11.179) and (11.190), we have for the case of the harmonic oscillator,

$$Z_{classical} = Z_{KIN} \times Z_{POT} = (2\pi\, mkT)^{1/2} \times \left(\frac{2\pi\, kT}{m\omega^2}\right)^{1/2} = \frac{2\pi\, kT}{\omega}. \tag{11.194}$$

So,

$$Z = \frac{Z_{classical}}{h} = \frac{2\pi\, kT}{h\omega} = \frac{kT}{\hbar\omega}. \tag{11.195}$$

Thus, by the equation (11.173),

$$U_1 = kT^2 \frac{\partial}{\partial T}(\ln Z) = kT^2 \frac{\partial}{\partial T} \ln\left(\frac{kT}{\hbar\omega}\right) = kT, \tag{11.196}$$

in accordance with the result (11.193) for the same quantity, obtained in a different way. A notable feature of this classical result is that the internal energy of a single harmonic oscillator does not depend on the oscillator frequency ω. Thus, for a classical system composed of N mutually noninteracting oscillators with whatever frequencies, the net vibration internal energy is simply given by

$$U_{VIB} = N U_1 = NkT. \tag{11.197}$$

Note that the equation (11.197) applies even if the N oscillators have different frequencies ω. By virtue of the classical result (11.193), any information about the actual oscillators, frequencies are completely washed out in the classical limit.

11.6.3 Classical Limit of Quantum Partition Function

The partition function Z employed in our discussions so far was an approximation to the exact quantum-mechanical partition function

$$Z_{QM} = \sum_{\alpha} \exp\left(-\frac{E_\alpha}{kT}\right). \tag{11.198}$$

The sum in (11.198) is done over all states α, and if there are degenerate states, then a term is written for each of the degenerate states. In this way, we do not need to employ the degeneracy factor g_α, simply because the states with the same energy are, by convention, repeated in the sum in (11.198). The energy levels are obtained by solving the Schrödinger equation, as discussed in the Part One of this book. For example, for the one-dimensional harmonic oscillator discussed in Section 11.6.2, "Harmonic and Anharmonic Oscillators," the energy levels are given by the equation (4.20). In general, by equations (11.198) and (8.22), the one-particle internal energy is

$$U_1(T) = kT^2 \frac{\partial}{\partial T} \ln Z_{QM} = kT^2 \frac{1}{Z_{QM}} \frac{\partial Z_{QM}}{\partial T}$$

$$= kT^2 \frac{1}{Z_{QM}} \frac{\partial}{\partial T} \left[\sum_\alpha \exp\left(-\frac{E_\alpha}{kT}\right) \right]$$

$$= kT^2 \frac{1}{Z_{QM}} \sum_\alpha \exp\left(-\frac{E_\alpha}{kT}\right) \frac{E_\alpha}{kT^2}$$

$$= \sum_\alpha E_\alpha \frac{1}{Z_{QM}} \exp\left(-\frac{E_\alpha}{kT}\right) = \sum_\alpha E_\alpha P_\alpha, \qquad (11.199)$$

that is,

$$U_1(T) = \sum_\alpha E_\alpha P_\alpha, \qquad (11.200)$$

with

$$P_\alpha = \frac{1}{Z_{QM}} \exp\left(-\frac{E_\alpha}{kT}\right). \qquad (11.201)$$

The equation (11.201) suggests that the quantity P_α is the probability of finding the quantum-mechanical system in the state with energy E_α. In accordance with the quantities P_α being a set of probabilities, are the facts that these quantities are obviously positive [see (11.201)], and that their sum is

$$\sum_\alpha P_\alpha = \frac{1}{Z_{QM}} \sum_\alpha \exp\left(-\frac{E_\alpha}{kT}\right) = \frac{Z_{QM}}{Z_{QM}} = 1. \qquad (11.202)$$

Thus, one can write the equation (11.200) as an average,

$$U_1(T) = <E_\alpha> = \sum_\alpha E_\alpha P_\alpha. \qquad (11.203)$$

The major feature of the result (11.201) is

$$P_\alpha \sim \exp\left(-\frac{E_\alpha}{kT}\right). \qquad (11.204)$$

That is, the probability to find the system in the state α is proportional to the quantum-mechanical Boltzmann factor $\exp[-E_\alpha/kT]$. In the limit $T \to 0$, the probability of finding the system in the ground state (minimum energy state, with energy E_0)

approaches 1, whereas the probabilities of all other, excited states (with energies $E_\alpha > E_0$) approach 0. Indeed, by the equation (11.201), we have

$$\frac{P_\alpha(T)}{P_0(T)} = \exp\left(-\frac{E_\alpha - E_0}{kT}\right) \to 0, \tag{11.205}$$

for $T \to 0$. This, combined with the condition,

$$1 = \sum_\alpha P_\alpha = P_0 \left[1 + \sum_{\alpha \neq 0} \frac{P_\alpha(T)}{P_0(T)}\right], \tag{11.206}$$

implies that, for $T \to 0$,

$$P_0 = \left[1 + \sum_{\alpha \neq 0} \exp\left(-\frac{E_\alpha - E_0}{kT}\right)\right]^{-1} \to 1, \tag{11.207}$$

whereas $P_\alpha(T) \to 0$ for *all* excited states with $\alpha \neq 0$. Thus, by the equation (11.203),

$$U_1(T) = < E_\alpha > = \sum_\alpha E_\alpha P_\alpha = E_0 P_0 + \sum_{\alpha \neq 0} E_\alpha P_\alpha \to E_0, \tag{11.208}$$

that is, the internal energy approaches the ground-state energy for $T \to 0$.

In view of the above discussions, it is clear that the exact quantum-mechanical treatment is essential at low enough temperatures, when one needs to handle the sum over the discrete quantum states carefully. This situation, however, changes with increasing temperatures. For large enough temperatures, the terms in the quantum-mechanical partition function (11.198) will vary slowly with increasing energy. This suggests that, at large enough temperatures, the sum over all states α in the equation (11.198) may be replaced by an integral. To see this, let us consider the one-dimensional system with the Hamiltonian $H_1(x, p)$. In Section 11.6.2, "Harmonic and Anharmonic Oscillators," we already wrote the approximate partition function for this system, in the form of the integral

$$Z = \int \int \frac{dx\,dp}{h} \exp\left(-\frac{H_1(x, p)}{kT}\right). \tag{11.209}$$

In the equation (11.209), there is a division by the Planck constant, yet, as illustrated in Section 11.6.2, "Harmonic and Anharmonic Oscillators," this feature does not affect any of the interesting quantities, such as the internal energy $U_1(T) = < E_\alpha >$. Let us now try to understand the approximate result in the equation (11.209) in terms of the exact quantum-mechanical partition function in the equation (11.198). For this purpose, it is useful to think of the classical physics as of quantum mechanics taken in the limit of zero Planck constant $h \to 0$. Indeed, in this limit, the Heisenberg uncertainty relation would tell us that

$$\Delta x \cdot \Delta p = h \to 0. \tag{11.210}$$

meaning that one would be able to simultaneously measure the position x and momentum p with infinite precisions, i.e., $\Delta x \to 0$ and $\Delta p \to 0$, just as in classical physics. In this limit, one can think of quantum states as of small cells in the two-dimensional (x,p) phase space, each having the

$$cell\ area = \Delta x \cdot \Delta p = h. \tag{11.211}$$

If h is small (i.e., $\Delta x \to 0$ and $\Delta p \to 0$), each of these cells would be a state with energy $E_{cell} = H_1(x,p)$. Thus, for a small Planck constant, the exact quantum-mechanical partition function (11.198), would be approximated by

$$Z_{QM} \approx \sum_{cells} \exp\left(-\frac{E_{cell}}{kT}\right). \tag{11.212}$$

For a small h, the number of cells within the phase-space area element $dx\,dp$ around the point (x,p) is

$$\frac{dx\,dp}{h}, \tag{11.213}$$

and the energy of each of these states around the point (x,p) is

$$E_{cell} = H_1(x,p). \tag{11.214}$$

By the equation (11.213), we can replace the sum of cells by the integral

$$\sum_{cells} \Rightarrow \int\int \frac{dx\,dp}{h}. \tag{11.215}$$

By equations (11.212), (11.214), and (11.215),

$$Z_{QM} \approx \sum_{cells} \exp\left(-\frac{E_{cell}}{kT}\right)$$

$$\approx \int\int \frac{dx\,dp}{h} \exp\left(-\frac{H_1(x,p)}{kT}\right). \tag{11.216}$$

Thus, we arrive at the important conclusion, that in the classical limit (formally for $h \to 0$), one has the approximation

$$Z_{QM} = \sum_{\alpha} \exp\left(-\frac{E_\alpha}{kT}\right)$$

$$\approx Z = \int \int \frac{dx\,dp}{h} \exp\left(-\frac{H_1(x,p)}{kT}\right). \tag{11.217}$$

This proves the validity of using the approximate partition function in the equation (11.209) to study the classical limit in the one-dimensional case discussed in Section 11.6.2, "Harmonic and Anharmonic Oscillators." For the three dimensional case, we had, in Section 11.6.1, "General Maxwell–Boltzmann Distribution,"

$$Z = \int_{\vec{r}} \int_{\vec{p}} \frac{d^3r\,d^3p}{h^3} \exp\left(-\frac{H_1(\vec{r},\vec{p})}{kT}\right). \tag{11.218}$$

This result can be derived in the classical limit along the same lines as the one for the one-dimensional case. The only difference is that here the phase-space is six-dimensional, and that, by the uncertainty relations, we have

$$\Delta x \cdot \Delta p_x = h, \quad \Delta y \cdot \Delta p_y = h, \quad \Delta z \cdot \Delta p_z = h, \tag{11.219}$$

so each cell occupies the phase-space volume

$$\text{cell volume} = \Delta x \cdot \Delta p_x \times \Delta y \cdot \Delta p_y \times \Delta z \cdot \Delta p_z = h^3. \tag{11.220}$$

Hence, in the equation (11.218), we have the division by h^3 to count the number of cells in the classical phase-space volume element

$$d\Gamma_{Cl} = dx\,dy\,dz\,dp_x\,dp_y\,dp_z = d^3r\,d^3p. \tag{11.221}$$

It should be stressed that the above conclusions about the classical limit do not depend on the form of the Hamiltonian. Rather, they are all based on the equations such as (11.210) or (11.219), which determine the cell volume in the appropriate phase space. The classical limit of the N-particle problem with whatever form of the Hamiltonian $H_N = H_N(\vec{r}_1, \vec{p}_1, \vec{r}_2, \vec{p}_2, ..., \vec{r}_N, \vec{p}_N)$, can be handled in exactly the same way. In this case, the cell volume is $h^{d \cdot N}$, for the particles moving in a d-dimensional space. The classical limit of the exact quantum mechanical N-particle partition function is thus given by

$$Z_N = C \left[\prod_{a=1}^{N} \int_{\vec{r}_a} \int_{\vec{p}_a} \frac{d^d r_a\,d^d p_a}{h^d} \right] \exp\left(-\frac{H_N}{kT}\right), \tag{11.222}$$

or

$$Z_N = \frac{C}{h^{d \cdot N}} \left[\prod_{a=1}^{N} \int_{\vec{r}_a} \int_{\vec{p}_a} d^d r_a d^d p_a \right] \exp\left(-\frac{H_N}{kT}\right). \tag{11.223}$$

Here, the pre-factor $C = 1$, if the particles are not identical to each other, i.e., if one deals with N different particles. If one deals with N identical particles, then, by the discussions of Section 8.1, "Free Energy and Partition Function," $C = 1/N!$. By a similar discussion for a mixture of N_1 particles of one kind with N_2 particles of another kind, one finds $C = 1/(N_1!N_2!)$, etc. In Chapter 13, we will use the equation (11.223) to discuss nonideal classical gases of N interacting monatomic molecules in $d = 3$, described by the Hamiltonian

$$H_N = \sum_{a=1}^{N} \frac{p_a^2}{2m} + U(\vec{r}_1, \vec{r}_2, \dots \vec{r}_N), \tag{11.224}$$

where the second term is a potential energy term dependent on all particle positions.

Finally, we note that the classical limit commonly coincides with the high-temperature limit, as illustrated in the following chapters. Typically, there exists a certain system dependent characteristic "quantum" temperature T^*, such that for the temperatures $T \gg T^*$ one can ignore quantum effects and treat thermodynamics classically by means of partitions functions such as those in equations (11.209), (11.218), or (11.223). Frequently, a more complex situation is realized, with several different quantum temperatures. This is the case, for example, for diatomic gases discussed in Chapter 12.

11.7 Problems with Solutions

Problem 1

An ideal monatomic gas, with given specific heat at constant volume c_V, undergoes a quasi-static change of state from the state (T_I, V_I) to the state (T_F, V_F).

(a) Derive the expression for the entropy change of the gas.
(b) Is the result for the entropy change dependent on the path taken between the two states?

Solution

(a) Using the first law of thermodynamics, we can write

$$dQ = T dS = dU + p dV = m c_V dT + p dV, \tag{11.225}$$

or

$$dS = m c_V \frac{dT}{T} + \frac{p}{T} dV. \tag{11.226}$$

Using now the thermal-state equation for the ideal monatomic gas

$$\frac{p}{T} = m \frac{R_g}{V}, \tag{11.227}$$

we obtain

$$dS = m c_V \frac{dT}{T} + m R_g \frac{dV}{V}. \tag{11.228}$$

The entropy change in this process is then given by

$$\Delta S = m c_V \int_{T_I}^{T_F} \frac{dT}{T} + m R_g \int_{V_I}^{V_F} \frac{dV}{V}, \tag{11.229}$$

or

$$\Delta S = m c_V \ln \frac{T_F}{T_I} + m R_g \ln \frac{V_F}{V_I}. \tag{11.230}$$

(b) The result for the entropy change (11.230) is path-independent. Although the differential dQ is an inexact differential, when it is multiplied by its integration factor $1/T$, we obtain an exact differential $dS = dQ/T$. Thus, our result for the entropy change is path-independent. This feature reflects the fundamental fact that the entropy is a state function, $S(N, V, T)$, so the entropy change can be expressed purely in terms of the initial and final state (N, V, T). In particular, the above result in the equation (11.230) can be obtained also from the expression for the ideal gas entropy $S(N, V, T)$ given by (11.81), simply by taking the difference $\Delta S = S(N, V_F, T_F) - S(N, T_I, V_I)$ and using the elementary properties of the ln-function.

Problem 2

A closed cylinder of total length $L = 0.8$ m is divided into two compartments by a very light piston initially fixed at $l_1 = 0.3$ m from the left cylinder base, as shown in the Fig. 11.2.

In the left compartment, there is 1 mole of a monatomic ideal gas with pressure $p_1 = 5 \times 10^5$ N/m^2, while in the right compartment there is some unspecified quantity of another monatomic ideal gas with pressure $p_2 = 10^5$ N/m^2. The cylinder is submerged into the water, and the entire system is thermally insulated at the uniform initial temperature $T_I = 298$ K. If we release the piston, it moves to a new equilibrium

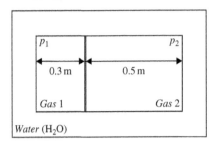

Figure 11.2 A closed cylinder submerged into the water.

position. Assume that the walls of the cylinder are perfectly rigid, so the cylinder has a constant volume. Also, assume that the volume occupied by the water is constant, i.e., that the water confining container has perfectly rigid walls.

(a) Calculate the final temperature of the water.
(b) Calculate the final position of the piston.
(c) Calculate the increase of the total entropy of the system.

Solution

(a) Both gases are ideal monatomic gases and their thermal equation of state is given by

$$pV = \frac{m}{M}RT. \tag{11.231}$$

Thus, for 1 mole of the first gas ($n_1 = 1$), we obtain the initial volume

$$V_{1I} = \frac{RT_I}{p_{1I}} = \frac{8.314 \times 298}{5 \times 10^5} = 0.00485\,\text{m}^3. \tag{11.232}$$

On the other hand, the geometry of the system gives that the volume of the second gas is

$$V_{2I} = \frac{5}{3}V_{1I} = 0.00808\,\text{m}^3. \tag{11.233}$$

From the equation of state, we find the number of moles of the second gas

$$n_2 = \frac{p_{2I}V_{2I}}{RT_{2I}} = \frac{10^5 \times 808 \times 10^{-5}}{8.314 \times 298} = \frac{1}{3}. \tag{11.234}$$

This defines the initial state of the system. In the final state of the system, the pressures of both gases are the same and the volume of the first gas must be three times the volume of the second gas. The final temperature of both gases is also equal ($T_{1F} = T_{2F} = T_F$).

The next question to ask is, what is the final temperature of this entire system made of two gases and the water? To answer this, let us apply the first law of thermodynamics to the *entire* system, $\Delta Q = \Delta U + L$. The entire system is thermally insulated. Thus, $\Delta Q = 0$. Also, the entire system has a constant volume fixed by the outer rigid walls of the water container (see the Fig. 11.2). So, the entire system does zero work on *its environment*, and thus $L = 0$. So, $\Delta U = \Delta Q - L = 0 - 0 = 0$, i.e., the total internal energy of the entire system is constant in this process.

Next, our entire system here is composed of the two ideal gases and water, so the net internal energy can be written as

$$U = \frac{3}{2} N_1 k T_1 + \frac{3}{2} N_2 k T_2 + U_w (T_w, V_w) \tag{11.235}$$

for any temperatures and volumes of the three parts of this entire system. A notable feature of the above internal energy is that it does *not* depend on the volumes of the ideal gases (which do change in this process). This feature (applying to ideal gases only) gives rise to the above simple form for the entire system's internal *energy*, which is just a function of three variables (T_1, T_2, T_w) that can (in principle) change in this system (recall that the water volume is fixed). Thus, the result found above $\Delta U = 0$, i.e., $U_F = U_I$, can (by means of the above result for the entire system's internal *energy*) be written as

$$\frac{3}{2} (N_1 + N_2) k T_I + U_w (T_I, V_w) = \frac{3}{2} (N_1 + N_2) k T_F + U_w (T_F, V_w). \tag{11.236}$$

From the above equation, it is obvious that $T_F = T_I$.

(b) The final position of the piston is determined by the ideal-gas law $pV = nRT$, applied to both gases (at the same final pressure and temperature) yielding $V_1/V_2 = n_1/n_2$, so

$$\frac{V_1}{V_2} = \frac{l_1}{l_2} = 3, \tag{11.237}$$

or

$$\frac{l_1}{l_2} + 1 = \frac{l_1 + l_2}{l_2} = \frac{L}{l_2} = 4 \Rightarrow l_2 = \frac{L}{4} = 0.2\,\text{m}. \tag{11.238}$$

Thus, we finally obtain the required result $l_1 = 3/4 L = 0.6$ m.

(c) The entropy change of the water is equal to zero, since both its temperature and its volume remain unchanged. The entropy change of the entire system is therefore equal to the entropy change of the two gases. For the second gas we have

$$\Delta S_2 = \int \frac{dQ_2}{T} = \int_{V_{2I}}^{V_{2F}} \frac{p_2 \, dV_2}{T_2} = R \int_{V_{2I}}^{V_{2F}} \frac{dV_2}{3V_2}, \tag{11.239}$$

or

$$\Delta S_2 = \frac{1}{3}R\ln\frac{V_{2F}}{V_{2I}}. \tag{11.240}$$

For the first gas we have

$$\Delta S_1 = \int\frac{dQ_1}{T} = R\ln\frac{V_{1F}}{V_{1I}}. \tag{11.241}$$

Thus, we obtain

$$\Delta S = \Delta S_2 + \Delta S_1 = \frac{1}{3}R\ln\frac{V_{2F}}{V_{2I}} + R\ln\frac{V_{1F}}{V_{1I}}, \tag{11.242}$$

or

$$\Delta S = \frac{1}{3}R\ln\frac{2}{5} + R\ln\frac{6}{3} = 3.22\,\frac{J}{K}. \tag{11.243}$$

Problem 3

An ideal monatomic gas of volume V is subject to an infinitesimal quasi-static process. Derive the result for the heat transferred in this process and show that it equals to

$$dQ = \frac{c_V}{nR}V\,dp + \frac{c_P}{nR}p\,dV, \tag{11.244}$$

where $n = \frac{N}{N_{AV}} = \frac{m}{M}$ is the number of moles of the gas.

Solution

For an ideal monatomic gas we have $pV = nRT$ and thereby

$$p\,dV + V\,dp = nR\,dT \Rightarrow dT = \frac{1}{nR}(p\,dV + V\,dp). \tag{11.245}$$

Using (11.245), we obtain the change of the internal energy of the gas dU due to the temperature change dT as follows:

$$dU = c_V\,dT = \frac{c_V}{nR}(p\,dV + V\,dp). \tag{11.246}$$

Using now the first law of thermodynamics ($dU = dQ - p\,dV$), we obtain

$$dQ = dU + p\,dV = \frac{c_V}{nR}(p\,dV + V\,dp) + p\,dV, \tag{11.247}$$

or

$$dQ = \frac{c_V}{nR} V\, dp + \frac{c_V + nR}{nR} p\, dV. \tag{11.248}$$

On the other hand, for ideal monatomic gas, we have $C_P = C_V + nR$ and the result (11.248) finally becomes

$$dQ = \frac{c_V}{nR} V\, dp + \frac{c_P}{nR} p\, dV, \tag{11.249}$$

which is identical to the required result (11.244).

Problem 4

Calculate the fraction of hydrogen molecules at zero altitude and at $T = 300\,\text{K}$ with sufficient speed to escape the gravitational field of the Earth.

Solution

The so-called *escape velocity* is the initial speed v_e of a massive particle required to go from an initial point r in a gravitational potential field to infinity $r \to \infty$ with a residual speed equal to zero $v_f = 0$. Thus, the final mechanical energy of the particle is equal to zero. From the energy conservation law, we can write

$$\frac{mv_e^2}{2} - \frac{GMm}{r} = 0, \tag{11.250}$$

where $G = (6.67428 \pm 0.00067) \times 10^{-11}\,\text{Nm}^2\,\text{kg}^{-2}$ is the gravitational constant, $M = 5.9742 \times 10^{24}\,\text{kg}$ is the mass of the Earth and m is the mass of a hydrogen molecule. Thus, we obtain

$$v_e = \sqrt{\frac{2GM}{r}}. \tag{11.251}$$

At zero altitude ($r = R_E = 6371\,\text{km}$) we have

$$v_e = \sqrt{\frac{2GM}{R_E}} = \sqrt{2 \frac{GM}{R_E^2} R_E} = \sqrt{2gR_E} \approx 11000\,\frac{\text{m}}{\text{s}}, \tag{11.252}$$

where $g = 9.81\,\text{ms}^{-2}$. The probability density for speeds of molecules of an ideal monatomic gas is given by (11.33), i.e.,

$$d\omega(v) = \left(\frac{m}{2\pi kT}\right)^{3/2} 4\pi v^2 dv \exp\left(-\frac{mv^2}{2kT}\right) = \nu(v)dv. \tag{11.253}$$

The fraction of molecules having speeds larger than the escape velocity is then given by

$$f = \left(\frac{m}{2\pi kT}\right)^{3/2} 4\pi \int_{v_e}^{\infty} v^2 dv \exp\left(-\frac{mv^2}{2kT}\right). \qquad (11.254)$$

Introducing here a new variable w as follows:

$$v = \sqrt{\frac{2kT}{m}} w \Rightarrow dv = \sqrt{\frac{2kT}{m}} dw, \qquad (11.255)$$

we obtain

$$f = \left(\frac{m}{2\pi kT}\right)^{3/2} 4\pi \left(\frac{2kT}{m}\right)^{3/2} \int_{6.9647}^{\infty} w^2 e^{-w^2} dw, \qquad (11.256)$$

where

$$\sqrt{\frac{m}{2kT}} v_e \approx 6.9647. \qquad (11.257)$$

Using suitable tables or by numerical integration, we obtain

$$f = \frac{4}{\sqrt{\pi}} \int_{6.9647}^{\infty} w^2 e^{-w^2} dw \approx 1.5 \times 10^{-20}, \qquad (11.258)$$

where we see that a very tiny fraction of all molecules have a sufficient speed to escape the gravitational field of the Earth.

Problem 5

Calculate the average reciprocal speed $< v^{-1} >$ of the molecules of mass m in an ideal monatomic gas.

Solution

Using the probability density for speeds of molecules of an ideal monatomic gas, given by (11.33), i.e.,

$$d\omega(v) = \left(\frac{m}{2\pi kT}\right)^{3/2} 4\pi v^2 dv \exp\left(-\frac{mv^2}{2kT}\right) = v(v)dv, \qquad (11.259)$$

we obtain

$$< v^{-1} > = \left(\frac{m}{2\pi kT}\right)^{3/2} 4\pi \int_{v_e}^{\infty} v \, dv \exp\left(-\frac{mv^2}{2kT}\right).$$ (11.260)

Introducing here a new variable w as follows:

$$v = \sqrt{\frac{2kT}{m}} w \Rightarrow dv = \sqrt{\frac{2kT}{m}} dw,$$ (11.261)

we obtain

$$< v^{-1} > = \left(\frac{m}{2\pi kT}\right)^{3/2} 4\pi \frac{2kT}{m} \int_0^{\infty} w e^{-w^2} dw,$$ (11.262)

or

$$< v^{-1} > = 2\left(\frac{m}{2\pi kT}\right)^{1/2} \int_0^{\infty} e^{-w^2} d(w^2),$$ (11.263)

or finally

$$< v^{-1} > = 2\left(\frac{m}{2\pi kT}\right)^{1/2} = \left(\frac{2m}{\pi kT}\right)^{1/2}.$$ (11.264)

Problem 6

In an absorption refrigerator, the cooling is obtained by mixing two gases at constant pressure. Calculate the maximum reduction of the temperature that can be achieved by mixing the same amounts of two ideal monatomic gases with pressure p and temperature T to a mixture with total pressure p.

Solution

The maximum reduction of the temperature is calculated using the second law of thermodynamics $\Delta S \geq 0$. The entropy of an ideal monatomic gas is given by the result (11.81), i.e.,

$$S = Nk \ln\frac{V}{N} + \frac{3}{2}Nk\left[1 + \ln\left(\frac{2\pi mkT}{h^2}\right)\right] + kN.$$ (11.265)

Using the thermal equation of state $pV = NkT$, we can rewrite (11.265) as follows:

$$S = Nk \ln\frac{kT}{p} + \frac{3}{2}Nk\left[1 + \ln\left(\frac{2\pi mkT}{h^2}\right)\right] + kN,$$ (11.266)

or

$$S = \frac{5}{2}Nk \ln T - Nk \ln p + \frac{3}{2}Nk \left[1 + \frac{2}{3}\ln k + \ln\left(\frac{2\pi mk}{h^2}\right) \right] + kN. \qquad (11.267)$$

In case of an ideal monatomic gas, the specific heat at constant pressure per constituent particle is equal to $c_P = 5/2k$, and we can rewrite (11.267) as follows:

$$S(p,T) = N c_P \ln T - Nk \ln p + C, \qquad (11.268)$$

where we have collected all constant terms into a constant C.

In the initial state, we have two equal quantities of two ideal monatomic gases ($N_1 = N_2 = N$) with the same pressure p and same temperature T. Thus, the total entropy $S(p,T)$ as a sum of entropies of the two gases is just double the entropy of any of the two gases

$$S_I = S_1 + S_2 = 2N c_P \ln T - 2Nk \ln p + 2C. \qquad (11.269)$$

In the final state, after the mixing of the two gases the total pressure of the mixture is still equal to p, and each of the gases contributes with equal partial pressure ($p_1 = p_2 = p/2$). The temperature has changed to a new temperature T_N. Since the two components in the mixture still have the same pressure ($p/2$) and the same temperature T_N, their contribution to the total entropy of the mixture is still the same. Thus, the final state entropy is just double the entropy of any of the two components

$$S_F = 2N c_P \ln T_N - 2Nk \ln\left(\frac{p}{2}\right) + 2C. \qquad (11.270)$$

The entropy change in the mixing process is then

$$\Delta S = S_F - S_I = 2N c_P \left(\frac{T_N}{T}\right) - 2Nk \ln\left(\frac{1}{2}\right) \geq 0. \qquad (11.271)$$

From the result (11.271), with $c_P = 5/2k$, we obtain

$$c_P \left(\frac{T_N}{T}\right) \geq k \ln\left(\frac{1}{2}\right) \Rightarrow T_N \geq T\left(\frac{1}{2}\right)^{2/5} \approx 0.76T. \qquad (11.272)$$

The maximum reduction of the temperature is obtained when the entropy change is zero ($\Delta S = 0$) and in such a case $T_N \approx 0.76T$, or $\Delta T_{max} = T_N - T \approx 0.24T$.

Problem 7

Joule–Kelvin cooling describes the temperature change of a gas or liquid when it is forced through a heat-insulated valve so that no heat is exchanged with the environment. This process is sometimes called a *throttling process*. During Joule–Kelvin cooling, the enthalpy of the substance remains constant.

Consider an ideal monatomic gas (e.g., a noble gas like Helium or Neon at room temperature) that is throttled through a heat-insulated valve so that its pressure is reduced from $p_1 = 2 \times 10^5$ Pa to $p_2 = 5 \times 10^4$ Pa. If the initial temperature of the gas was equal to $T_I = 300$ K, calculate the new temperature T_N.

Solution

The enthalpy for an ideal monatomic gas is given by the result (11.84), i.e.,

$$H = U + pV = \frac{3}{2}NkT + NkT = \frac{5}{2}NkT. \tag{11.273}$$

Now, we see that the enthalpy is not dependent on the pressure of the gas and depends only on its temperature. The condition that the enthalpy remains constant in the Joule–Kelvin cooling process ($H = $ Constant) therefore implies that the temperature of the gas remains constant. Thus, the new temperature is $T_N = T_I = 300$ K.

Problem 8

Consider an ideal gas inside a vertical cylinder with the height L and the base area A_B. So, for the N molecules inside the cylinder, their z-co-ordinates are within the range $0 < z < L$. The molecules move in an external potential $V(\vec{r}) = V(z)$ acting on *all* molecules. Note that this potential depends on the z-co-ordinate only.

Use general thermodynamic principles to calculate the local pressure exerted by the gas on the upper base of the cylinder. Show that this pressure can be expressed as

$$p(z = L) = n(z = L) \cdot kT, \tag{11.274}$$

where $n(z = L)$ is the local gas density at $z = L$.

Remark: The result (11.274) is to be expected from the statement in the equation (11.167). So by doing this problem, we will effectively confirm the unproved statement in the equation (11.167).

Conceptual hint: Think of the upper cylinder's base (with the area A_B) as a piston at the position $z = L$.

Solution

The force and the pressure exerted on the piston can be obtained from the general result (8.42) by noting that $V = A_B \cdot L$. Thus, we have

$$p(z = L) = -\frac{1}{A_B}\left(\frac{\partial F_N}{\partial L}\right)_T = \frac{kT}{A_B}\left(\frac{\partial \ln Z_N}{\partial L}\right)_T. \tag{11.275}$$

Here, the gas free energy is given by $F_N = -kT \ln Z_N$, with the N-particle partition function Z_N as in the equation (11.107), i.e.,

$$Z_N = \frac{Z^N}{N!} \Rightarrow \ln Z_N = N \ln Z - \ln(N!). \tag{11.276}$$

By the equations (11.275) and (11.276), we thus have

$$p(z = L) = \frac{NkT}{A_B} \left(\frac{\partial \ln Z}{\partial L} \right)_T. \tag{11.277}$$

Using here (11.179),

$$Z_{\text{classical}} = Z_{\text{KIN}} \cdot Z_{\text{POT}} \Rightarrow \ln(Z) = \ln(Z_{\text{KIN}}) + \ln(Z_{\text{POT}}). \tag{11.278}$$

By noting that Z_{KIN} in the equation (11.143) is a function of temperature only and does not depend on L, by equations (11.277) and (11.278), we have

$$p(z = L) = \frac{NkT}{A_B} \left(\frac{\partial \ln Z_{\text{POT}}}{\partial L} \right)_T, \tag{11.279}$$

or

$$p(z = L) = \frac{NkT}{A_B} \frac{1}{Z_{\text{POT}}} \left(\frac{\partial Z_{\text{POT}}}{\partial L} \right)_T. \tag{11.280}$$

Here, by the equation (11.144),

$$Z_{\text{POT}} = \int_{\vec{r}} d^3 r \exp\left(-\frac{V(\vec{r})}{kT} \right) = \int \int dx \, dy \int_0^L dz \exp\left(-\frac{V(z)}{kT} \right)$$

$$= A_B \int_0^L dz \exp\left(-\frac{V(z)}{kT} \right). \tag{11.281}$$

Thus, the Z_{POT} is a function of L given by

$$Z_{\text{POT}} = A_B \int_0^L dz \exp\left(-\frac{V(z)}{kT} \right). \tag{11.282}$$

Using here the basic calculus fact that

$$\frac{\partial}{\partial L} \int_0^L dz f(z) = f(z = L), \tag{11.283}$$

we obtain

$$\frac{\partial Z_{\text{POT}}}{\partial L} = \frac{\partial}{\partial L} \left[A_B \int_0^L dz \exp\left(-\frac{V(z)}{kT}\right) \right] = A_B \exp\left(-\frac{V(z=L)}{kT}\right). \tag{11.284}$$

By (11.284), the equation (11.280) yields

$$\begin{aligned} p(z=L) &= \frac{NkT}{A_B} \frac{1}{Z_{\text{POT}}} \left(\frac{\partial Z_{\text{POT}}}{\partial L}\right)_T \\ &= \frac{NkT}{A_B} \frac{1}{Z_{\text{POT}}} \exp\left(-\frac{V(z=L)}{kT}\right). \end{aligned} \tag{11.285}$$

That is,

$$p(z=L) = \frac{NkT}{A_B} \frac{1}{Z_{\text{POT}}} \exp\left(-\frac{V(z=L)}{kT}\right). \tag{11.286}$$

On the other hand, for $V(\vec{r}) = V(z)$, by the Boltzmann formula (11.154), we have

$$n(z) = \frac{N}{Z_{\text{POT}}} \exp\left(-\frac{V(z)}{kT}\right), \tag{11.287}$$

yielding

$$n(z=L) = \frac{N}{Z_{\text{POT}}} \exp\left(-\frac{V(z=L)}{kT}\right). \tag{11.288}$$

From the equations (11.286) and (11.188), it is obvious that the local pressure in (11.286) can be expressed as

$$p(z=L) = n(z=L) \cdot kT. \tag{11.289}$$

Thus, we have proved the equation (11.274), and in this way, we also confirmed the statement made in the equation (11.167).

12 Ideal Diatomic Gases

In this chapter, we consider the statistical thermodynamics of very dilute (ideal) gases of diatomic molecules, such as, for example, CO or HCl molecules. For such gases, the contributions from the rotational and vibrational degrees of freedom of the gas particles should be taken into account in the calculations of the thermodynamic quantities such as the experimentally significant specific heats. For simplicity, we will ignore the presence of the spins of the two nuclei constituting the molecule. Also, we will consider the nuclei as two distinguishable particles, as appropriate for the heteronuclear molecules, such as CO or HCl, and comment on the situation for the homonuclear molecules, such as O_2 or N_2. As discussed below, a single diatomic molecule partition function can be represented as the product of three terms,

$$Z = Z_{TR} \times Z_{ROT} \times Z_{VIB}. \tag{12.1}$$

Here, Z_{TR} originates from the molecule's center of mass translation, and it has the same form as for monatomic molecules discussed in Chapter 11, "Ideal Monatomic Gases." On the other side, the terms Z_{ROT} and Z_{VIB} in the equation (12.1) originate, respectively, from the molecules' rotations and vibrations. To derive the equation (12.1), we recall the discussions of Chapters 3 through 5, by which a diatomic molecule made of two atoms with masses m_1 and m_2 can be envisioned as composite system consisting of a particle with total mass $m_{molecule} = m_1 + m_2$ (moving along with molecule's center of mass) combined with an internally moving subsystem executing rotational and vibrational motions. Energy states of this composite system are labeled by the six quantum numbers: n_x, n_y, n_z (describing the translational motion of the molecules center of mass), l, m (describing the molecules rotational motion, see Chapter 5, "Kinetic Energy of Rotations"), and n (describing bond vibrations of the molecule, see Chapter 4, "Energy of Vibrations"). The energy of this state is, by the discussions of Chapters 3 through 5, given by

$$E\left(n_x, n_y, n_z, l, m, n\right) = E_{TR}\left(n_x, n_y, n_z\right) + E_{ROT}(l, m) + E_{VIB}(n), \tag{12.2}$$

with

$$E_{TR}\left(n_x, n_y, n_z\right) = \frac{h^2}{8mV^{2/3}}\left(n_x^2 + n_y^2 + n_z^2\right), \tag{12.3}$$

$$E_{ROT}(l, m) = \frac{\hbar^2}{2I}l(l+1), \quad E_{VIB}(n) = \hbar\omega\left(n + \frac{1}{2}\right). \tag{12.4}$$

Introductory Statistical Thermodynamics

The corresponding molecule's partition function Z is a sum over all possible states, i.e., over all possible values of the quantum numbers n_x, n_y, n_z, l, m, and n, of the form

$$Z = \sum_{n_x=0}^{\infty} \sum_{n_y=0}^{\infty} \sum_{n_z=0}^{\infty} \sum_{l=0}^{\infty} \sum_{m=-l}^{+l} \sum_{n=0}^{\infty} \exp\left[-\frac{E\left(n_x, n_y, n_z, l, m, n\right)}{kT}\right]. \tag{12.5}$$

Using the standard manipulations with multiple sums, and the above additive form of $E\left(n_x, n_y, n_z, l, m, n\right)$, in which the contributions from translations, rotations, and vibrations simply add to each other, one can easily see that the partition function Z indeed has the multiplicative form displayed in (12.1), with

$$Z_{TR} = \sum_{n_x=0}^{\infty} \sum_{n_y=0}^{\infty} \sum_{n_z=0}^{\infty} \exp\left[-\frac{E_{TR}\left(n_x, n_y, n_z\right)}{kT}\right], \tag{12.6}$$

$$Z_{ROT} = \sum_{l=0}^{\infty} \sum_{m=-l}^{+l} \exp\left[-\frac{E_{ROT}(l, m)}{kT}\right] = \sum_{l=0}^{\infty} \sum_{m=-l}^{+l} \exp\left[-\frac{\hbar^2 l(l+1)}{2IkT}\right]$$

$$= \sum_{l=0}^{\infty} (2l+1) \exp\left[-\frac{\hbar^2 l(l+1)}{2IkT}\right], \tag{12.7}$$

and

$$Z_{VIB} = \sum_{n=0}^{\infty} \exp\left[-\frac{E_{VIB}(n)}{kT}\right] = \sum_{n=0}^{\infty} \exp\left[-\frac{\hbar\omega}{kT}\left(n + \frac{1}{2}\right)\right]. \tag{12.8}$$

By the equation (12.1),

$$\ln Z = \ln Z_{TR} + \ln Z_{ROT} + \ln Z_{VIB}. \tag{12.9}$$

Combining the above relation with the equation (8.22), we readily obtain that the internal energy of an ideal diatomic gas has the additive form

$$U = U_{TR} + U_{ROT} + U_{VIB}, \tag{12.10}$$

with

$$U_X = NkT^2 \frac{\partial}{\partial T}(\ln Z_X)_V, \tag{12.11}$$

where $X \in \{TR, ROT, VIB\}$. Obviously, the diatomic gas internal energy per unit mass and the gas specific heat also have the additive forms

$$u = u_{TR} + u_{ROT} + u_{VIB} \tag{12.12}$$

and

$$c_V = c_{V_{TR}} + c_{V_{ROT}} + c_{V_{VIB}}. \tag{12.13}$$

We note that the above translational partition function Z_{TR} is identical to the one discussed in the Chapter 11, and displayed in the equation (11.26). It can be thus directly adopted here,

$$Z_{TR} = V \left(\frac{2\pi m_{molecule} kT}{h^2} \right)^{3/2}, \tag{12.14}$$

with $m_{molecule} = m_1 + m_2$, the total mass of the diatomic molecule. In this chapter, we thus focus our attention on the here derived rotational Z_{ROT} and the vibrational Z_{VIB} partition functions, and their corresponding contributions to the specific internal energies and specific heats in the equations (12.12) and (12.13). From the above expressions for Z_{ROT} and Z_{VIB}, it is clear that these two partition functions do not depend on the system volume V. This is physically unsurprising, in view of the fact that these two partition functions originate from the molecule's microscopic internal motions, which are unaffected by the macroscopic quantities such as the gas container volume V. On the other hand, the translational motion of the molecule's center of mass does sense the magnitude of V, and because of this, the above-displayed translational partition function Z_{TR} does depend on the gas volume. There is an important consequence of these facts: molecular vibrations and rotations do not affect the pressure exerted by an ideal diatomic gas on the walls of its confining container. In other words, the thermal equation of state of an ideal diatomic (and, for that mater, polyatomic) gas has exactly the same form as for the monatomic gas, that is,

$$pV = NkT, \tag{12.15}$$

where N stands for total number of molecules, no matter whether they are monatomic, diatomic, or, in general, polyatomic with whatever number of atoms. This fact is easily derived from the general formula for an ideal gas free energy F in (8.1), by calculating the gas pressure p via the general formula (8.42) that expresses p in terms of the derivative of F with respect to V. Using these formulas with $\ln Z = \ln Z_{TR} + \ln Z_{ROT} + \ln Z_{VIB}$ and recalling that Z_{ROT} and Z_{VIB} do not depend on the system volume V, whereas Z_{TR} has the same form as for monatomic gases, one easily deduces that the pressure of a polyatomic gas must be given by the same law we had before for monatomic gases. This law can then be written as

$$p = nkT, \tag{12.16}$$

with $n = N/V$, the molecular number density. A notable feature of this result is that the pressure does not depend on how large individual molecules are. That is, the pressure does not depend on the molecular mass. Rather, for the pressure of dilute molecular gases, the only decisive factor is the number of molecules per unit volume, $n = N/V$, and, of course, the gas temperature T.

Finally, we note that, because the Z_{TR} has the same form as for monatomic gases, one easily deduces, by (12.11), that the translational contribution to the internal energy of a polyatomic gas must be given by the same law we had before for monatomic gases, that is,

$$U_{TR} = \frac{3}{2}NkT. \tag{12.17}$$

This yields the specific internal energy due to translations given by

$$u_{TR} = \frac{U_{TR}}{m_{gas}} = \frac{3}{2}R_g T, \tag{12.18}$$

indicating that the translational contribution to the specific heat at constant volume is the constant,

$$c_{V_{TR}} = \left(\frac{\partial u_{TR}}{\partial T}\right)_V = \frac{3}{2}R_g, \tag{12.19}$$

for any dilute (ideal) polyatomic gas.

12.1 Rotations of Gas Particles

Let us now discuss the thermodynamic effects of the molecular rotational motion. The partition function associated with diatomic molecule rotations calculated in the introductory part of this chapter can be written as

$$Z_{ROT} = \sum_{l=0}^{\infty}(2l+1)\exp\left[-\frac{T_{ROT}}{T}l(l+1)\right], \tag{12.20}$$

where we introduced a characteristic quantum temperature associated with rotations, defined by

$$T_{ROT} = \frac{\hbar^2}{2Ik}. \tag{12.21}$$

Because of the (relative) smallness of the Planck constant, this temperature is typically small for realistic diatomic gases. For example, for the HCl molecule, $T_{ROT} = 15.2\,K$, whereas for the N_2 molecule, $T_{ROT} = 2.86\,K$. In practical applications, one is thus typically interested in the high-temperature regime $T \gg T_{ROT}$. In this regime, the terms in the above-written partition function sum change slowly with increasing l, so the sum can be calculated accurately by replacing it with an integral over l,

$$Z_{ROT} \approx \int_0^{\infty} dl(2l+1)\exp\left[-\frac{T_{ROT}}{T}l(l+1)\right]. \tag{12.22}$$

Let us introduce a variable λ as follows:

$$\lambda = l(l+1) = l^2 + l \quad \Rightarrow \quad d\lambda = (2l+1)dl. \tag{12.23}$$

Substituting (12.23) into (12.22), we obtain

$$Z_{\text{ROT}} \approx \int_0^\infty d\lambda \exp\left(-\frac{T_{\text{ROT}}}{T}\lambda\right) = \frac{T}{T_{\text{ROT}}}. \tag{12.24}$$

As noted in the introductory part of this chapter, the rotational contribution to the internal energy of the system is

$$U_{\text{ROT}} = NkT^2 \frac{\partial}{\partial T}[\ln(Z_{\text{ROT}})], \tag{12.25}$$

or

$$U_{\text{ROT}} \approx NkT^2 \frac{\partial}{\partial T}\left[\ln\left(\frac{T}{T_{\text{ROT}}}\right)\right] = NkT, \tag{12.26}$$

in the high-temperature limit $T \gg T_{\text{ROT}}$. Note that the above result for U_{ROT} does not depend on the value of the Planck constant. In view of this, the high-temperature limit represents also the classical limit. In this respect, it is illuminating to note that in the formal limit of zero Planck constant (reducing quantum to classical mechanics), the above-defined T_{ROT} would also go to zero, meaning that the classical results for U_{ROT} in (12.26) would hold at any temperature (as has to be expected, since quantum mechanics reduces to classical mechanics for zero Planck constant). In reality, the Planck constant is not zero, yet it is relatively small, giving rise to typically small values of T_{ROT}, as mentioned above.

In the high-temperature limit, the specific internal energy due to rotations is given by

$$u_{\text{ROT}} = \frac{U_{\text{ROT}}}{m_{\text{gas}}} = R_g T, \tag{12.27}$$

indicating that the rotational contribution to the specific heat at constant volume is a constant,

$$c_{V_{\text{ROT}}} = \left(\frac{\partial u_{\text{ROT}}}{\partial T}\right)_V = R_g. \tag{12.28}$$

We note that equations (12.26) through (12.28) apply also to polyatomic molecules gases, in the classical limit. In this limit, nuclear spins and the effects of indistinguishability of identical nuclei in homonuclear diatomic molecules (as discussed in Section 6.1, "Introduction and Definitions") disappear. Thus, in the classical high-temperature limit, the equations (12.26) through (12.28) apply both to heteronuclear (e.g., HCl, CO) and homonuclear (e.g., N_2, O_2) diatomic gases.

12.2   Vibrations of Gas Particles

Let us now discuss the thermodynamic effects of the molecular vibrational motion. The partition function associated with diatomic molecule vibrations calculated in the introductory part of this chapter can be written as

$$Z_{\text{VIB}} = \sum_{n=0}^{\infty} \exp\left[-\frac{T_{\text{VIB}}}{T}\left(n+\frac{1}{2}\right)\right], \tag{12.29}$$

where we introduced a characteristic quantum temperature associated with the vibrations, defined by

$$T_{\text{VIB}} = \frac{\hbar\omega}{k}. \tag{12.30}$$

Due to relatively large angular frequencies ω, this temperature is typically large for realistic diatomic gases (in spite the "smallness" of the Planck constant). For example, for the HCl molecule, $T_{\text{VIB}} = 4140$ K, whereas for the N_2 molecule, $T_{\text{VIB}} = 3340$ K. In practical situations, we are normally interested in the temperatures $T < T_{\text{VIB}}$. Thus, one is frequently interested in the low-temperature behavior of the vibrational partition function Z_{VIB}. On the other side, much like in the discussions we had in the Section 12.1, "Rotations of Gas Particles," the high-temperature limit $T \gg T_{\text{VIB}}$ is the classical limit in which we can replace the sum in Z_{VIB} by an integral to obtain

$$Z_{\text{VIB}} \approx \int_0^{\infty} dn \exp\left[-\frac{T_{\text{VIB}}}{T}\left(n+\frac{1}{2}\right)\right] = \exp\left(-\frac{T_{\text{VIB}}}{2T}\right)\int_0^{\infty} dn \exp\left(-\frac{T_{\text{VIB}}}{T}n\right)$$

$$= \exp\left(-\frac{T_{\text{VIB}}}{2T}\right)\frac{T}{T_{\text{VIB}}} \approx \frac{T}{T_{\text{VIB}}}, \tag{12.31}$$

where we used the fact that $\exp(x) \approx 1$ for $x \ll 1$. Thus, in the high-temperature limit, $T \gg T_{\text{VIB}}$,

$$Z_{\text{VIB}} \approx \frac{T}{T_{\text{VIB}}}. \tag{12.32}$$

With T_{VIB} as in (12.30), we see that the equation (12.32) agrees with the classical prediction for the single harmonic oscillator partition function in the equation (11.195) of Section 11.6.2, "Harmonic and Anharmonic Oscillators." As noted in the introductory part of this chapter, the vibrational contribution to the internal energy of the system is

$$U_{\text{VIB}} = NkT^2 \frac{\partial}{\partial T}[\ln(Z_{\text{VIB}})] = NkT^2 \frac{\partial}{\partial T}\left[\ln\left(\frac{T}{T_{\text{VIB}}}\right)\right] = NkT \tag{12.33}$$

in the high-temperature limit $T \gg T_{\text{VIB}}$. Note that the above result for U_{VIB} does not depend on the value of the Planck constant. In view of this, the high-temperature limit represents also the classical limit. In fact, the above result has been anticipated before in Sections 11.4.6, "Equipartition Theorem," and 11.6.2, "Harmonic and Anharmonic Oscillators," on the basis of equipartition theorem applied to the system of N classical harmonic oscillators.

In the high-temperature limit, the specific internal energy due to vibrations is given by

$$u_{\text{VIB}} = \frac{U_{\text{VIB}}}{m_{\text{gas}}} = R_g T, \tag{12.34}$$

indicating that the vibrational contribution to the specific heat at constant volume approaches the constant,

$$c_{V_{\text{VIB}}} = \left(\frac{\partial u_{\text{VIB}}}{\partial T} \right)_V = R_g, \tag{12.35}$$

in the high-temperature limit $T \gg T_{\text{VIB}}$. However, as noted above, one is typically interested in $T < T_{\text{VIB}}$. Fortunately, by using the well-known result for the geometric series

$$\sum_{l=0}^{\infty} a^l = \frac{1}{1-a}, \quad |a| < 1, \tag{12.36}$$

the vibrational partition function Z_{VIB} can be calculated exactly at any temperature T as follows:

$$\begin{aligned} Z_{\text{VIB}} &= \exp\left(-\frac{T_{\text{VIB}}}{2T} \right) \sum_l \left[\exp\left(-\frac{T_{\text{VIB}}}{T} \right) \right]^l \\ &= \frac{\exp\left(-\frac{T_{\text{VIB}}}{2T} \right)}{1 - \exp\left(-\frac{T_{\text{VIB}}}{T} \right)} = \frac{\exp\left(\frac{T_{\text{VIB}}}{2T} \right)}{\exp\left(\frac{T_{\text{VIB}}}{T} \right) - 1}. \end{aligned} \tag{12.37}$$

As noted in the introductory part of this chapter, the vibrational contribution to the internal energy of the system is

$$U_{\text{VIB}} = NkT^2 \frac{\partial}{\partial T} [\ln (Z_{\text{VIB}})]. \tag{12.38}$$

Substituting here the exact Z_{VIB} from (12.37), we obtain the exact U_{VIB} as follows:

$$
\begin{aligned}
U_{VIB} &= NkT^2 \frac{\partial}{\partial T} \left[\ln \left(\frac{\exp\left(\frac{T_{VIB}}{2T}\right)}{\exp\left(\frac{T_{VIB}}{T}\right) - 1} \right) \right] \\
&= NkT^2 \frac{\partial}{\partial T} \left[\frac{T_{VIB}}{2T} - \ln \left(\exp\left(\frac{T_{VIB}}{T}\right) - 1 \right) \right] \\
&= NkT^2 \left[-\frac{T_{VIB}}{2T^2} - \frac{\exp\left(\frac{T_{VIB}}{T}\right)}{\exp\left(\frac{T_{VIB}}{T}\right) - 1} \left(-\frac{T_{VIB}}{T^2} \right) \right] \\
&= NkT_{VIB} \left[\frac{\exp\left(\frac{T_{VIB}}{T}\right)}{\exp\left(\frac{T_{VIB}}{T}\right) - 1} - \frac{1}{2} \right],
\end{aligned}
\tag{12.39}
$$

or, per unit mass,

$$
u_{VIB} = \frac{U_{VIB}}{m_{molecule}} = R_g T_{VIB} \left[\frac{\exp\left(\frac{T_{VIB}}{T}\right)}{\exp\left(\frac{T_{VIB}}{T}\right) - 1} - \frac{1}{2} \right].
\tag{12.40}
$$

We note that, in the high-temperature limit $T_{VIB}/T \ll 1$, in the equation (12.39), we can approximate $\exp(T_{VIB}/T)$ by $1 + T_{VIB}/T$. With this, one easily regains from the equation (12.40) the previously obtained classical result $U_{VIB} = NkT$, valid for $T \gg T_{VIB}$. Again, we recall that this result was expected on the basis of equipartition theorem. Using the identity,

$$
\frac{\exp(x)}{\exp(x) - 1} - \frac{1}{2} = \frac{1}{2} + \frac{1}{\exp(x) - 1},
\tag{12.41}
$$

and the relation (12.13), the equation (12.39) can be written as

$$
U_{VIB} = N U_{osc},
\tag{12.42}
$$

where

$$
U_{osc} = \frac{\hbar\omega}{2} + \frac{\hbar\omega}{\exp\left(\frac{\hbar\omega}{kT}\right) - 1}
\tag{12.43}
$$

is the internal energy of a single harmonic oscillator. In the low-temperature limit $T \ll T_{VIB}$, or equivalently $\hbar\omega \gg kT$, by the equation (12.43), we have

$$
U_{osc} \approx \frac{\hbar\omega}{2} + \hbar\omega \exp\left(-\frac{\hbar\omega}{kT}\right) \rightarrow \frac{\hbar\omega}{2} \quad \text{for} \quad T \rightarrow 0,
\tag{12.44}
$$

meaning that, in the zero-temperature limit, the internal energy reduces to the quantum-mechanical ground-state energy of the harmonic oscillator. In this limit, the probabilities of all excited states ($n = 1, 2, 3, \ldots$) [with the energies $E_n = (n + 1/2)\hbar\omega$] approach zero, and the system approaches its ground state with $n = 0$. On the other hand, in the high-temperature limit $T \gg T_{VIB}$, or equivalently $\hbar\omega \ll kT$, by using $\exp(x) \approx 1 + x + x^2/2$, for $x = \hbar\omega/kT \ll 1$, one finds by the equation (12.43) that $U_{osc} \approx kT$. This means that the single-oscillator internal energy approaches its classical value predicted by the equipartition theorem (see Section 11.6.2, "Harmonic and Anharmonic Oscillators"). So, the high-temperature limit is also the classical limit. In view of the form of the energy levels for the harmonic oscillator in the equation (4.21),

$$E_n = \frac{\hbar\omega}{2} + n \cdot \hbar\omega, \tag{12.45}$$

it is illuminating to write the equation (12.43) as

$$U_{osc} =< E_n >= \frac{\hbar\omega}{2} + <n> \cdot \hbar\omega, \tag{12.46}$$

with

$$<n>= \frac{1}{\exp\left(\frac{\hbar\omega}{kT}\right) - 1}. \tag{12.47}$$

The equations (12.46) and (12.47) best relate the oscillator thermodynamics to our discussions of photons or phonons in Chapter 17, "Photon Gas in Equilibrium," and Chapter 18, "Other Examples of Boson Systems," respectively. Indeed, the n-th state of the oscillator may be interpreted as a state with n photons (or phonons) present, each having the energy $\hbar\omega$, such that the total energy is $n \cdot \hbar\omega$. The average number of these photons (or phonons) $< n >$ is given by the equation (12.47), so they contribute to the internal energy of the system by the amount equal to $< n > \cdot\hbar\omega$ corresponding to the second term in the equation (12.46). The first term in the equation (12.46), emerging from the oscillator ground-state energy

$$E_{n=0} = \frac{\hbar\omega}{2}, \tag{12.48}$$

is frequently called the *vacuum energy* (or zero-point energy), as it corresponds to the zero-temperature situation with no photons (or phonons) present. This seemingly strange contribution to the internal energy may be of interest in some physical problems (see Chapters 17 and 18).

Let us now return to our N diatomic molecules ideal gas and calculate the vibration energy contribution to the gas specific heat. It is obtained from u_{VIB} using the definition

$$c_{V_{VIB}} = \left(\frac{\partial u_{VIB}}{\partial T}\right)_V. \tag{12.49}$$

Substituting (12.40) into (12.49), we can calculate

$$c_{V_{\text{VIB}}} = R_g T_{\text{VIB}} \frac{\partial}{\partial T} \left[\frac{1}{1 - \exp\left(-\frac{T_{\text{VIB}}}{T}\right)} \right] = R_g T_{\text{VIB}} \frac{(-1)\exp\left(-\frac{T_{\text{VIB}}}{T}\right)\left(-\frac{T_{\text{VIB}}}{T^2}\right)}{\left[1 - \exp\left(-\frac{T_{\text{VIB}}}{T}\right)\right]^2},$$

$$(12.50)$$

or finally

$$c_{V_{\text{VIB}}}(T) = R_g \left(\frac{T_{\text{VIB}}}{T}\right)^2 \frac{\exp\left(\frac{T_{\text{VIB}}}{T}\right)}{\left[\exp\left(\frac{T_{\text{VIB}}}{T}\right) - 1\right]^2}. \tag{12.51}$$

From the equation (12.51), we see that the vibrational contribution to the specific heat is not constant, but rather a function of temperature. At low temperatures, the vibrational specific heat tends to zero, i.e., we have

$$c_{V_{\text{VIB}}}(T \to 0) \to R_g \left(\frac{T_{\text{VIB}}}{T}\right)^2 \exp\left(-\frac{T_{\text{VIB}}}{T}\right) \to 0. \tag{12.52}$$

Thus, the vibrational effects in ideal polyatomic gases are significant only at higher temperatures. In the high-temperature limit, we can put $\exp(T_{\text{VIB}}/T) \to 1$ in the numerator of the equation (12.21) and $\exp(T_{\text{VIB}}/T) \to 1 + T_{\text{VIB}}/T$ in the denominator of the equation (12.21). Thus, we obtain the classical limit for the vibrational contribution to the specific heat as follows:

$$c_{V_{\text{VIB}}}(T) = R_g \left(\frac{T_{\text{VIB}}}{T}\right)^2 \frac{1}{\left(\frac{T_{\text{VIB}}}{T}\right)^2} = R_g, \quad T \gg T_{\text{VIB}}. \tag{12.53}$$

This result is in accordance with the previously discussed result for this quantity in the classical limit, whence the result $U_{\text{VIB}} = NkT$ applies.

Finally, let us consider the diatomic molecules ideal gas at realistic temperatures above the typically low rotational temperature T_{ROT}. By using the equation (12.13) (additivity of specific heats) and the results for the translational, rotational, and vibrational specific heats discussed in this chapter, the total specific heat at constant volume for an ideal diatomic gas is given by

$$c_V(T) = R_g \left\{ \frac{5}{2} + \left(\frac{T_{\text{VIB}}}{T}\right)^2 \frac{\exp\left(\frac{T_{\text{VIB}}}{T}\right)}{\left[\exp\left(\frac{T_{\text{VIB}}}{T}\right) - 1\right]^2} \right\}, \tag{12.54}$$

which in the classical limit ($T \gg T_{\text{VIB}}$) becomes $c_V = \frac{7}{2} R_g$.

Unlike diatomic molecules, polyatomic molecules with more than two nuclei (e.g., CH_4 or NH_3, or polymers) have more than one vibrational mode. Different vibrational modes have different frequencies and, by the equation (12.13), different characteristic quantum temperatures. For the case of polyatomic molecules, the last term in (12.54) is simply replaced by a sum of such terms, over all modes. Thus, in the classical high-temperature limit, the specific heat at constant volume is given by $c_V = (5/2 + N_{VIB})R_g$, where N_{VIB} is the total number of the vibrational modes of the polyatomic molecule.

12.3 Problems with Solutions

Problem 1

Two identical mutually noninteracting particles of mass m are subject to an external harmonic potential, such that the Hamiltonian of the system is given by

$$H = \frac{p_1^2}{2m} + \frac{p_2^2}{2m} + \frac{m\omega^2}{2}\left(x_1^2 + x_2^2\right). \tag{12.55}$$

(a) Derive the expression for the energy levels of the system.
(b) Derive the expression for the partition function of the system, if the two particles are bosons.
(c) Derive the expression for the partition function of the system, if the two particles are fermions.

Solution

(a) The total energy of the two uncoupled harmonic oscillators is given by

$$E_{n_1 n_2} = \hbar\omega\left(n_1 + \frac{1}{2}\right) + \hbar\omega\left(n_2 + \frac{1}{2}\right) \tag{12.56}$$

or

$$E_{n_1 n_2} = \hbar\omega\left(n_1 + n_2 + 1\right). \tag{12.57}$$

(b) The partition function of the system, if the two particles are bosons, is given by

$$Z_B = \sum_{n_1 \leq n_2} \exp\left(-\frac{E_{n_1 n_2}}{kT}\right), \tag{12.58}$$

or

$$Z_B = \sum_{n_1 \leq n_2} \exp\left[-\frac{\hbar\omega}{kT}\left(n_1 + n_2 + 1\right)\right], \tag{12.59}$$

where both particles are allowed to be in the same state ($n_1 = n_2$). By dividing the
sum into two parts, one with $n_1 = n_2$ and the other with $n_1 < n_2$, we can write

$$Z_B = \sum_{n_1=n_2=n} \exp\left[-\frac{\hbar\omega}{kT}(n_1 + n_2 + 1)\right] + \sum_{n_1<n_2} \exp\left[-\frac{\hbar\omega}{kT}(n_1 + n_2 + 1)\right],$$

$$(12.60)$$

or

$$Z_B = \sum_{n=0}^{\infty} \exp\left[-\frac{\hbar\omega}{kT}(2n + 1)\right] + \sum_{n_1<n_2} \exp\left[-\frac{\hbar\omega}{kT}(n_1 + n_2 + 1)\right], \quad (12.61)$$

or

$$Z_B = \exp\left(-\frac{\hbar\omega}{kT}\right) \sum_{n=0}^{\infty}\left[\exp\left(-\frac{2\hbar\omega}{kT}\right)\right]^n + \exp\left(-\frac{\hbar\omega}{kT}\right)$$

$$\times \sum_{n_1=0}^{\infty}\left[\exp\left(-\frac{\hbar\omega}{kT}\right)\right]^{n_1} \sum_{n_2=n_1+1}^{\infty}\left[\exp\left(-\frac{\hbar\omega}{kT}\right)\right]^{n_2}, \quad (12.62)$$

or

$$Z_B = \frac{\exp\left(-\frac{\hbar\omega}{kT}\right)}{1 - \exp\left(-\frac{2\hbar\omega}{kT}\right)} + \exp\left(-\frac{\hbar\omega}{kT}\right)$$

$$\times \sum_{n_1=0}^{\infty}\left[\exp\left(-\frac{\hbar\omega}{kT}\right)\right]^{n_1} \sum_{j=0}^{\infty}\left[\exp\left(-\frac{\hbar\omega}{kT}\right)\right]^{j+n_1+1}, \quad (12.63)$$

or

$$Z_B = \frac{\exp\left(-\frac{\hbar\omega}{kT}\right)}{1 - \exp\left(-\frac{2\hbar\omega}{kT}\right)} + \exp\left(-\frac{2\hbar\omega}{kT}\right)$$

$$\times \sum_{n_1=0}^{\infty}\left[\exp\left(-\frac{\hbar\omega}{kT}\right)\right]^{2n_1} \sum_{j=0}^{\infty}\left[\exp\left(-\frac{\hbar\omega}{kT}\right)\right]^{j}, \quad (12.64)$$

or

$$Z_B = \frac{\exp\left(-\frac{\hbar\omega}{kT}\right)}{1 - \exp\left(-\frac{2\hbar\omega}{kT}\right)} + \exp\left(-\frac{2\hbar\omega}{kT}\right)$$

$$\times \frac{1}{1 - \exp\left(-\frac{2\hbar\omega}{kT}\right)} \times \frac{1}{1 - \exp\left(-\frac{\hbar\omega}{kT}\right)}, \quad (12.65)$$

or

$$Z_B = \frac{\exp\left(-\frac{\hbar\omega}{kT}\right)}{1 - \exp\left(-\frac{2\hbar\omega}{kT}\right)} \times \left[1 + \frac{\exp\left(-\frac{\hbar\omega}{kT}\right)}{1 - \exp\left(-\frac{\hbar\omega}{kT}\right)}\right]. \tag{12.66}$$

Thus, we finally obtain for the two bosons

$$Z_B = \frac{\exp\left(-\frac{\hbar\omega}{kT}\right)}{1 - \exp\left(-\frac{2\hbar\omega}{kT}\right)} \cdot \frac{1}{1 - \exp\left(-\frac{\hbar\omega}{kT}\right)}. \tag{12.67}$$

(c) The partition function of the system, if the two particles are fermions, is given by

$$Z_B = \sum_{n_1 < n_2} \exp\left(-\frac{En_1n_2}{kT}\right), \tag{12.68}$$

or

$$Z_B = \sum_{n_1 < n_2} \exp\left[-\frac{\hbar\omega}{kT}(n_1 + n_2 + 1)\right], \tag{12.69}$$

where the two particles are now not allowed to be in the same state ($n_1 \neq n_2$ and $n_1 < n_2$). Thus, we can write

$$Z_B = \exp\left(-\frac{\hbar\omega}{kT}\right)\sum_{n_1=0}^{\infty}\left[\exp\left(-\frac{\hbar\omega}{kT}\right)\right]^{n_1} \times \sum_{n_2=n_1+1}^{\infty}\left[\exp\left(-\frac{\hbar\omega}{kT}\right)\right]^{n_2}, \tag{12.70}$$

or

$$Z_B = \exp\left(-\frac{\hbar\omega}{kT}\right)\sum_{n_1=0}^{\infty}\left[\exp\left(-\frac{\hbar\omega}{kT}\right)\right]^{n_1} \times \sum_{j=0}^{\infty}\left[\exp\left(-\frac{\hbar\omega}{kT}\right)\right]^{j+n_1+1}, \tag{12.71}$$

or

$$Z_B = \exp\left(-\frac{2\hbar\omega}{kT}\right)\sum_{n_1=0}^{\infty}\left[\exp\left(-\frac{\hbar\omega}{kT}\right)\right]^{2n_1} \times \sum_{j=0}^{\infty}\left[\exp\left(-\frac{\hbar\omega}{kT}\right)\right]^{j}. \tag{12.72}$$

Thus, we finally obtain for the two fermions

$$Z_B = \frac{\exp\left(-\frac{2\hbar\omega}{kT}\right)}{1 - \exp\left(-\frac{2\hbar\omega}{kT}\right)} \cdot \frac{1}{1 - \exp\left(-\frac{\hbar\omega}{kT}\right)}. \tag{12.73}$$

Problem 2

Consider a thermodynamic system with two modes of vibration with vibration frequencies ω and 2ω at a fixed temperature T. Determine the probability that the energy of the system is lower than $4\hbar\omega$.

Solution

The two modes of vibration have the energy spectra

$$E_{n_1} = \hbar\omega\left(n_1 + \frac{1}{2}\right), \quad E_{n_2} = 2\hbar\omega\left(n_2 + \frac{1}{2}\right). \tag{12.74}$$

The total energy of the system is given by

$$E_{n_1 n_2} = \hbar\omega\left(n_1 + 2n_2 + \frac{3}{2}\right), \tag{12.75}$$

and the ground-state energy ($n_1 = n_2 = 0$) is

$$E_{00} = \frac{3}{2}\hbar\omega. \tag{12.76}$$

The energy levels below $4\hbar\omega$ are

$$E_{00} = \frac{3}{2}\hbar\omega, \ E_{10} = \frac{5}{2}\hbar\omega, \ E_{01} = E_{20} = \frac{7}{2}\hbar\omega. \tag{12.77}$$

Thus, the probability that $E_{n_1 n_2} < 4\hbar\omega$ is the sum of probabilities for each of these four states, i.e.,

$$P_{E<4\hbar\omega} = P_{00} + P_{10} + P_{01} + P_{20}. \tag{12.78}$$

The probability of an arbitrary state of the system $P_{n_1 n_2}$ is the probability of the simultaneous occurrence of the two modes of vibration, given by the following product

$$P_{n_1 n_2} = P_{n_1} \cdot P_{n_2} = \frac{1}{Z_1}\exp\left(-\frac{E_{n_1}}{kT}\right) \cdot \frac{1}{Z_2}\exp\left(-\frac{E_{n_2}}{kT}\right). \tag{12.79}$$

On the other hand, for a single mode of vibration of frequency ω, the partition function Z is given by

$$Z = \sum_{n=0}^{\infty}\exp\left(-\frac{E_n}{kT}\right) = \sum_{n=0}^{\infty}\exp\left[-\frac{\hbar\omega}{kT}\left(n + \frac{1}{2}\right)\right], \tag{12.80}$$

or

$$Z = \exp\left(-\frac{\hbar\omega}{2kT}\right) \sum_{n=0}^{\infty} \left[\exp\left(-\frac{\hbar\omega}{kT}\right)\right]^n,$$

(12.81)

or

$$Z = \frac{\exp\left(-\frac{\hbar\omega}{2kT}\right)}{1 - \exp\left(-\frac{\hbar\omega}{kT}\right)}.$$

(12.82)

Thus, we have

$$\frac{1}{Z_1} = \left[1 - \exp\left(-\frac{\hbar\omega}{kT}\right)\right] \exp\left(\frac{\hbar\omega}{2kT}\right)$$

(12.83)

and

$$\frac{1}{Z_2} = \left[1 - \exp\left(-\frac{2\hbar\omega}{kT}\right)\right] \exp\left(\frac{\hbar\omega}{kT}\right).$$

(12.84)

Substituting the results (12.83) and (12.84) into (12.79), we obtain

$$P_{n_1 n_2} = \exp\left[-\frac{\hbar\omega}{kT}\left(n_1 + \frac{1}{2}\right)\right] \exp\left[-\frac{\hbar\omega}{kT}(2n_2 + 1)\right]$$

$$\times \left[1 - \exp\left(-\frac{\hbar\omega}{kT}\right)\right] \exp\left(\frac{\hbar\omega}{2kT}\right) \times \left[1 - \exp\left(-\frac{2\hbar\omega}{kT}\right)\right] \exp\left(\frac{\hbar\omega}{kT}\right),$$

(12.85)

or

$$P_{n_1 n_2} = \exp\left[-\frac{\hbar\omega}{kT}(n_1 + 2n_2)\right] \left[1 - \exp\left(-\frac{\hbar\omega}{kT}\right)\right] \times \left[1 - \exp\left(-\frac{2\hbar\omega}{kT}\right)\right].$$

(12.86)

Thus, the final result for the probability that the energy of the system is lower than $4\hbar\omega$ is

$$P_{E<4\hbar\omega} = P_{00} + P_{10} + P_{01} + P_{20},$$

(12.87)

with $P_{n_1 n_2}$ defined by (12.86).

13 Nonideal Gases

13.1 Partition Function for Nonideal Gases

Thus far, we have considered very dilute, ideal gases with intermolecular separations much larger than the range of intermolecular interactions. In accordance with this, the interaction potential $U(\vec{r}_1, \vec{r}_2, \ldots, \vec{r}_N)$ between the constituent particles was simply set to be zero. This is a good approximation for many realistic gases under normal conditions ($p \sim 1$ atm, $T \sim$ room temperature). However, even under normal conditions, the interparticle interactions may produce possibly small yet experimentally measurable effects in realistic, nonideal gases. In this chapter, we would like to understand these effects in the framework of statistical mechanics. Importantly, interparticle interactions are under certain conditions capable of producing qualitatively significant effects such as the condensation of a gas into a liquid or freezing of a liquid into a solid. Understanding these phase transitions was and, to some extent, still one of the most challenging problems of statistical mechanics. In order to analyze the effects of interparticle interactions, we start with the total energy of the N-particle system

$$E_N = \sum_{a=1}^{N} \frac{p_a^2}{2m} + U(\vec{r}_1, \vec{r}_2, \ldots, \vec{r}_N), \tag{13.1}$$

where $\vec{r}_a = (x_a, y_a, z_a)$ is the position vector of the constituent particle labeled by index a. The classical partition function for the N-particle system is given by (11.222), i.e.,

$$Z_N = \frac{1}{N!} \int_\Gamma \exp\left(-\frac{E_N}{kT}\right) d\Gamma, \tag{13.2}$$

where the dimensionless infinitesimal element of the $6N$-dimensional phase space of the entire system, with N constituent particles, is given by

$$d\Gamma = \left[\prod_{a=1}^{N} \frac{4\pi p_a^2 dp_a}{h^3}\right]\left[\prod_{a=1}^{N} dx_a dy_a dz_a\right]. \tag{13.3}$$

Substituting (13.1) into (13.2), we obtain

$$Z_N = \frac{1}{N!} \int_\Gamma \exp\left(-\sum_{a=1}^{N} \frac{p_a^2}{2mkT} - \frac{U}{kT}\right) d\Gamma, \tag{13.4}$$

or

$$Z_N = Z_{0N} \times Z_{IN} = \frac{Z_0^N}{N!} \times Z_{IN}. \tag{13.5}$$

In the equation (13.5), the quantity $Z_{0N} = Z_0^N/N!$ is the partition function for the ideal gas without particle interactions ($U = 0$), defined by

$$Z_{0N} = \frac{V^N}{N!} \int\limits_{\Gamma_p} \exp\left(-\sum_{a=1}^{N} \frac{p_a^2}{2mkT}\right) \prod_{a=1}^{N} \frac{4\pi p_a^2 dp_a}{h^3}, \tag{13.6}$$

or

$$Z_{0N} = \frac{1}{N!} \left[\frac{V}{h^3} \int\limits_0^\infty \exp\left(-\frac{p_a^2}{2mkT}\right) 4\pi p_a^2 dp_a \right]^N = \frac{Z_0^N}{N!}, \tag{13.7}$$

where

$$Z_0 = \frac{V}{h^3} \int\limits_0^\infty \exp\left(-\frac{p_a^2}{2mkT}\right) 4\pi p_a^2 dp_a = V \left(\frac{2\pi mkT}{h^2}\right)^{3/2}. \tag{13.8}$$

The quantity Z_{IN} in (13.5) is the interaction partition function defined by

$$Z_{IN} = \frac{1}{V^N} \int\limits_{\Gamma_V} \exp\left(-\frac{U}{kT}\right) \prod_{a=1}^{N} dx_a dy_a dz_a. \tag{13.9}$$

Introducing the volume elements $dV_a = dx_a dy_a dz_a$ for each particle, we obtain

$$Z_{IN} = \frac{1}{V^N} \int\limits_{\Gamma_V} \exp\left(-\frac{U}{kT}\right) dV_1 dV_2 \ldots dV_N. \tag{13.10}$$

From the equation (13.10), we see that for the ideal gas model, in which the U is set to zero, we have $Z_{IN} = 1$. Thus, in the absence of interparticle interactions, the N-particle partition function in the equation (13.5) reduces to the ideal gas partition function in the equation (13.11).

13.2 Free Energy of Nonideal Gases

Substituting the result (13.5) into the general definition of the free energy (8.14), we can write

$$F = -kT \ln Z_{0N} - kT \ln Z_{IN}, \tag{13.11}$$

or

$$F = -kT \ln \frac{Z_0^N}{N!} - kT \ln Z_{IN}, \tag{13.12}$$

or

$$F = -NkT \ln Z_0 + kT \ln N! - kT \ln Z_{IN}. \tag{13.13}$$

Using here the Stirling formula (7.8), we obtain

$$F = -NkT \ln Z_0 + NkT \ln N - NkT - kT \ln Z_{IN}, \tag{13.14}$$

or

$$F = -NkT \ln \frac{Z_0}{N} - NkT - kT \ln Z_{IN}. \tag{13.15}$$

Thus, the free energy of nonideal gases with particle interactions is given by

$$F = F_0 + F_I = -NkT \ln \frac{Z_0}{N} - NkT - kT \ln Z_{IN}, \tag{13.16}$$

where

$$F_0 = -NkT \ln \left[\frac{V}{N} \left(\frac{2\pi mkT}{h^2} \right)^{3/2} \right] - NkT \tag{13.17}$$

is the free energy of the ideal gas and $F_I = -kT \ln Z_{IN}$ is the contribution to the free energy emerging from the interactions of the constituent particles. For this contribution, by the equation (13.10), we have

$$F_I = -kT \ln \left[\frac{1}{V^N} \int_{\Gamma_V} \exp\left(-\frac{U}{kT} \right) dV_1 dV_2 \ldots dV_N \right]. \tag{13.18}$$

From the definition (13.18), we see that for $U = 0$, we have

$$F_I = -kT \ln \left[\frac{V^N}{V^N} \right] = 0. \tag{13.19}$$

Using (13.19), we can add unity and subtract V^N/V^N within the logarithm on the right-hand side of the equation (13.18), to obtain

$$F_I = -kT \ln \left\{ 1 + \frac{1}{V^N} \int_{\Gamma_V} \left[\exp\left(-\frac{U}{kT} \right) - 1 \right] dV_1 dV_2 \ldots dV_N \right\}. \tag{13.20}$$

By the equation (13.20), it manifests that F_I vanishes in the absence of interactions whence $U = 0$.

13.3   Free Energy of Particle Interactions

In dilute (small-density) nonideal gases, effects of the interactions between the constituent particles are weak. In such gases, the interactions are significant only during the infrequent (small probability) events when some two particles come close to each other. Moreover, in dilute gases, we can neglect the clustered configurations involving three or more particles coming close to each other. Thus, in dealing with dilute nonideal gases, it is sufficient to focus on the effects of the smallest possible clusters, namely the pairs that can be formed out of N particles. It is easy to see that there are a total of $N(N-1)/2$ pairs that can be formed out of N particles. For example, for $N=4$ particles, we have $4(4-1)/2 = 6$ possible pairs, i.e., (12), (13), (14), (23), (24), and (34). The dominant contribution to the integral in (13.20) comes from the $N(N-1)/2$ configurations involving pairs of any two particles: the configuration with the (12) pair with all other particles not in pairs (i.e., not interacting), then the configuration with the (13) pair with all other particles not in pairs (i.e., not interacting), etc. Next, because the particles are statistically indistinguishable, all these $N(N-1)/2$ configurations will contribute equally to the integral in (13.20). In view of these observations, for dilute gases, the integral in (13.20) can be calculated approximately as follows:

$$
I = \int \left[\exp\left(-\frac{U}{kT}\right) - 1 \right] dV_1 dV_2 \dots dV_N
$$

$$
= \frac{1}{2} N(N-1) \int \left[\exp\left(-\frac{U_{12}}{kT}\right) - 1 \right] dV_1 dV_2 \int dV_3 \dots dV_N. \tag{13.21}
$$

Since the interaction potential $U_{12} = U_{12}(\vec{r}_1, \vec{r}_2)$ is the function of the co-ordinates of the two interacting particles only, we can integrate over the co-ordinates of all the other particles, as indicated in the equation (13.21). Furthermore, we can use the approximation $N(N-1) \approx N^2$ valid for $N \gg 1$. Thus, we obtain

$$
I = \frac{N^2 V^{N-2}}{2} \int_{V_1, V_2} \left[\exp\left(-\frac{U_{12}}{kT}\right) - 1 \right] dV_1 dV_2. \tag{13.22}
$$

Substituting (13.22) into (13.20), we obtain

$$
F_I = -kT \ln \left\{ 1 + \frac{N^2}{2V^2} \int_{V_1, V_2} \left[\exp\left(-\frac{U_{12}}{kT}\right) - 1 \right] dV_1 dV_2 \right\}. \tag{13.23}
$$

For weak interactions (with $U_{12} = U_{12}(\vec{r}_1, \vec{r}_2) \ll kT$), and/or dilute gases (with a small particle number density N/V), the first, unity term under the logarithm in the equation (13.23) dominates over the second term therein. Thus, we can use the approximation $\ln(1+x) \approx x$, valid for $x \ll 1$, to obtain

$$
F_I = -\frac{N^2 kT}{V} \frac{1}{2V} \int_{V_1, V_2} \left[\exp\left(-\frac{U_{12}}{kT}\right) - 1 \right] dV_1 dV_2, \tag{13.24}
$$

or

$$F_I = \frac{N^2 kT}{V} B(T),$$ (13.25)

where

$$B(T) = \frac{1}{2V} \int\limits_{V_1, V_2} \left[1 - \exp\left(-\frac{U_{12}}{kT}\right)\right] dV_1 dV_2.$$ (13.26)

Let us now assume that the interaction potential $U_{12} = U_{12}(r)$ is a function of the relative distance $r = |\vec{r}_2 - \vec{r}_1|$ between the two interacting particles only. In such a case, we can express the coordinates of the two interacting particles $\vec{r}_1$ and $\vec{r}_2$ in terms of the co-ordinates of their mutual center of mass $\vec{r}_{CM}$ and the co-ordinates of their relative distance $\vec{r}$. Thus, we can write

$$\vec{r}_{CM} = \frac{m_1 \vec{r}_1 + m_2 \vec{r}_2}{m_1 + m_2} = \frac{\vec{r}_1 + \vec{r}_2}{2}, \quad \vec{r} = \vec{r}_2 - \vec{r}_1,$$ (13.27)

or

$$\vec{r}_1 = \vec{r}_{CM} - \frac{\vec{r}}{2}, \quad \vec{r}_2 = \vec{r}_{CM} + \frac{\vec{r}}{2}.$$ (13.28)

Using the equations (13.28), we can formally write

$$dV_1 dV_2 = \frac{\partial(\vec{r}_1, \vec{r}_2)}{\partial(\vec{r}_{CM}, \vec{r})} dV_{CM} dV,$$ (13.29)

where

$$\frac{\partial(\vec{r}_1, \vec{r}_2)}{\partial(\vec{r}_{CM}, \vec{r})} = \begin{vmatrix} \frac{\partial \vec{r}_1}{\partial \vec{r}_{CM}} & \frac{\partial \vec{r}_1}{\partial \vec{r}} \\ \frac{\partial \vec{r}_2}{\partial \vec{r}_{CM}} & \frac{\partial \vec{r}_2}{\partial \vec{r}} \end{vmatrix} = \begin{vmatrix} 1 & -\frac{1}{2} \\ 1 & +\frac{1}{2} \end{vmatrix} = 1.$$ (13.30)

Thus, we have $dV_1 dV_2 = dV_{CM} dV$, and the equation (13.26) can be rewritten as

$$B(T) = \frac{1}{2} \int\limits_V \left[1 - \exp\left(-\frac{U_{12}}{kT}\right)\right] dV \frac{1}{V} \int\limits_V dV_{CM},$$ (13.31)

or, as $\frac{1}{V} \int_V dV_{CM} = 1$,

$$B(T) = \frac{1}{2} \int\limits_0^\infty \left[1 - \exp\left(-\frac{U_{12}}{kT}\right)\right] 4\pi r^2 dr.$$ (13.32)

The equation (13.25) for F_I in combination with the result (13.32), giving the so-called *second virial coefficient* $B(T)$, are the central results of the above-described perturbative calculation of the nonideal gas free energy. For ideal gases, in the absence of interparticle interactions, $B(T) = 0$, and the first-order perturbative correction to the gas free energy displayed in (13.25) vanishes. In order to calculate $B(T)$, we need to know the detailed form of the interparticle interaction potential $U_{12} = U_{12}(r)$. However, from some sound physical arguments, one can derive a very illuminating approximation for $B(T)$ in the equation (13.32) even without specifying the detailed form of the interaction potential $U_{12} = U_{12}(r)$. Indeed, realistic molecules are much like hard, impenetrable balls with the radius r_0 (hard-core radius). The centers of two such balls cannot be brought to each other to a distance r smaller than $2r_0$. This corresponds to a positive repulsive $U_{12}(r)$, which is essentially infinite for r less than $2r_0$. Thus, in the equation (13.32), we can set,

$$\exp\left(-\frac{U_{12}}{kT}\right) \to 0, \quad r \le 2r_0. \tag{13.33}$$

On the other hand for $r > 2r_0$, one typically has a relatively weak negative (attractive) potential. Indeed, realistic molecules generally attract each other via the van der Waals interactions with $U_{12}(r) \sim -1/r^6$, for a large r. In the region $r > 2r_0$, for a weak U_{12}, we can use the approximation $1 - \exp(-x) \approx x$, to obtain

$$1 - \exp\left(-\frac{U_{12}}{kT}\right) \approx \frac{U_{12}}{kT}, \quad r > 2r_0. \tag{13.34}$$

Using the approximations (13.33) and (13.34), we can rewrite the result (13.32) as follows:

$$B(T) = \frac{1}{2} \int_0^{2r_0} \left[1 - \exp\left(-\frac{U_{12}}{kT}\right)\right] 4\pi r^2 dr$$

$$+ \frac{1}{2} \int_{2r_0}^{\infty} \left[1 - \exp\left(-\frac{U_{12}}{kT}\right)\right] 4\pi r^2 dr, \tag{13.35}$$

or

$$B(T) \approx \frac{1}{2} \int_0^{2r_0} 4\pi r^2 dr + \frac{1}{2} \int_{2r_0}^{\infty} \frac{U_{12}}{kT} 4\pi r^2 dr = 2\pi \int_0^{2r_0} r^2 dr + \frac{2\pi}{kT} \int_{2r_0}^{\infty} U_{12} r^2 dr$$

$$= 16\pi \frac{r_0^3}{3} - \frac{1}{T}\left(-\frac{2\pi}{k}\int_{2r_0}^{\infty} U_{12} r^2 dr\right) = b - \frac{a}{T}. \tag{13.36}$$

Since the interaction potential is negative $U_{12} < 0$ for $r > 2r_0$, the integral

$$a = -\frac{2\pi}{k} \int\limits_{2r_0}^{\infty} U_{12} r^2 dr > 0 \tag{13.37}$$

is a positive quantity. In (13.36), we also introduced the quantity

$$b = 16\pi \frac{r_0^3}{3} = 4 \times 4\pi \frac{r_0^3}{3}, \tag{13.38}$$

i.e., the b is four times the volume of a constituent particle. Substituting (13.36) into (13.25), we obtain

$$F_I = \frac{N^2 kT}{V} \left(b - \frac{a}{T} \right) = \frac{N^2 k}{V} (bT - a), \tag{13.39}$$

or finally

$$F_I = NkT \frac{Nb}{V} - \frac{N^2 ka}{V}. \tag{13.40}$$

Equation (13.40) gives our final result for the free-energy contribution coming from interactions in a dilute nonideal gas.

13.4 van der Waals Equation

The total free energy of a nonideal gas is given by the equation (13.16) with F_I as in equation (13.40), yielding

$$F = F_0 + F_I = -NkT \ln \left[\frac{1}{N} \left(\frac{2\pi mkT}{h^2} \right)^{3/2} \right]$$

$$- NkT \left(\ln V - \frac{Nb}{V} \right) - \frac{N^2 ka}{V}. \tag{13.41}$$

For a dilute nonideal gas, the total volume per particle V/N is much larger than four times the volume of a single constituent particle $b = 4 \times 4\pi r_0^3/3$. Thus, we have $b \ll V/N$ or

$$\frac{Nb}{V} \ll 1. \tag{13.42}$$

Using again the approximation $\ln(1 + x) \approx x$ for $x \ll 1$, we have

$$\ln(V - Nb) = \ln V \left(1 - \frac{Nb}{V}\right)$$

$$= \ln V + \ln\left(1 - \frac{Nb}{V}\right) \approx \ln V - \frac{Nb}{V}. \tag{13.43}$$

Substituting (13.43) into (13.41), we obtain

$$F = -NkT \ln\left[\frac{1}{N}\left(\frac{2\pi mkT}{h^2}\right)^{3/2}\right] - NkT \ln(V - Nb) - \frac{N^2 ka}{V}. \tag{13.44}$$

The thermal-state equation for nonideal gases is now obtained using (8.42) as follows:

$$p = -\left(\frac{\partial F}{\partial V}\right)_T = \frac{NkT}{V - Nb} - \frac{N^2 ka}{V^2}. \tag{13.45}$$

After rearranging the terms in the equation (13.45), we finally obtain the so-called van der Waals thermal-state equation for nonideal gases in the form

$$\left(p + \frac{N^2 ka}{V^2}\right)(V - Nb) = NkT. \tag{13.46}$$

In the absence of interactions, i.e., with no long-range attraction and with zero hard-core radius $r_0 = 0$, by equations (13.37) and (13.38), the constants a and b in (13.46) vanish, and the van der Waals equation reduces to the familiar ideal gas law.

13.5   Caloric-State Equation for Nonideal Gases

The van der Waals equation is the thermal-state equation for nonideal gases. The caloric-state equation for nonideal gases is obtained from the general result (8.22), in the following form

$$U = kT^2 \frac{\partial}{\partial T}(\ln Z_N)_V, \tag{13.47}$$

or, in terms of the free energy $F = -kT \ln Z_N$,

$$U = -kT^2 \left(\frac{\partial}{\partial T}\frac{F}{kT}\right)_V. \tag{13.48}$$

Using here the equation (13.16), i.e, that $F = F_0 + F_I$, we have

$$U = U_0 + U_I, \tag{13.49}$$

with U_0 being the ideal gas contribution,

$$U_0 = -kT^2 \left(\frac{\partial}{\partial T} \frac{F_0}{kT} \right)_V = \frac{3}{2} NkT,$$ (13.50)

whereas the interactions contribute the term,

$$U_I = -kT^2 \left(\frac{\partial}{\partial T} \frac{F_I}{kT} \right)_V.$$ (13.51)

By the above equation and equation (13.40), one obtains

$$U_I = -kT^2 \left[\frac{\partial}{\partial T} \frac{1}{kT} \left(NkT \; \frac{Nb}{V} - \frac{N^2 ka}{V} \right) \right]_V = -\frac{N^2 ka}{V}.$$ (13.52)

Thus, the caloric equation for dilute nonideal gases has the form

$$U = U_0 + U_I = \frac{3}{2} NkT - \frac{N^2 ka}{V},$$ (13.53)

where the second term is the energy contribution coming from interactions. Note that $U_I/U_0 \sim N/V =$ particle number density, so for dilute (low-density) gases, the ideal gas contribution U_0 dominates over the interaction contribution U_I.

13.6 Specific Heats for Nonideal Gases

From the caloric equation for nonideal gases (13.53), we obtain the internal energy per unit mass as follows:

$$u = \frac{3}{2} R_g T - R_g \frac{Na}{V},$$ (13.54)

where we used the relation $Nk = mR_g$. The specific heat at constant volume, defined by (9.55), is then calculated as follows:

$$c_V = \left(\frac{\partial u}{\partial T} \right)_V = \frac{3}{2} R_g.$$ (13.55)

Thus, the specific heat at constant volume for dilute nonideal gases (13.55) is the same as in the case of the ideal gas (11.95). The specific heat at constant pressure c_P is a more complex function of temperature and volume of the system and is not constant as in the case of ideal gas. However, using the van der Waals thermal-state equation, we can derive the relation between the two specific heats of a nonideal gas, described by the van der Waals equation. Let us recall the general result (9.86) for a gas of an arbitrary mass m, i.e.,

$$m(c_p - c_V) = TV \frac{\alpha^2}{\gamma}, \tag{13.56}$$

and the definitions of the parameters α and γ, given by (9.26) and (9.41), respectively, i.e.,

$$\alpha = \frac{1}{V} \left(\frac{\partial V}{\partial T} \right)_P, \quad \gamma = -\frac{1}{V} \left(\frac{\partial V}{\partial p} \right)_T. \tag{13.57}$$

From the identity (see Section 9.4.3, "Isothermal Expansion")

$$\frac{\partial(V,p)}{\partial(T,p)} \frac{\partial(T,p)}{\partial(T,V)} \frac{\partial(T,V)}{\partial(V,p)} = 1, \tag{13.58}$$

we obtain

$$\left(\frac{\partial V}{\partial T} \right)_P = -\frac{\partial(V,T)}{\partial(p,T)} \frac{\partial(p,V)}{\partial(T,V)} = -\left(\frac{\partial V}{\partial p} \right)_T \left(\frac{\partial p}{\partial T} \right)_V, \tag{13.59}$$

or

$$\frac{1}{V} \left(\frac{\partial V}{\partial T} \right)_P = -\frac{1}{V} \left(\frac{\partial V}{\partial p} \right)_T \left(\frac{\partial p}{\partial T} \right)_V. \tag{13.60}$$

Using here (13.59), we have

$$\alpha = \gamma \left(\frac{\partial p}{\partial T} \right)_V. \tag{13.61}$$

Thus, we obtain

$$m(c_p - c_V) = TV\gamma \left[\left(\frac{\partial p}{\partial T} \right)_V \right]^2, \tag{13.62}$$

or

$$m(c_p - c_V) = -T \left(\frac{\partial V}{\partial p} \right)_T \left[\left(\frac{\partial p}{\partial T} \right)_V \right]^2. \tag{13.63}$$

The relation (13.63) is an alternative general relation between the two specific heats, suitable for the analysis of nonideal gases. Using the thermal-state equation for nonideal gases (13.45), we have

$$\left(\frac{\partial p}{\partial T} \right)_V = \frac{Nk}{V - Nb}, \tag{13.64}$$

and

$$\left(\frac{\partial V}{\partial p}\right)_T = \left[\left(\frac{\partial p}{\partial V}\right)_T\right]^{-1} = -\left[\frac{NkT}{(V-Nb)^2} - \frac{2N^2ka}{V^3}\right]^{-1}. \tag{13.65}$$

Substituting (13.64) and (13.65) into (13.63), we obtain

$$m(c_p - c_V) = T\left[\frac{NkT}{(V-Nb)^2} - \frac{2N^2ka}{V^3}\right]^{-1}\left(\frac{Nk}{V-Nb}\right)^2. \tag{13.66}$$

Using here $Nk = mR_g$ and rearranging, we finally obtain

$$c_p - c_V = \frac{R_g}{1 - 2Na(V-Nb)^2/(TV^3)}. \tag{13.67}$$

From the result (13.67), we see that, when the interaction potential tends to zero $U_{12} \to 0$ ($a \to 0$), the relation between the two specific heats approaches the result for an ideal gas (11.97), i.e.,

$$c_p - c_V = R_g. \tag{13.68}$$

Combining the results (13.55) and (13.67), we obtain the result for the specific heat at constant pressure in the form

$$c_p = \frac{3}{2}R_g + \frac{R_g}{1 - 2Na(V-Nb)^2/(TV^3)}. \tag{13.69}$$

13.7 Problems with Solutions

Problem 1

Consider 1 mole ($Nk = R$) of a nonideal gas described by the van der Waals equation of state

$$\left(p + \frac{A}{V^2}\right)(V - B) = RT, \tag{13.70}$$

where $A = N^2ka$ and $B = Nb$. If the gas expands freely with no heat exchanged with the surroundings from the volume V to volume $2V$, determine the temperature change ΔT of the gas as a function of the initial volume V, the constant specific heat at constant volume c_V, and the two constants A and B in the thermal equation of state.

Solution

The gas is expanding freely, which means that it does not perform any work to the surroundings ($\Delta L = 0$), and it does not exchange any heat with the surroundings ($\Delta Q = 0$). From the first law of thermodynamics ($\Delta U = \Delta Q - \Delta L$), we conclude that the internal energy of the gas remains constant ($\Delta U = 0$) during the expansion. Using the differential formulation of the first law of thermodynamics (8.39), we can write

$$dU = TdS - pdV \Rightarrow \left(\frac{\partial U}{\partial V}\right)_T = T\left(\frac{\partial S}{\partial V}\right)_T - p. \tag{13.71}$$

Using now the identity (9.80), i.e.,

$$\frac{\partial(T,S)}{\partial(p,V)} = \frac{\partial(S,T)}{\partial(V,p)} = 1, \tag{13.72}$$

we can calculate

$$\left(\frac{\partial S}{\partial V}\right)_T = \frac{\partial(S,T)}{\partial(V,T)} = \frac{\partial(V,p)}{\partial(V,T)} = \left(\frac{\partial p}{\partial T}\right)_V. \tag{13.73}$$

Substituting (13.73) into (13.71) and using the equation of state (13.70), yields

$$\left(\frac{\partial U}{\partial V}\right)_T = T\left(\frac{\partial p}{\partial T}\right)_V - p = \frac{RT}{V-B} - p = \frac{A}{V^2}. \tag{13.74}$$

Integrating the equation (13.74), we obtain

$$U(T,V) = -\frac{A}{V} + f(T), \tag{13.75}$$

where the unknown function $f(T)$ is determined from the definition of the molar heat capacity (since we have here 1 mole) at constant volume c_V, i.e.,

$$c_V = \left(\frac{\partial U}{\partial T}\right)_V = f'(T) \Rightarrow f(T) = c_V T + \text{Constant}. \tag{13.76}$$

Thus, the constant internal energy of the nonideal gas is given by

$$U(T,V) = -\frac{A}{V} + c_V T + \text{Constant}. \tag{13.77}$$

Unsurprisingly, the equation (13.77) is (after setting the Constant $= 0$) identical to equation (13.53). The initial state (T_1, V) and the final state (T_2, $2V$) have the same internal energy, and we can write

$$-\frac{A}{V} + c_V T_1 = -\frac{A}{2V} + c_V T_2. \tag{13.78}$$

The temperature change ΔT of the gas is now calculated as follows:

$$\Delta T = T_2 - T_1 = -\frac{A}{2c_V V}. \tag{13.79}$$

Problem 2

Consider 1 mole ($Nk = R$) of a nonideal gas described by the van der Waals equation of state

$$\left(p + \frac{A}{V^2}\right)(V - B) = RT, \tag{13.80}$$

where $A = N^2 ka$ and $B = Nb$. If the gas expands at constant temperature ($T = $ Constant) from the volume $V_1 = 3B$ to the volume $V_2 = 6B$ (B is the constant from the above equation of state), calculate the heat developed in the process.

Solution

Using again the identity (9.80), i.e.,

$$\frac{\partial(T,S)}{\partial(p,V)} = \frac{\partial(S,T)}{\partial(V,p)} = 1, \tag{13.81}$$

we can use the equation of state (13.81) to calculate

$$\left(\frac{\partial S}{\partial V}\right)_T = \frac{\partial(S,T)}{\partial(V,T)} = \frac{\partial(V,p)}{\partial(V,T)} = \left(\frac{\partial p}{\partial T}\right)_V = \frac{R}{V - B}. \tag{13.82}$$

At constant temperature, we then have

$$dS = \frac{R}{V - B} dV. \tag{13.83}$$

The heat developed in the process can be calculated using

$$Q = \int_{3B}^{6B} T dS = \int_{3B}^{6B} \frac{RT}{V - B} dV = RT \ln\left(\frac{6B - B}{3B - B}\right), \tag{13.84}$$

and the final result is given by

$$Q = RT \ln \frac{5}{2}. \tag{13.85}$$

Problem 3

Consider 1 mole ($Nk = R$) of a nonideal gas described by the van der Waals equation of state

$$\left(p + \frac{A}{V^2}\right)(V - B) = RT,$$ (13.86)

where $A = N^2 ka$ and $B = Nb$. The curves in $p - V$ plane, for various values of T, have a maximum and a minimum at the two points where $(\partial p/\partial V)_T = 0$. At some critical temperature $T = T_C$, the maximum and the minimum merge into a single critical point (p_C, V_C, T_C), where both $(\partial p/\partial V)_T = 0$ and $(\partial^2 p/\partial V^2)_T = 0$.

(a) Express A and B in terms of T_C and V_C.
(b) Express p_C in terms of T_C and V_C.
(c) Write the van der Waals equation in terms of the dimensionless variables $p_1 = p/p_C$, $V_1 = V/V_C$, and $T_1 = T/T_C$.

Solution

(a) From the equation of state (13.86), we can write

$$p = \frac{RT}{V - B} - \frac{A}{V^2}.$$ (13.87)

Thus, we obtain

$$\left(\frac{\partial p}{\partial V}\right)_{T=T_C} = -\frac{RT_C}{(V_C - B)^2} + \frac{2A}{V_C^3} = 0,$$ (13.88)

and

$$\left(\frac{\partial^2 p}{\partial V^2}\right)_{T=T_C} = \frac{2RT_C}{(V_C - B)^3} - \frac{6A}{V_C^4} = 0.$$ (13.89)

Hence, we have

$$\frac{2A}{V_C^3} = \frac{RT_C}{(V_C - B)^2},$$ (13.90)

and

$$\frac{6A}{V_C^4} = \frac{2RT_C}{(V_C - B)^3}.$$ (13.91)

Dividing (13.91) by (13.90), we obtain

$$\frac{3}{V_C} = \frac{2}{V_C - B} \Rightarrow 3(V_C - B) = 2V_C \Rightarrow B = \frac{V_C}{3}.$$ (13.92)

Using (13.92), we further obtain

$$A = \frac{RT_C V_C^3}{2(V_C - B)^2} = \frac{RT_C V_C^3}{2\frac{4}{9}V_C^2} = \frac{9}{8}RT_C V_C. \tag{13.93}$$

Using (13.92) and (13.93), the equation of state (13.86) becomes

$$\left(p + \frac{9RT_C V_C}{8V^2}\right)\left(V - \frac{V_C}{3}\right) = RT. \tag{13.94}$$

(b) From the equation (13.94) at the critical point (p_C, V_C, T_C), we have

$$\left(p_C + \frac{9RT_C}{8V_C}\right)\frac{2V_C}{3} = RT_C, \tag{13.95}$$

or

$$p_C = \frac{3}{2}\frac{RT_C}{V_C} - \frac{9RT_C}{8V_C} = \frac{3}{8}\frac{RT_C}{V_C}. \tag{13.96}$$

(c) Substituting (13.96) into (13.94), the equation of state becomes

$$\left(p + 3p_C\frac{V_C^2}{V^2}\right)(3V - V_C) = 3RT, \tag{13.97}$$

or

$$\left(\frac{p}{p_C} + 3\frac{V_C^2}{V^2}\right)\left(3\frac{V}{V_C} - 1\right) = \frac{3RT_C}{p_C V_C}\frac{T}{T_C}, \tag{13.98}$$

or

$$\left(p_1 + \frac{3}{V_1^2}\right)(3V_1 - 1) = \frac{3RT_C}{p_C V_C}T_1. \tag{13.99}$$

Using here (13.96), we finally obtain

$$\left(p_1 + \frac{3}{V_1^2}\right)(3V_1 - 1) = 8T_1. \tag{13.100}$$

Problem 4

The free expansion of a gas is a process where the internal energy of the gas remains constant. Consider 1 mole ($Nk = R$) of a nonideal gas described by the van der Waals equation of state

$$\left(p + 3p_C \frac{V_C^2}{V^2}\right)(3V - V_C) = 3RT, \tag{13.101}$$

where $p_C = ka/27b^2$ and $V_C = 3Nb$. Calculate the change of temperature $\Delta T = T_2 - T_1$ of this nonideal gas, when it is expanded freely from V_1 to V_2.

Solution

In order to calculate the temperature change $\Delta T = T_2 - T_1$, we need to derive the general result for $(\partial T/\partial V)_U$ of the freely expanding gas. From the first law of thermodynamics

$$dU = TdS - pdV, \tag{13.102}$$

we obtain

$$\left(\frac{\partial U}{\partial V}\right)_T = T\left(\frac{\partial S}{\partial V}\right)_T - p. \tag{13.103}$$

On the other hand, we have

$$\left(\frac{\partial S}{\partial V}\right)_T = \frac{\partial(T, s)}{\partial(T, V)} = \frac{\partial(p, V)}{\partial(T, V)} = \left(\frac{\partial p}{\partial T}\right)_V, \tag{13.104}$$

and

$$\left(\frac{\partial U}{\partial V}\right)_T = T\left(\frac{\partial p}{\partial T}\right)_V - p. \tag{13.105}$$

Now, we can use

$$\left(\frac{\partial U}{\partial V}\right)_T = \frac{\partial(U, T)}{\partial(V, T)} = \frac{\partial(U, T)}{\partial(U, V)} \frac{\partial(U, V)}{\partial(V, T)}$$

$$= -\left(\frac{\partial T}{\partial V}\right)_U \left(\frac{\partial U}{\partial T}\right)_V = -\left(\frac{\partial T}{\partial V}\right)_U c_V, \tag{13.106}$$

to obtain

$$\left(\frac{\partial T}{\partial V}\right)_U = -\frac{1}{c_V}\left(\frac{\partial U}{\partial V}\right)_T = -\frac{1}{c_V}\left[T\left(\frac{\partial p}{\partial T}\right)_V - p\right]. \tag{13.107}$$

From the equation of state (13.101), we can write

$$p = \frac{3RT}{3V - V_C} - 3pc\frac{V_C^2}{V^2},$$
(13.108)

and

$$\left(\frac{\partial p}{\partial T}\right)_V = \frac{3R}{3V - V_C}.$$
(13.109)

Thus, we have

$$T\left(\frac{\partial p}{\partial T}\right)_V - p = \frac{3RT}{3V - V_C} - \frac{3RT}{3V - V_C} + 3pc\frac{V_C^2}{V^2},$$
(13.110)

or

$$T\left(\frac{\partial p}{\partial T}\right)_V - p = 3pc\frac{V_C^2}{V^2}.$$
(13.111)

Substituting (13.111) into (13.107), we obtain

$$\left(\frac{\partial T}{\partial V}\right)_U = -\frac{3pc}{c_V}\frac{V_C^2}{V^2}.$$
(13.112)

Thus, for $U = $ Constant, we have

$$dT = -\frac{3pc}{c_V}\frac{V_C^2}{V^2}dV,$$
(13.113)

or

$$\int_{T_1}^{T_2} dT = -\frac{3pcV_C^2}{c_V}\int_{V_1}^{V_2}\frac{dV}{V^2}.$$
(13.114)

Finally, we have

$$\Delta T = \frac{3pcV_C^2}{c_V}\left(\frac{1}{V_2} - \frac{1}{V_1}\right) < 0.$$
(13.115)

From the result (13.115), we see that the temperature of the gas decreases ($\Delta T < 0$) in the course of the free expansion from volume V_1 to a larger volume V_2.

Problem 5

The inversion curve for a nonideal gas, described by the van der Waals equation of state, is obtained from the condition that the so-called Joule–Kelvin coefficient is equal to zero, i.e.,

$$\left(\frac{\partial T}{\partial p}\right)_H = 0, \tag{13.116}$$

where $H = U + pV$ is the enthalpy of the gas. Derive the inversion curve for a van der Waals gas, expressed in terms of the dimensionless variables $p_1 = p/p_C$, $V_1 = V/V_C$, and $T_1 = T/T_C$. Find $p_1 = p_1(T_1)$ along the inversion curve.

Solution

The Joule–Kelvin coefficient can be calculated as

$$\left(\frac{\partial T}{\partial p}\right)_H = \frac{\partial(T,H)}{\partial(p,H)} = -\frac{\partial(T,H)}{\partial(T,p)}\frac{\partial(p,T)}{\partial(p,H)}, \tag{13.117}$$

or

$$\left(\frac{\partial T}{\partial p}\right)_H = -\left(\frac{\partial H}{\partial p}\right)_T\left[\left(\frac{\partial H}{\partial T}\right)_P\right]^{-1} = -\frac{1}{c_P}\left(\frac{\partial H}{\partial p}\right)_T. \tag{13.118}$$

Using now the differential of enthalpy $dH = TdS + Vdp$, we have

$$\left(\frac{\partial H}{\partial p}\right)_T = T\left(\frac{\partial S}{\partial p}\right)_T + V. \tag{13.119}$$

On the other hand, using

$$\left(\frac{\partial S}{\partial p}\right)_T = \frac{\partial(T,S)}{\partial(T,p)} = -\frac{\partial(V,p)}{\partial(T,p)} = -\left(\frac{\partial V}{\partial T}\right)_p, \tag{13.120}$$

we obtain

$$\left(\frac{\partial H}{\partial p}\right)_T = -T\left(\frac{\partial V}{\partial T}\right)_p + V. \tag{13.121}$$

Substituting (13.121) into (13.118), we obtain

$$\left(\frac{\partial T}{\partial p}\right)_H = \frac{V}{c_P}\left[\frac{T}{V}\left(\frac{\partial V}{\partial T}\right)_p - 1\right]. \tag{13.122}$$

The condition (13.116), with (13.122), becomes

$$\frac{T}{V}\left(\frac{\partial V}{\partial T}\right)_p - 1 = 0, \tag{13.123}$$

or

$$\frac{T}{V}\left(\frac{\partial V}{\partial T}\right)_p = -\frac{T}{V}\left(\frac{\partial p}{\partial T}\right)_V \left[\left(\frac{\partial p}{\partial V}\right)_T\right]^{-1} = 1, \tag{13.124}$$

or

$$-\frac{V}{T}\left(\frac{\partial p}{\partial V}\right)_T \left[\left(\frac{\partial p}{\partial T}\right)_V\right]^{-1} = 1, \tag{13.125}$$

or

$$-\frac{V_1}{T_1}\left(\frac{\partial p_1}{\partial V_1}\right)_{T_1} \left[\left(\frac{\partial p_1}{\partial T_1}\right)_{V_1}\right]^{-1} = 1. \tag{13.126}$$

Using the result (13.100), we can write

$$p_1 = \frac{8T_1}{3V_1 - 1} - \frac{3}{V_1^2}, \tag{13.127}$$

and

$$\left[\left(\frac{\partial p_1}{\partial T_1}\right)_{V_1}\right]^{-1} = \left[\frac{8}{3V_1 - 1}\right]^{-1} = \frac{3V_1 - 1}{8}, \tag{13.128}$$

and

$$\left(\frac{\partial p_1}{\partial V_1}\right)_{T_1} = -\frac{24T_1}{(3V_1 - 1)^2} + \frac{6}{V_1^3}. \tag{13.129}$$

Substituting (13.128) and (13.129) into (13.127), we obtain

$$-\frac{V_1}{T_1}\frac{3V_1 - 1}{8}\left[-\frac{24T_1}{(3V_1 - 1)^2} + \frac{6}{V_1^3}\right] = 1, \tag{13.130}$$

or

$$\frac{3V_1}{3V_1 - 1} - \frac{3(3V_1 - 1)}{4T_1 V_1^2} = 1, \tag{13.131}$$

or

$$\frac{12T_1V_1^3 - 3(3V_1 - 1)^2}{4T_1V_1^2 \ (3V_1 - 1)} = 1,$$

(13.132)

or

$$12T_1V_1^3 - 3(3V_1 - 1)^2 = 4T_1V_1^2(3V_1 - 1),$$

(13.133)

or

$$3(3V_1 - 1)^2 = 4T_1V_1^2,$$

(13.134)

or

$$\frac{3V_1 - 1}{V_1} = 2\sqrt{\frac{T_1}{3}}.$$

(13.135)

The solution of this equation is given by

$$V_1 = \frac{1}{3 - 2\sqrt{T_1/3}},$$

(13.136)

such that we have

$$3V_1 - 1 = \frac{2\sqrt{T_1/3}}{3 - 2\sqrt{T_1/3}}.$$

(13.137)

Substituting (13.136) and (13.137) into (13.127), we obtain

$$p_1 = 8T_1\frac{3 - 2\sqrt{T_1/3}}{2\sqrt{T_1/3}} - 3\left(3 - 2\sqrt{T_1/3}\right)^2,$$

(13.138)

or

$$p_1 = -27 + 24\sqrt{3}\sqrt{T_1} - 12T_1,$$

(13.139)

or

$$p_1 = 9 - 12\left(3 - 2\sqrt{3}\sqrt{T_1} + T_1\right).$$

(13.140)

Thus, we finally obtain the inversion curve in the following form

$$p_1 = 9 - 12\left(\sqrt{T_1} - \sqrt{3}\right)^2.$$

(13.141)

Problem 6

Consider 1 mole ($Nk = R$) of a nonideal gas described by the van der Waals equation of state

$$\left(p + 3p_C\frac{V_C^2}{V^2}\right)(3V - V_C) = 3RT,$$ (13.142)

where $p_C = ka/27b^2$, $V_C = 3Nb$, and $T_C = 8a/27b$ and determine the critical point, where both $(\partial p/\partial V)_T = 0$ and $(\partial^2 p/\partial V^2)_T = 0$.

(a) Show that the pressure function $p(V, T)$ can be expanded about the critical point (p_C, V_C, T_C) as follows:

$$p = p_C + c_1(T - T_C) + c_2(T - T_C)(V - V_C) + c_3(V - V_C)^3$$ (13.143)

and determine the coefficients c_1, c_2, and c_3.

(b) Show that, for $T = T_C$, we have

$$p - p_C = c_3(V - V_C)^\delta,$$ (13.144)

and determine δ.

(c) When p lies on the phase boundary line near the critical point, defined by $p - p_C = c_1(T - T_C)$, show that we have

$$V - V_C = c_4(T_C - T)^\beta,$$ (13.145)

and determine c_4 and β.

Solution

(a) From the equation of state (13.142), we can write

$$p(V, T) = \frac{3RT}{3V - V_C} - 3p_C\frac{V_C^2}{V^2}.$$ (13.146)

Developing this function in Taylor series about the critical point, we obtain

$$p(V, T) = p(V_C, T_C) + \left(\frac{\partial p}{\partial V}\right)_C (V - V_C) + \left(\frac{\partial p}{\partial T}\right)_C (T - T_C)$$

$$+ \frac{1}{2}\left(\frac{\partial^2 p}{\partial V^2}\right)_C (V - V_C)^2 + \left(\frac{\partial^2 p}{\partial V \partial T}\right)_C (T - T_C)(V - V_C)$$

$$+ \frac{1}{2}\left(\frac{\partial^2 p}{\partial T^2}\right)_C (T - T_C)^2 + \frac{1}{6}\left(\frac{\partial^3 p}{\partial V^3}\right)_C (V - V_C)^3 + \cdots.$$

(13.147)

In the equation (13.147), we observe that $p(V, T)$ is a linear function of temperature and only the first-order partial derivative with respect to temperature is not equal to zero. The second and all higher order partial derivatives of the function $p(V, T)$ with respect to temperature vanish. Furthermore, at the critical point, the first-order and the second-order derivatives of the function $p(V, T)$ with respect to volume vanish by definition of the critical point. Neglecting the higher order terms, we obtain the required result

$$p = p_C + c_1(T - T_C) + c_2(T - T_C)(V - V_C) + c_3(V - V_C)^3 \qquad (13.148)$$

where the coefficients c_1, c_2, and c_3 are identified to be

$$c_1 = \left(\frac{\partial p}{\partial T}\right)_C, \quad c_2 = \left(\frac{\partial^2 p}{\partial V \partial T}\right)_C, \quad c_3 = \frac{1}{6}\left(\frac{\partial^3 p}{\partial V^3}\right)_C. \qquad (13.149)$$

From the equation (13.146), we can calculate the following derivatives

$$\left(\frac{\partial p}{\partial T}\right)_V = \frac{3R}{3V - V_C} \Rightarrow \left(\frac{\partial p}{\partial T}\right)_C = \frac{3R}{2V_C} \qquad (13.150)$$

$$\left(\frac{\partial^2 p}{\partial V \partial T}\right)_C = \left[-\frac{9R}{(3V - V_C)^2}\right]_C = -\frac{9R}{4V_C^2}, \qquad (13.151)$$

and

$$\left(\frac{\partial p}{\partial V}\right)_T = -\frac{9RT}{(3V - V_C)^2} + \frac{6p_C V_C^2}{V^3}, \qquad (13.152)$$

$$\left(\frac{\partial^2 p}{\partial V^2}\right)_T = \frac{54RT}{(3V - V_C)^3} - \frac{18p_C V_C^2}{V^4}, \qquad (13.153)$$

$$\left(\frac{\partial^3 p}{\partial V^3}\right)_T = -\frac{486RT}{(3V - V_C)^4} + \frac{72p_C V_C^2}{V^5}. \qquad (13.154)$$

At the critical point, we have $RT_C = 8p_C V_C/3$, and the result (13.154) becomes

$$\left(\frac{\partial^3 p}{\partial V^3}\right)_C = -\frac{486}{16V_C^4}\frac{8p_C V_C}{3} + \frac{72p_C V_C^2}{V_C^5} = -\frac{9p_C}{V_C^3}. \qquad (13.155)$$

Thus, the values of the coefficients c_1, c_2, and c_3 are the following

$$c_1 = \frac{3R}{2V_C}, \quad c_2 = -\frac{9R}{4V_C^2}, \quad c_3 = -\frac{3p_C}{2V_C^3}. \qquad (13.156)$$

(b) The result (13.148), for $T = T_C$, becomes

$$p - p_C = c_3(V - V_C)^3 \Rightarrow \delta = 3. \qquad (13.157)$$

(c) With $p - p_C = c_1(T - T_C)$, the result (13.148) becomes

$$0 = c_2(T - T_C)(V - V_C) + c_3(V - V_C)^3, \tag{13.158}$$

or

$$(V - V_C) = \sqrt{\frac{c_2}{c_3}}(T_C - T)^{1/2}. \tag{13.159}$$

Thus, we finally obtain

$$c_4 = \sqrt{\frac{c_2}{c_3}}, \quad \beta = \frac{1}{2}. \tag{13.160}$$

Problem 7

Consider the result for the free energy of the van der Waals model

$$F = -NkT \ln\left[\frac{1}{N}\left(\frac{2\pi mkT}{h^2}\right)^{3/2}\right] - NkT \ln(V - Nb) - \frac{N^2 ka}{V}. \tag{13.161}$$

(a) Show that the free-energy density can be written as

$$f = -nkT\left\{\ln\left[\left(\frac{2\pi mkT}{h^2}\right)^{3/2}\left(\frac{1}{n} - b\right)\right] + \frac{na}{T}\right\}, \tag{13.162}$$

where $n = N/V$ is the particle number density, and $f = F/V$.

(b) What is the physical significance of the condition $d^2 f/dn^2 = 0$? Show that it gives the equation

$$n(1 - bn)^2 = \frac{T}{2a}. \tag{13.163}$$

(c) Find the maximum of the expression on the left-hand side of the equation (13.163) with respect to n, and show that it determines the critical temperature $T_C = 8a/27b$.

Solution

(a) From the equation (13.161), we obtain

$$f = \frac{F}{V} = -\frac{N}{V}kT \ln\left[\left(\frac{2\pi mkT}{h^2}\right)^{3/2}\right] - \frac{N}{V}kT \ln\frac{V - Nb}{N} - \frac{N^2}{V^2}ka, \tag{13.164}$$

or

$$f = -nkT \ln \left[\left(\frac{2\pi mkT}{h^2} \right)^{3/2} \right] - nkT \ln \left(\frac{1}{n} - b \right) nkT \frac{na}{T}. \qquad (13.165)$$

Thus, we get the required result

$$f = -nkT \left\{ \ln \left[\left(\frac{2\pi mkT}{h^2} \right)^{3/2} \left(\frac{1}{n} - b \right) \right] + \frac{na}{T} \right\}. \qquad (13.166)$$

(b) The pressure of a nonideal gas is obtained from the free energy as follows:

$$p = - \left(\frac{\partial F}{\partial V} \right)_T = \frac{NkT}{V - Nb} - \frac{N^2 ka}{V^2}. \qquad (13.167)$$

The usual two conditions for the critical point are

$$\left(\frac{\partial p}{\partial V} \right)_{T,N} = \left(\frac{\partial^2 F}{\partial V^2} \right)_{T,N} = -\frac{NkT}{(V - Nb)^2} + \frac{2N^2 ka}{V^3} = 0, \qquad (13.168)$$

$$\left(\frac{\partial^2 p}{\partial V^2} \right)_{T,N} = \left(\frac{\partial^3 F}{\partial V^3} \right)_{T,N} = \frac{2NkT}{(V - Nb)^3} - \frac{6N^2 ka}{V^4} = 0, \qquad (13.169)$$

or

$$\frac{T}{(V - Nb)^2} = \frac{2Na}{V^3} \Rightarrow n(1 - bn)^2 = \frac{T}{2a}, \qquad (13.170)$$

$$\frac{T}{(V - Nb)^3} = \frac{3Na}{V^4} \Rightarrow n(1 - bn)^3 = \frac{T}{3a}. \qquad (13.171)$$

The first equation of the critical point (13.171) is identical to the equation that will be derived in the following from the condition $d^2 f/dn^2 = 0$. Thus, the physical significance of the condition $(\partial^2 f/\partial n^2)_T = 0$ is that it is an analog of the usual condition $(\partial^2 F/\partial V^2)_{T,N} = 0$. Similarly, the condition $(\partial^3 f/\partial n^3)_T = 0$ could also be expected to be an analog of the condition $(\partial^3 F/\partial V^3)_{T,N} = 0$. Thus, the two conditions $d^2 f/dn^2 = 0$ and $d^3 f/dn^3 = 0$ are the conditions for the critical point in this density-based formulation of the theory. Let us now show that the condition $d^2 f/dn^2 = 0$ indeed leads to the result (13.171). We rewrite the result (13.166) as follows:

$$f = -nkT \ln \left(\frac{2\pi mkT}{h^2} \right)^{3/2} + nkT \ln n$$

$$- nkT \ln(1 - bn) + n^2 ka. \qquad (13.172)$$

Now, we can calculate the first derivative df/dn as follows:

$$\frac{df}{dn} = -kT \ln \left(\frac{2\pi mkT}{h^2}\right)^{3/2} + kT \ln n + kT$$

$$- kT \ln(1 - bn) + \frac{nbkT}{1 - bn} - 2nka. \tag{13.173}$$

The second derivative d^2f/dn^2 is then

$$\frac{d^2f}{dn^2} = \frac{kT}{n} + \frac{2bkT}{1 - bn} + \frac{nb^2kT}{(1 - bn)^2} - 2ka, \tag{13.174}$$

or

$$\frac{d^2f}{dn^2} = kT\frac{(1 - bn)^2 + 2bn(1 - bn) + b^2n^2}{n(1 - bn)^2} - 2ka, \tag{13.175}$$

or

$$\frac{d^2f}{dn^2} = kT\frac{1 - 2bn + n^2b^2 + 2bn - 2b^2n^2 + n^2b^2}{n(1 - bn)^2} - 2ka. \tag{13.176}$$

Thus, the condition for the critical point becomes

$$\frac{d^2f}{dn^2} = \frac{kT}{n(1 - bn)^2} - 2ka = 0, \tag{13.177}$$

or

$$n(1 - bn)^2 = \frac{T}{2a}, \tag{13.178}$$

which is identical to the result (13.170) as required.

(c) Let us now examine the condition $d^3f/dn^3 = 0$. Using (13.177), it becomes

$$\frac{d^3f}{dn^3} = -\frac{kT}{[n(1 - bn)^2]^2} \frac{d}{dn}[n(1 - bn)^2] = 0. \tag{13.179}$$

It is easily seen that the condition $d^3f/dn^3 = 0$ is equivalent to the condition that the left-hand side of the equation (13.170), i.e., $n(1 - bn)^2$, has a maximum. Thus, we obtain

$$\frac{d}{dn}[n(1 - bn)^2] = (1 - bn)^2 + 2n(1 - bn)(-b) = 0, \tag{13.180}$$

or

$$(1 - bn)(1 - bn - 2bn) = (1 - bn)(1 - 3bn) = 0. \tag{13.181}$$

Since $n \neq 1/b$, we obtain the critical density $n_C = 1/3b$. Substituting the result for critical density n_C into (13.178), we obtain

$$\frac{1}{3b}\left(1 - b\frac{1}{3b}\right)^2 = \frac{T_C}{2a} \Rightarrow T_C = \frac{8a}{27b}. \tag{13.182}$$

Thus, we have obtained the correct result for the critical temperature.

Problem 8

An unknown substance has the thermal equation of state

$$pV = AT^3, \tag{13.183}$$

and the caloric equation of state

$$U(T, V) = BT^n \ln\left(\frac{V}{V_0}\right) + f(T), \tag{13.184}$$

where A, B, n, and V_0 are constants, and $f(T)$ is an unspecified function of temperature T. Using the general thermodynamic relations, determine the integer n and the constant B, as a function of constant A.

Solution

Using the general result (13.74) obtained in Problem 1, i.e.,

$$\left(\frac{\partial U}{\partial V}\right)_T = T\left(\frac{\partial S}{\partial V}\right)_T - p = T\left(\frac{\partial p}{\partial T}\right)_V - p, \tag{13.185}$$

we obtain

$$\left(\frac{\partial U}{\partial V}\right)_T = \frac{3AT^3}{V} - p = \frac{2AT^3}{V}. \tag{13.186}$$

On the other hand, from the caloric-state equation (13.184), we have

$$\left(\frac{\partial U}{\partial V}\right)_T = \frac{BT^n}{V}. \tag{13.187}$$

Comparison of the two results (13.186) and (13.187) shows that we must have $n = 3$ and $B = 2A$.

14 Quasi-Static Thermodynamic Processes

In this chapter, we will discuss a number of practically interesting thermodynamic processes involving gases. Our discussions are based on a straightforward application of the results obtained in previous chapters. For concreteness, we will frequently tacitly assume one deals with an ideal gas with constant (temperature and density independent) specific heats. This assumption is true only in the classical limit and if the intermolecular interactions can be ignored. Yet, some of the results of this chapter are valid more generally, even for nonideal gases, and in the presence of quantum effects. If so, this fact will be emphasized in the text.

14.1 Isobaric Process

The isobaric process is a process that takes place at the constant pressure ($p = $ Constant, $dp = 0$). The differential of enthalpy per unit mass for an isobaric process, obtained from the definition (9.19), is given by

$$dh = d(u + pv) = du + pdv + vdp = du + pdv. \tag{14.1}$$

Using here the first law of thermodynamics, i.e., $du = dq - pdv$, we further obtain

$$dh = dq = c_P dT. \tag{14.2}$$

In the p-V plane, an isobaric process is represented by a straight line parallel to the V-axis, as shown in Fig. 14.1.

The quantity of heat exchanged in the process is then given by

$$q_{12} = c_P (T_2 - T_1). \tag{14.3}$$

The work performed during an isobaric process is obtained as

$$l_{12} = \int_1^2 pdv = p \int_1^2 dv = p(v_2 - v_1). \tag{14.4}$$

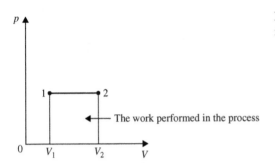

Figure 14.1 The isobaric process in the p-V plane.

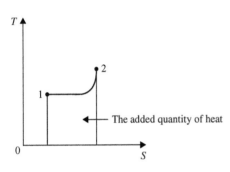

Figure 14.2 The isobaric process in the T-S plane.

The infinitesimal change in the internal energy is now given by

$$du = dq - pdv = c_p dT - pdv,$$ (14.5)

and the total change in the internal energy during an isobaric process is obtained as follows:

$$u_{12} = q_{12} - l_{12} = c_p(T_2 - T_1) - p(v_2 - v_1).$$ (14.6)

The infinitesimal quantity of heat exchanged in an isobaric process, expressed in terms of an infinitesimal change of entropy, is then given by

$$dq = Tds = c_p dT \Rightarrow ds = c_p \frac{dT}{T}.$$ (14.7)

Integrating the equation (14.7), we obtain

$$s_2 - s_1 = c_p \ln \frac{T_2}{T_1} \Rightarrow T_2 = T_1 \exp\left(\frac{s_2 - s_1}{c_p}\right).$$ (14.8)

In the T-S plane, an isobaric process is represented by an exponential function, defined by the result (14.8), as shown in Fig. 14.2.

We stress that the added amount of heat during this process (and any other process) is generally equal to the area under the curve representing the process in the T-S plane.

This is a simple consequence of the general thermodynamic relation $dQ = TdS$. In particular, the above ignored quantum effects and interparticle interactions will certainly modify the form of, say, the isobaric process curve in the S-T plane. Yet, the amount of the added heat would still be equal to the area under the process curve in the T-S plane. This seemingly simple observation shows that the very concept of entropy transcends any detailed mechanical (or quantum mechanical) description of any physical system. It is thus no surprise that basic thermodynamic ideas and laws have survived the advent of quantum mechanics.

14.2 Isochoric Process

The isochoric process is a process that takes place at the constant volume (V = Constant, $dV = 0$). In the p-V plane, an isochoric process is represented by a straight line parallel to the p-axis, as shown in Fig. 14.3.

The performed mechanical work in an isochoric process is obviously equal to zero

$$l_{12} = \int_1^2 pdv \equiv 0. \tag{14.9}$$

Thus, by the first law of thermodynamics, the change in the internal energy of the system is equal to the heat added to the system,

$$du = dq = c_V dT \Rightarrow u_{12} = q_{12} = c_V (T_2 - T_1). \tag{14.10}$$

The infinitesimal quantity of heat exchanged in an isochoric process, expressed in terms of an infinitesimal change of entropy, is then given by

$$dq = Tds = c_V dT \Rightarrow ds = c_V \frac{dT}{T}. \tag{14.11}$$

Integrating the equation (14.11), we obtain

$$s_2 - s_1 = c_V \ln \frac{T_2}{T_1} \Rightarrow T_2 = T_1 \exp\left(\frac{s_2 - s_1}{c_V}\right). \tag{14.12}$$

Figure 14.3 The isochoric process in the p-V plane.

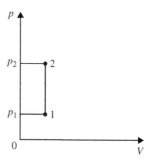

Similarly to the case of an isobaric process, an isochoric process is represented in the T-S plane by an exponential function, defined by the result (14.12), as shown in Fig. 14.4.

It should be noted that the isochoric curve in the T-S plane is steeper than the isobaric curve in the T-S plane, since $c_V < c_P$.

14.3 Isothermal Process

The isothermal process is a process that takes place at the constant temperature ($T = $ Constant, $dT = 0$). In the T-S plane, an isothermal process is represented by a straight line parallel to the S-axis, as shown in Fig. 14.5.

The exchanged quantity of heat in an isothermal process is given by

$$dq = T ds \Rightarrow q_{12} = T (s_2 - s_1). \tag{14.13}$$

For an ideal gas, the internal energy is a function of its temperature but not of its volume (in the absence of the quantum effects discussed in Part Four of this book). Thus, for such a system, the change of its internal energy during an isothermal process is zero. As an example, for an ideal monatomic gas, we have $u = 3/2 R_g T$ (in classical limit). When $T = $ Constant, we have $u = $ Constant ($du = 0$). Thus, we may write

$$du = c_V dT = 0 \Rightarrow du = dq - dl = 0. \tag{14.14}$$

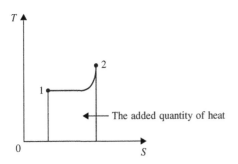

Figure 14.4 The isochoric process in the T-S plane.

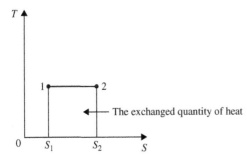

Figure 14.5 The isothermal process in the T-S plane.

Thus, for ideal classical gases, the work performed during an isothermal process is equal to the exchanged quantity of heat, and we have

$$l_{12} = q_{12} = T(s_2 - s_1).$$ (14.15)

For an isothermal process in the p-V plane, by the thermal-state equation $T = T(p, v)$, we have

$$dT = \left(\frac{\partial T}{\partial p}\right)_V dp + \left(\frac{\partial T}{\partial v}\right)_P dv = 0,$$ (14.16)

or, by Jacobian identities of Section 9.4.3, "Isothermal Expansion,"

$$dp = -\left(\frac{\partial p}{\partial T}\right)_V \left(\frac{\partial T}{\partial v}\right)_P dv = -\frac{\partial(p, v)}{\partial(T, v)}\frac{\partial(T, p)}{\partial(v, p)} dv$$

$$= \frac{\partial(p, T)}{\partial(v, T)} dv = \left(\frac{\partial p}{\partial v}\right)_T dv$$ (14.17)

as, of course, expected for $T = $ Constant, from the thermal-state equation written in the form $p(T, V)$. Obviously, for any fluid, the function $p = p(v)$ describing an isothermal process in the p-V plane is just the thermal-state equation $p = p(v, T)$ with $T = $ Constant. For example, for ideal gases (in the absence of quantum effects discussed in Part Four), an isothermal curve is given by

$$p = p(v) = \frac{R_g T}{v} = \frac{\text{Constant}}{v}, \quad T = \text{Constant},$$ (14.18)

even for polyatomic ideal gases, as discussed in Chapter 12, "Ideal Diatomic Gases." On the other hand, for nonideal gases, isothermal curves can be frequently well described by approximate state equations such as the van der Waals equation (13.45). The function $p = p(v)$ is shown in Fig. 14.6.

Figure 14.6 The isothermal process in the p-V plane.

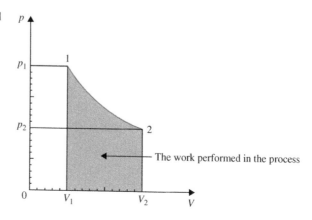

The work performed in the process

The work performed by an ideal classical gas during an isothermal process is easily calculated as follows:

$$l_{12} = \int_1^2 p\,dv = \int_1^2 \frac{R_g T}{v}\,dv = R_g T \ln \frac{v_2}{v_1}. \tag{14.19}$$

14.4 Adiabatic Process

The adiabatic process is a process that takes place at the constant entropy ($S = $ Constant, $dS = 0$). For any physical system, in the T-S plane, an adiabatic process is represented by a straight line parallel to the T-axis, as shown in Fig. 14.7. The condition that the entropy of the system is constant implies that the system is thermally insulated, i.e., that there is no exchange of heat with the surroundings. Thus we can write

$$dq = T\,ds = 0. \tag{14.20}$$

Using the first law of thermodynamics for an ideal gas, with the result (14.20), we obtain

$$du = c_V dT = -p\,dv = -\frac{R_g T}{v}\,dv, \tag{14.21}$$

or

$$\frac{dT}{T} + \frac{R_g}{c_V}\frac{dv}{v} = \frac{dT}{T} + \frac{c_P - c_V}{c_V}\frac{dv}{v} = 0, \tag{14.22}$$

where we have used the result $c_P - c_V = R_g$ valid for the classical ideal gas. Introducing here the ratio of the two specific heats of an ideal gas

$$\kappa = c_P/c_V, \tag{14.23}$$

we can rewrite the equation (14.22) as follows:

$$\frac{dT}{T} + (\kappa - 1)\frac{dv}{v} = 0 \Rightarrow d\left(\ln Tv^{\kappa-1}\right) = 0, \tag{14.24}$$

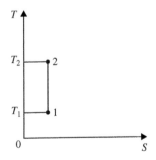

Figure 14.7 The adiabatic process in the T-S plane.

or

$$Tv^{\kappa-1} = \text{Constant.} \tag{14.25}$$

On the other hand, since $T = pv/R_g$, we also have

$$dT = \frac{v}{R_g}dp + \frac{p}{R_g}dv \Rightarrow \frac{dT}{T} = \frac{v}{R_g T}dp + \frac{p}{R_g T}dv, \tag{14.26}$$

or

$$\frac{dT}{T} = \frac{dp}{p} + \frac{dv}{v}. \tag{14.27}$$

Combining the equations (14.24) and (14.27), we further obtain

$$\frac{dp}{p} + \frac{dv}{v} + (\kappa - 1)\frac{dv}{v} = \frac{dp}{p} + \kappa\frac{dv}{v} = 0. \tag{14.28}$$

Thus, we finally obtain the equation of the adiabatic curve in the p-V plane as follows:

$$d\left(\ln pv^{\kappa}\right) = 0 \Rightarrow pv^{\kappa} = \text{Constant}, \tag{14.29}$$

or

$$p = p(v) = \frac{\text{Constant}}{v^{\kappa}}, \quad S = \text{Constant.} \tag{14.30}$$

This curve in the p-V plane is similar to the one obtained in the case of an isothermal process, but the adiabatic curve is steeper since, generally, $\kappa = c_p/c_v > 1$ (e.g., for monatomic gases $\kappa = 5/3$). There is a simple physical reason for this: during an adiabatic expansion (volume increase), the positive work done $L > 0$ must be accompanied by a decrease of internal energy

$$\Delta U = \Delta Q - L = -L < 0. \tag{14.31}$$

Since U is an increasing function of T only (for classical ideal gases), the temperature T must decrease during an adiabatic volume increase. This is possible only if the adiabatic curve is steeper than any member of the family of isothermal curves in the p-V plane. The adiabatic curve in the p-V plane is shown in Fig. 14.8.

The work performed in an adiabatic process for an ideal gas is calculated using the result

$$pv^{\kappa} = p_1 v_1^{\kappa} = \text{Constant} \Rightarrow p = \frac{p_1 v_1^{\kappa}}{v^{\kappa}}. \tag{14.32}$$

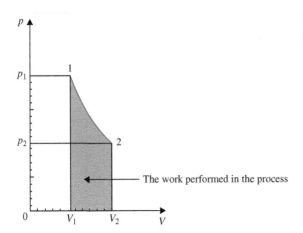

Figure 14.8 The adiabatic process in the p-V plane.

The work performed in the process

Thus, we obtain

$$l_{12} = \int_1^2 p\,dv = p_1 v_1^\kappa \int_1^2 \frac{dv}{v^\kappa} = \frac{p_1 v_1^\kappa}{1-\kappa}\left(v_2^{1-\kappa} - v_1^{1-\kappa}\right), \tag{14.33}$$

or finally

$$l_{12} = \frac{p_1 v_1^\kappa}{\kappa - 1}\left(v_1^{1-\kappa} - v_2^{1-\kappa}\right). \tag{14.34}$$

14.5 Polytropic Process

Polytropic processes described here are more general in character than the processes discussed thus far, in which one of the thermodynamic variables (p, v, T, s) is kept constant. A polytropic process is represented in the p-V plane by the following equation

$$pv^n = \text{Constant}, \tag{14.35}$$

where n is an arbitrary constant. All the other quasi-static processes described thus far are special cases of the polytropic process, i.e.,

1. For $n = 0$, the polytropic process is the isobaric process.
2. For $n \to \infty$, the polytopic process is the isochoric process.
3. For $n = 1$, the polytropic process is the isothermal process.
4. For $n = \kappa$, the polytropic process is the adiabatic process.

The polytropic process concept is motivated by the following interesting mathematical problem: in the p-V diagram, find the shape of the process curve along which the added heat is given by

$$dq = c_n dT,$$ (14.36)

where c_n is an arbitrarily given constant, the so-called *polytropic specific heat*. Let us solve this problem for ideal classical gases. Using now the caloric-state equation $dq = du + pdv$, with $du = c_V dT$, we obtain for an ideal gas

$$dq = c_n dT = c_V dT + pdv = c_V dT + \frac{R_g T}{v} dv.$$ (14.37)

Using $R_g = c_P - c_V$ and rearranging, we obtain

$$(c_n - c_V) \frac{dT}{T} = (c_P - c_V) \frac{dv}{v},$$ (14.38)

or

$$\frac{dT}{T} = \frac{(c_P - c_V)}{(c_n - c_V)} \frac{dv}{v}.$$ (14.39)

Using here the general result (14.27), we have

$$\frac{dT}{T} = \frac{dp}{p} + \frac{dv}{v} = \frac{(c_P - c_V)}{(c_n - c_V)} \frac{dv}{v},$$ (14.40)

or

$$\frac{dp}{p} + \left(1 - \frac{c_P - c_V}{c_n - c_V}\right) \frac{dv}{v} = 0,$$ (14.41)

or

$$\frac{dp}{p} + \frac{c_n - c_P}{c_n - c_V} \frac{dv}{v} = 0 \Rightarrow \frac{dp}{p} + n \frac{dv}{v} = 0,$$ (14.42)

where we introduced the following notation

$$n = \frac{c_n - c_P}{c_n - c_V}.$$ (14.43)

The equation (14.42) is readily integrated. Thus, for ideal classical gases, we obtain the equation of the polytropic curve in the p-V plane of the form

$$d\left(\ln pv^n\right) = 0 \Rightarrow pv^n = \text{Constant}.$$ (14.44)

The work performed in a polytropic process for an ideal gas is calculated using the result

$$pv^n = p_1 v_1^n = \text{Constant} \Rightarrow p = \frac{p_1 v_1^n}{v^n}.$$ (14.45)

Thus, we obtain

$$l_{12} = \int_1^2 p dv = p_1 v_1^n \int_1^2 \frac{dv}{v^n} = \frac{p_1 v_1^n}{n-1} \left(v_1^{1-n} - v_2^{1-n} \right). \tag{14.46}$$

14.6 Cyclic Processes: Carnot Cycle

The processes in which the final state of the system coincides with its initial state are called the *cyclic processes*. The simplest example of a cyclic process in the T-S plane is the so-called *Carnot cycle*, and it is shown in Fig. 14.9.

In the Carnot cycle, the quantity of heat Q_2 is added to the system from the surroundings during the isothermal expansion $a - b$ at the constant higher temperature T_2. During the adiabatic expansion $b - c$, the system performs work, and its temperature is reduced to the lower temperature T_1 without heat exchange with the surroundings. During the isothermal compression $c - d$ at the constant lower temperature T_1, the quantity of heat Q_1 is removed from the system and transferred to the surroundings. Finally, during the adiabatic compression $d - a$, the work is done on the system and its temperature is increased back to the higher temperature T_2 without heat exchange with the surroundings.

In a cyclic process, there is no change in the internal energy of the system $\Delta U = 0$. The reason for this is that the final state of the system, after a complete cycle, is the same as the initial state of the cycle. Using now the first law of thermodynamics, we have

$$\Delta U_C = \Delta Q_C - \Delta L_C = 0 \Rightarrow \Delta L_C = \Delta Q_C. \tag{14.47}$$

The meaning of the equation (14.47) is that the positive work extracted from the system is equal to the *net* heat added to the system during the Carnot cycle. Since there is no heat exchange with the surroundings during the two adiabatic processes, the *net* heat added to the system during the Carnot cycle is equal to the difference between the heat added to the system Q_2 (during the isothermal process at $T = T_2$) and the heat removed from the system Q_1 (during the isothermal process at $T = T_1$), i.e.,

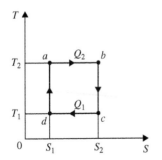

Figure 14.9 The Carnot cycle in the T-S plane.

$\Delta Q_C = Q_2 - Q_1$. On the other hand, for these two isothermal processes depicted in the T-S diagram, we can write down following the two relations as follows:

$$Q_2 = T_2 \Delta S, \quad Q_1 = T_1 \Delta S, \tag{14.48}$$

where $\Delta S = S_2 - S_1$. Thus, we obtain

$$\Delta L_C = \Delta Q_C = Q_2 - Q_1 = (T_2 - T_1) \Delta S. \tag{14.49}$$

The thermal efficiency of any cyclic process is defined as the ratio between the extracted work from the system ΔL_C and the heat added to the system Q_2, i.e.,

$$\eta = \frac{\Delta L_C}{Q_2} = \frac{Q_2 - Q_1}{Q_2}, \tag{14.50}$$

i.e.,

$$\eta = 1 - \frac{Q_1}{Q_2}. \tag{14.51}$$

In the special case of the Carnot cycle, we can use the results (14.48), to obtain

$$\eta = 1 - \frac{T_1}{T_2}. \tag{14.52}$$

Thus, the thermal efficiency of the Carnot cycle is fully determined by the constant temperatures T_1 and T_2 of the two isothermal processes.

The major feature of any cyclic process is that there is no change of internal energy U of the system after completing the full cycle. It is because the system returns to its initial state. Thus, $\Delta U = 0$, and the first law of thermodynamics then predicts that the work done by the system is equal to the net heat added to the system,

$$\Delta L = \Delta Q_{net}. \tag{14.53}$$

For any cyclic process, the net added heat energy can be written as

$$\Delta Q_{net} = \Delta Q_2 - \Delta Q_1, \tag{14.54}$$

hence

$$\Delta L = \Delta Q_2 - \Delta Q_1. \tag{14.55}$$

A general cyclic process in the T-S plane is shown in Fig. 14.10.

Here, ΔQ_2 is the heat energy absorbed by the system, given by the integral

$$\Delta Q_2 = \int_{\text{upper}} T dS, \tag{14.56}$$

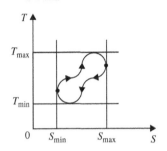

Figure 14.10 General cyclic process in the T-S plane.

done along the upper part of the loop in Fig. 14.10 (from S_{min} to S_{max}). On the other side, ΔQ_1 is the heat energy released from the system, given by the integral

$$\Delta Q_1 = \int_{lower} T dS, \tag{14.57}$$

done along the lower part of the loop in Fig. 14.10 (from S_{min} to S_{max}). It is, thus, obvious that the net work done by the system during any cyclic process,

$$\Delta L = \Delta Q_{net} = \Delta Q_2 - \Delta Q_1 = \int_{upper} T dS - \int_{lower} T dS, \tag{14.58}$$

is in general equal to the area enclosed by the cycle loop in the T-S diagram.

Any cyclic process is characterized by two important temperatures labeled in Fig. 14.10 by T_{max} and T_{min}, being a maximum and a minimum temperature occuring along the cycle, respectively. As $T > T_{min}$, we have

$$\Delta Q_1 = \int_{lower} T dS > \int_{S_{min}}^{S_{max}} T_{min} dS = (S_{max} - S_{min}) T_{min} \tag{14.59}$$

and, as, $T < T_{max}$, we have

$$\Delta Q_2 = \int_{upper} T dS < \int_{S_{min}}^{S_{max}} T_{max} dS = (S_{max} - S_{min}) T_{max}. \tag{14.60}$$

By the above two inequalities, we have

$$\frac{\Delta Q_2}{T_{max}} < S_{max} - S_{min} < \frac{\Delta Q_1}{T_{min}} \tag{14.61}$$

Thus, we obtain the inequality

$$\frac{\Delta Q_2}{T_{max}} < \frac{\Delta Q_1}{T_{min}} \tag{14.62}$$

implying the inequalities,

$$\frac{T_{min}}{T_{max}} < \frac{\Delta Q_1}{\Delta Q_2} \Rightarrow -\frac{T_{min}}{T_{max}} > -\frac{\Delta Q_1}{\Delta Q_2}, \tag{14.63}$$

or

$$-\frac{\Delta Q_1}{\Delta Q_2} < -\frac{T_{min}}{T_{max}}, \tag{14.64}$$

yielding the following fundamental result for the efficiency of any cycle as follows:

$$\eta = 1 - \frac{\Delta Q_1}{\Delta Q_2} < 1 - \frac{T_{min}}{T_{max}}. \tag{14.65}$$

By the above-mentioned result, in combination with the equation (14.52), we have derived a celebrated theorem of classical thermodynamics, stating that any cyclic process is less efficient than the Carnot's cyclic process that would operate between the same minimum and maximum temperatures, T_{min} and T_{max}. Thus, for given T_{min} and T_{max}, the Carnot's cycle is the most efficient cycle.

Finally, we would like to note the fact that the area enclosed by a cyclic process loop in the T-S plane is equal to the area enclosed by the corresponding loop in the p-V plane. Physically, this invariant area is just the work done by the system during the cyclic process. Mathematically, the invariance of this area under the mapping $(T, S) \leftrightarrow (p, V)$ is a direct reflection of the fact that the Jacobian of this mapping is 1, as stated in the equation (9.80). Thus, the equation (9.80) is directly related to the first law of thermodynamics applied to cyclic processes. Indeed, by integrating the first law of the thermodynamics

$$TdS = dU + pdV, \tag{14.66}$$

along any cycle loop, we have, in terms of loop integrals,

$$\oint TdS = \oint dU + \oint pdV, \tag{14.67}$$

or, as $\oint dU = \Delta U = 0$,

$$\oint TdS = \oint pdV = \Delta L. \tag{14.68}$$

The above two loop integrals are nothing else but the areas enclosed by the loops in the (T, S) and (p, V) planes,

$$\int\int_{loop} dTdS = \int\int_{loop} dpdV = \Delta L. \tag{14.69}$$

The above areas, equality holds for *any* choice of loop. This can be true only if the Jacobian of the mapping $(T, S) \leftrightarrow (p, V)$ is 1. Indeed, by the above equation,

$$\int\int_{\text{any loop}} dTdS = \int\int_{\text{any loop}} \frac{\partial(T, S)}{\partial(p, V)} dpdV = \int\int_{\text{any loop}} dpdV, \tag{14.70}$$

or

$$\int\int_{\text{any loop}} \left(\frac{\partial(T, S)}{\partial(p, V)} - 1\right) dpdV = 0. \tag{14.71}$$

Since the equation (14.71) holds for any choice of the loop, we must have

$$\frac{\partial(T, S)}{\partial(p, V)} = 1. \tag{14.72}$$

Hence, the equation (9.80) follows from the first law of thermodynamics applied to the cyclic processes.

14.7 Problems with Solutions

Problem 1

A gas undergoes a cyclic process consisting of two isochoric and two isobaric processes. The state of the gas in the p-V diagram changes from (p_1, V_1) over (p_1, V_2), (p_2, V_2), and (p_2, V_1) back to (p_1, V_1). Calculate the work done on the gas by external agents and the net head added to the gas during this cyclic process.

Solution

This cyclic process can be represented by the p-V diagram shown in Fig. 14.11.

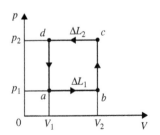

Figure 14.11 The cyclic process in the p-V plane.

The initial and the final states of the gas in a cyclic process are the same. Thus, the change in the internal energy is equal to zero ($\Delta U = 0$). Using here the first law of thermodynamics, we have

$$\Delta U = T\Delta S - p\Delta V = \Delta Q - \Delta L_S = 0. \tag{14.73}$$

In the equation (14.73), the work ΔL_S is by convention work done by the gas on the surroundings. The work done on the gas by the surroundings, i.e., by external agents, is the same with the opposite sign ($\Delta L_{EXT} = \Delta L_G = -\Delta L_S$), and we can write

$$\Delta U = \Delta Q + \Delta L_{EXT} = 0, \quad \Delta L_{EXT} = -p\Delta V. \tag{14.74}$$

Thus, the heat that the gas gains after the above cyclic process is equal to minus the work done on the gas, i.e.,

$$\Delta Q = -\Delta L_{EXT}. \tag{14.75}$$

In order to find the work done on the gas ΔL_{EXT}, we first note that the work done in an isochoric process ($dV = 0$) is equal to zero. The work done by external agents on the gas during the entire cyclic process is then equal to the sum of the work done on the gas during the two isobaric processes

$$\Delta L_{EXT} = -p_1\,(V_2 - V_1) - p_2\,(V_1 - V_2)$$
$$= (p_1 - p_2)(V_1 - V_2) = (p_2 - p_1)(V_1 - V_2) > 0. \tag{14.76}$$

So, a positive amount of work is done by the external agent on the gas. Finally, the net heat added to the gas during the above cyclic process is equal to

$$\Delta Q = -\Delta L_{EXT} = -(p_1 - p_2)(V_1 - V_2)$$
$$= -(p_2 - p_1)(V_2 - V_1) < 0. \tag{14.77}$$

From the equation (14.77), we see that this heat is negative. This fact is associated with the fact that the cycle loop is here oriented counter-clockwise in the p-V diagram. Such systems operate as time-reversed engines, i.e., as heat pumps. They can thus be used in home heating systems.

Problem 2

One mole of a monatomic ideal gas is subject to the cyclic process shown in Fig. 14.12.

Figure 14.12 The cyclic process in the p-V plane.

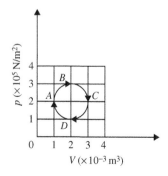

(a) Calculate the net work done by the gas during one cycle.
(b) Calculate the internal energy difference between states C and A.
(c) Calculate the heat absorbed by the gas moving between states A and C via the path ABC.

Solution

(a) The equation of the circle describing the cyclic process in Fig. 14.12 is given by

$$(\sigma - 2)^2 + (v - 2)^2 = 1, \quad \sigma = \frac{p}{p_0}, \quad v = \frac{V}{V_0}, \tag{14.78}$$

where $p_0 = 10^5 \, \text{N/m}^2$ and $V_0 = 10^{-3} \, \text{m}^3$. The work done by the gas in one cycle is defined by

$$\Delta L = \oint p \, dV = p_0 V_0 \oint \sigma \, dv. \tag{14.79}$$

Dividing this integral into two integrals, we can write

$$\Delta L = p_0 V_0 \left(\int_{ABC} \sigma \, dv + \int_{CDA} \sigma \, dv \right), \tag{14.80}$$

or

$$\Delta L = p_0 V_0 \left[\int_1^3 \left(2 + \sqrt{1 - (v - 2)^2} \right) dv + \int_3^1 \left(2 - \sqrt{1 - (v - 2)^2} \right) dv \right], \tag{14.81}$$

or

$$\Delta L = p_0 V_0 \left[2 \int_1^3 dv + \int_1^3 \sqrt{1 - (v - 2)^2} dv - 2 \int_1^3 dv + \int_1^3 \sqrt{1 - (v - 2)^2} dv \right]. \tag{14.82}$$

Thus, we obtain

$$\Delta L = 2 p_0 V_0 \int_1^3 \sqrt{1 - (v - 2)^2} dv. \tag{14.83}$$

Introducing here $x = v - 2$, we get

$$\Delta L = 2p_0 V_0 \int_{-1}^{1} \sqrt{1 - x^2}\, dx. \tag{14.84}$$

Introducing further $x = \sin\theta$, we have

$$\Delta L = 2p_0 V_0 \int_{-\pi/2}^{\pi/2} \cos^2\theta\, d\theta, \tag{14.85}$$

or

$$\Delta L = 2p_0 V_0 \int_{-\pi/2}^{\pi/2} \frac{1}{2}(1 + \cos 2\theta)\, d\theta = \pi p_0 V_0. \tag{14.86}$$

Thus, we finally obtain

$$\Delta L = \pi p_0 V_0 = 100\pi\, \text{J} = 314.16\, \text{J}. \tag{14.87}$$

There is a much simpler way to calculate the net work done by the gas during one cycle, if we remember that it is equal to the area of the cycle in the $p - V$ plane. Then, we can write

$$\frac{\Delta L}{p_0 V_0} = \pi r^2 = \pi, \tag{14.88}$$

or

$$\Delta L = \pi p_0 V_0 = 314.16\, \text{J}. \tag{14.89}$$

(b) In the state A, we have the pressure $p_A = 2 \times 10^5\, \text{N/m}^2$ and the volume $V_A = 10^{-3}\, \text{m}^3$. Using the equation of state, we obtain

$$T_A = \frac{p_A V_A}{R} = \frac{200\, \text{J}}{8.314\, \text{J/K}} = 24\, \text{K}. \tag{14.90}$$

In the state B, we have the pressure $p_C = 2 \times 10^5\, \text{N/m}^2$ and the volume $V_C = 3 \times 10^{-3}\, \text{m}^3$. Using the equation of state, we obtain

$$T_C = \frac{p_C V_C}{R} = \frac{600\, \text{J}}{8.314\, \text{J/K}} = 72\, \text{K}. \tag{14.91}$$

The internal energy difference between states C and A is then given by

$$\Delta U_{AC} = \int_A^C dU = c_V \int_{T_A}^{T_C} dT = c_V (T_C - T_A) = 598\, \text{J}. \tag{14.92}$$

(c) Using the first law of thermodynamics, we have

$$\Delta Q_{ABC} = \Delta U_{ABC} + \Delta L_{ABC} = \Delta U_{AC} + \Delta L_{ABC}. \tag{14.93}$$

The work done along the path ABC is given by

$$\Delta L_{ABC} = p_0 V_0 \int_1^3 \left(2 + \sqrt{1 - (v-2)^2}\right) dv, \tag{14.94}$$

or

$$\Delta L_{ABC} = p_0 V_0 \left[2 \int_1^3 dv + \int_1^3 \sqrt{1 - (v-2)^2} dv \right], \tag{14.95}$$

or finally

$$\Delta L_{ABC} = p_0 V_0 \left(4 + \frac{\pi}{2}\right) = 557 \, \text{J}. \tag{14.96}$$

Substituting (14.92) and (14.96) into (14.93), we obtain

$$\Delta Q_{ABC} = 598 \, \text{J} + 557 \, \text{J} = 1155 \, \text{J}. \tag{14.97}$$

Problem 3

One mole of an ideal diatomic gas undergoes a quasi-stationary change of state from A to C as shown in Fig. 14.13.

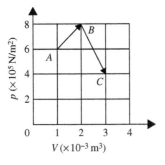

Figure 14.13 The quasi-static process in the p-V plane.

(a) Calculate the heat capacity at constant volume of this gas.
(b) Calculate the work done by the gas in the process ABC.
(c) Calculate the heat absorbed by the gas in the process ABC.
(d) Calculate the entropy change of the gas in the process ABC.

Solution

(a) Since we have one mole of the gas ($m = M$), we have $R_g = R$. Thus, for a diatomic gas with five $(3 + 2)$ degrees of freedom, we have

$$c_V = \left(\frac{\partial Q}{\partial T}\right)_V = \left(\frac{\partial U}{\partial T}\right)_V = \frac{5}{2}R = 20.785 \frac{J}{K}. \tag{14.98}$$

(b) The work done by the gas in the process ABC is given by

$$\Delta L_{ABC} = \int_{ABC} p\,dV = \int_{AB} p\,dV + \int_{BC} p\,dV. \tag{14.99}$$

Let us introduce the new variables

$$\sigma = \frac{p}{p_0}, \quad \nu = \frac{V}{V_0}, \tag{14.100}$$

where $p_0 = 10^5\,\text{N/m}^2$ and $V_0 = 10^{-3}\,\text{m}^3$. Thus, we can write

$$\Delta L_{ABC} = p_0 V_0 \left(\int_{AB} \sigma\,d\nu + \int_{BC} \sigma\,d\nu\right). \tag{14.101}$$

Along the path AB, we have

$$\sigma = \sigma_A + \frac{\sigma_B - \sigma_A}{\nu_B - \nu_A}(\nu - \nu_A) = 6 + \frac{8 - 6}{2 - 1}(\nu - 1), \tag{14.102}$$

or

$$\sigma = 2\nu + 4. \tag{14.103}$$

Similarly, along the path BC, we have

$$\sigma = \sigma_B + \frac{\sigma_C - \sigma_B}{\nu_C - \nu_B}(\nu - \nu_B) = 8 + \frac{4 - 8}{3 - 2}(\nu - 2), \tag{14.104}$$

or

$$\sigma = -4\nu + 16. \tag{14.105}$$

Substituting (14.103) and (14.105) into (14.101), we obtain

$$\Delta L_{ABC} = p_0 V_0 \left[\int_1^2 (2\nu + 4)\,d\nu + \int_2^3 (-4\nu + 16)\,d\nu\right], \tag{14.106}$$

or

$$\Delta L_{ABC} = 13 p_0 V_0 = 1300\,\text{J}. \tag{14.107}$$

(c) In the state A, we have the pressure $p_A = 6 \times 10^5 \, \text{N/m}^2$ and the volume $V_A = 10^{-3} \, \text{m}^3$. Using the equation of state, we obtain

$$T_A = \frac{p_A V_A}{R} = \frac{600 \, \text{J}}{8.314 \, \text{J/K}} = 72 \, \text{K}. \tag{14.108}$$

In the state B, we have the pressure $p_C = 4 \times 10^5 \, \text{N/m}^2$ and the volume $V_C = 3 \times 10^{-3} \, \text{m}^3$. Using the equation of state, we obtain

$$T_C = \frac{p_C V_C}{R} = \frac{1200 \, \text{J}}{8.314 \, \text{J/K}} = 144 \, \text{K}. \tag{14.109}$$

The internal energy difference between states C and A is then given by

$$\Delta U_{AC} = \int_A^C dU = c_V \int_{T_A}^{T_C} dT = c_V (T_C - T_A) = 1497 \, \text{J}. \tag{14.110}$$

Using the first law of thermodynamics, the heat absorbed by the gas in the process ABC is given by

$$\Delta Q_{ABC} = \Delta U_{AC} + \Delta L_{ABC} = 2797 \, \text{J}. \tag{14.111}$$

(d) The entropy change for 1 mole of an ideal gas in the reversible change of state is given by

$$\Delta S = c_V \ln \frac{T_F}{T_I} + R \ln \frac{V_F}{V_I}, \tag{14.112}$$

or

$$\Delta S_{AC} = c_V \ln \frac{T_C}{T_A} + R \ln \frac{V_C}{V_A} = c_V \ln 2 + R \ln 3. \tag{14.113}$$

Thus, we finally obtain

$$\Delta S_{AC} = 23.54 \, \frac{\text{J}}{\text{K}}. \tag{14.114}$$

Problem 4

An electric heat pump absorbs heat from the surrounding earth at the lower temperature $T_O = 273 \, \text{K}$ and then delivers heat to the interior of a building at the higher temperature $T_I = 298 \, \text{K}$, as shown in Fig. 14.14.

Calculate the maximum energy supplied to the building per 1 joule of electrical energy needed to operate the heat pump.

Figure 14.14 The cyclic process in the T-S plane.

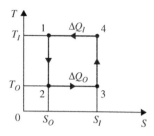

Solution

The maximum energy supplied to the building is obtained if we assume that our process is reversible, i.e.,

$$\Delta S = \frac{\Delta Q_O}{T_O} + \frac{\Delta Q_I}{T_I} = 0. \tag{14.115}$$

By convention, the heat supply to the interior of the building is given by $Q_S = -\Delta Q_I$, and we can write

$$\Delta S = \frac{Q_{23}}{T_O} - \frac{Q_S}{T_I} = 0. \tag{14.116}$$

The work performed by the electric motor of the heat pump is equal to the electric energy spent in the process, i.e.,

$$-\delta L = -\delta Q = Q_{41} + Q_{23} = -Q_S + Q_{23}. \tag{14.117}$$

Thus, we obtain

$$\delta L = Q_S - Q_{23} \Rightarrow Q_{23} = Q_S - \delta L, \tag{14.118}$$

and

$$\Delta S = \frac{Q_S - \delta L}{T_O} - \frac{Q_S}{T_I} = Q_S \left(\frac{1}{T_O} - \frac{1}{T_I} \right) - \frac{\delta L}{T_O} = 0. \tag{14.119}$$

From (14.119), we finally obtain

$$Q_S = \frac{T_I}{T_I - T_O} \delta L = \frac{298}{298 - 273} \delta L \approx 12\,\delta L. \tag{14.120}$$

Theoretically, for every joule of electric energy, we can expect 12 joules of heat supplied to the building.

Problem 5

Two identical bodies, with heat capacity at constant pressure equal to c_P, are used as heat reservoirs for a heat engine. Initially, the temperatures of the two bodies are T_1 and T_2, respectively. As a result of the operation of the heat engine, the bodies remain at a constant pressure and reach a common final temperature T_F.

(a) Calculate the work done by the engine in terms of the specified quantities c_P, T_1, T_2, and T_F.
(b) Using the second law of thermodynamics, derive the inequality relating the final temperature T_F to T_1 and T_2.
(c) For given T_1 and T_2, calculate the maximum work that can be obtained from the engine.

Solution

(a) At constant pressure, the amounts of heat energy released from the two reservoirs are

$$\Delta Q_1 = c_P (T_1 - T_F),$$
$$\Delta Q_2 = c_P (T_2 - T_F). \tag{14.121}$$

Because of the cyclic nature of heat engines' operation, the first law of thermodynamics is that the total heat energy released to an engine (after an integer number of cycles) is equal to the work done by the engine. Thus,

$$\Delta L = \Delta Q_1 + \Delta Q_2 = c_P (T_1 + T_2 - 2T_F). \tag{14.122}$$

(b) Because of the cyclic nature of heat engines' operation, the change of the entropy of the system is equal to the change of the entropy of the two reservoirs. It can be calculated as follows:

$$\Delta S = \int_{T_1}^{T_F} \frac{dQ_1}{T} + \int_{T_2}^{T_F} \frac{dQ_2}{T} = c_P \ln \frac{T_F}{T_1} + c_P \ln \frac{T_F}{T_2}. \tag{14.123}$$

Using the second law of thermodynamics, we can write

$$\Delta S = c_P \ln \frac{T_F^2}{T_1 T_2} \geq 0, \tag{14.124}$$

or

$$\frac{T_F^2}{T_1 T_2} \geq 1 \Rightarrow T_F \geq \sqrt{T_1 T_2}. \tag{14.125}$$

(c) The maximum work from the engine is obtained in case of a reversible process, i.e., when the entropy change is equal to zero $\Delta S = 0$. In such a case, we have an equality in the equation (14.125). Substituting (14.125) into (14.122), we obtain

$$\Delta L_{\max} = c_P \left(T_1 + T_2 - 2\sqrt{T_1 T_2} \right) = c_P \left(\sqrt{T_1} - \sqrt{T_2} \right)^2. \tag{14.126}$$

Problem 6

A gasoline engine can be described by the idealized cyclic process shown in Fig. 14.15.

Figure 14.15 The cyclic process in the p-V plane.

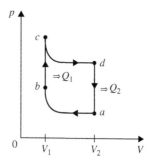

The process $a \to b$ describes the adiabatic compression of the air–gasoline mixture. The process $b \to c$ describes the rise in pressure at constant volume, due to the explosion of the air–gasoline mixture. The process $c \to d$ describes the adiabatic expansion of the mixture during which the engine performs the useful work.

Finally, the process $d \to a$ describes the cooling down of the remaining gas at constant volume. If this cycle is performed quasi-statically with a fixed amount of ideal gas, calculate the efficiency η for this process in terms of the quantities V_1, V_2, and $\kappa = c_P/c_V$.

Solution

The efficiency of a process is defined as the ratio between the work performed by the process $\Delta L = Q_1 - Q_2$ and the heat intake Q_1, i.e.,

$$\eta = \frac{Q_1 - Q_2}{Q_1} = 1 - \frac{Q_2}{Q_1}. \tag{14.127}$$

The exchanged quantity of heat during the isochoric process $d - a$ is given by

$$Q_2 = -U_{da} = -c_V \left(T_a - T_d \right) = c_V \left(T_d - T_a \right). \tag{14.128}$$

The exchanged quantity of heat during the isochoric process $b - c$ is given by

$$Q_1 = U_{bc} = c_V \left(T_c - T_b \right). \tag{14.129}$$

Substituting (14.128) and (14.129) into (14.127), we obtain

$$\eta = 1 - \frac{T_d - T_a}{T_c - T_b}.$$ (14.130)

On the other hand, for a reversible adiabatic process, we have

$$TV^{\kappa-1} = \text{Constant}.$$ (14.131)

Thus, we have

$$T_d = T_c \left(\frac{V_c}{V_d}\right)^{\kappa-1} = T_c \left(\frac{V_1}{V_2}\right)^{\kappa-1},$$ (14.132)

and

$$T_a = T_b \left(\frac{V_b}{V_a}\right)^{\kappa-1} = T_b \left(\frac{V_1}{V_2}\right)^{\kappa-1}.$$ (14.133)

Thus, we have

$$T_d - T_a = (T_c - T_b) \left(\frac{V_1}{V_2}\right)^{\kappa-1}.$$ (14.134)

Substituting (14.134) into (14.130), we finally obtain the efficiency of the cyclic process described by Fig. 14.15, in terms of the quantities V_1, V_2, and $\kappa = c_P/c_V$ as follows:

$$\eta = 1 - \left(\frac{V_1}{V_2}\right)^{\kappa-1}.$$ (14.135)

Problem 7

Determine the temperature $T(z)$ of the Earth's atmosphere as a function of altitude z using the following assumptions:

1. The atmosphere can be described as an ideal gas with molar mass of $M = 0.028\,\text{kg/mol}$ and the ratio between the specific heats $\kappa = C_P/C_V = 1.4$.
2. There is a mechanical equilibrium, i.e., we have $dp(z) = -n(z)mgdz$, where $n(z)$ is the number density of the constituent particles and $g = 9.81\,\text{m/s}^2$.
3. There is no heat transport in z-direction.

Solution

Using the thermal equation of state for an ideal gas, i.e.,

$$pV = NkT \Rightarrow p = \frac{N}{V}kT \Rightarrow p(z) = n(z)kT(z), \tag{14.136}$$

and the equilibrium condition $dp(z) = -n(z)mgdz$, we can write

$$\frac{dp(z)}{p(z)} = d[\ln p(z)] = -\frac{mgdz}{kT(z)}. \tag{14.137}$$

The third condition of the problem implies that the entropy per particle is independent on the altitude. In other words, the temperature change with the altitude can be described as an adiabatic change such that the result (14.30) is valid, and we have

$$pV^{\kappa} = C \Rightarrow p^{1-\kappa}(pV)^{\kappa} = C, \tag{14.138}$$

where C denotes a constant independent on any state variables. Using again the thermal equation of state for an ideal gas, we get

$$p^{1-\kappa}(NkT)^{\kappa} = C \Rightarrow p^{1-\kappa}T^{\kappa} = C, \tag{14.139}$$

or

$$pT^{\kappa/(1-\kappa)} = C \Rightarrow p = C \times T^{\kappa/(\kappa-1)}. \tag{14.140}$$

From the results (14.137) and (14.140), we obtain

$$d[\ln p(z)] = \frac{\kappa}{\kappa - 1}d[\ln T(z)] = -\frac{mgdz}{kT(z)}, \tag{14.141}$$

or

$$\frac{dT(z)}{T(z)} = -\frac{\kappa - 1}{\kappa}\frac{mgdz}{kT(z)}, \tag{14.142}$$

or

$$\frac{dT(z)}{dz} = -\frac{\kappa - 1}{\kappa}\frac{mg}{k}. \tag{14.143}$$

Assuming that at the surface of the Earth ($z = 0$), we have $T = T_0$, we obtain the temperature $T(z)$ of the Earth's atmosphere as a linear function of altitude z

$$T(z) = T_0 - \frac{\kappa - 1}{\kappa}\frac{mgz}{k}, \tag{14.144}$$

or

$$T(z) = T_0 - \frac{\kappa - 1}{\kappa} \frac{Mgz}{R}. \tag{14.145}$$

Using the actual values of the given numerical parameters $\kappa = 1.4$, $M = 0.028\,\text{kg/mol}$, $g = 9.81\,\text{m/s}^2$, and $R = 8.314\,\text{J/K mol}$, we obtain

$$T(z) = T_0 - 0.0094395 \left[\frac{K}{m} \right] \times z. \tag{14.146}$$

From the numerical result (14.146), we see that, in this model, the temperature of the Earth's atmosphere falls approximately 1 K every 100 m.

Problem 8

A cartridge of a volume $V_1 = 2 \times 10^{-7}\,\text{m}^3$ contains 6×10^{-3} moles of gun powder. When the rifle is fired, the temperature of the combustion gases is $T_1 = 4000\,\text{K}$, and the volume is assumed to be that of the cartridge. Estimate the speed of a bullet of mass equal to $m = 5 \times 10^{-3}\,\text{kg}$ leaving the rifle with the pipe volume of $V_2 = 2 \times 10^{-5}\,\text{m}^3$. The combustion gases are described as an ideal gas with $c_P/c_V = 1.5$ undergoing the reversible adiabatic expansion.

Solution

In the initial state, the combustion gases occupy the volume $V_1 = 2 \times 10^{-7}\,\text{m}^3$ at the temperature $T_1 = 4000\,\text{K}$, while in the final state, the combustion gases occupy the volume $V_2 = 2 \times 10^{-5}\,\text{m}^3$ at some unknown temperature T_2. In order to determine T_2, we can use the result (14.25), with $\kappa = 1.5$, to obtain

$$Tv^{\kappa-1} = Tv^{1/2} = \text{Constant} \Rightarrow T_2 = T_1 \sqrt{\frac{V_1}{V_2}}. \tag{14.147}$$

As $V_1/V_2 = 1/100$, the final temperature is 10 times lower than the initial temperature and we have $T_2 = 400\,\text{K}$. According to the first law of thermodynamics, for an ideal gas undergoing adiabatic change ($\Delta Q = 0$), we have

$$\Delta U = c_V \Delta T = \Delta Q - \Delta L = -\Delta L. \tag{14.148}$$

In this case, we have $c_P/c_V = 1.5$, which gives

$$\frac{c_P}{c_V} = \frac{c_V + nR}{c_V} = 1.5 \Rightarrow c_V = 2nR. \tag{14.149}$$

The work done by the expanding gas is then equal to

$$\Delta L = -c_V \Delta T = -2nR(T_2 - T_1) \approx 360\,\text{J}. \tag{14.150}$$

The work done by the expanding gas is converted into the kinetic energy of the bullet leaving the pipe of the rifle, and we can write

$$\frac{1}{2}mv^2 = \Delta L \Rightarrow v = \sqrt{\frac{2\Delta L}{m}} \approx 380\,\frac{m}{s}. \tag{14.151}$$

The speed of the bullet is thereby estimated at $v \approx 380\,\text{m/s}$, which is reasonably close to the actual bullet's speed.

Problem 9

The work required to change the surface area of a liquid drop by dA at a constant volume ($V = $ Constant) is equal to $dL = \gamma dA$, where $\gamma(T) = aT + b$ is the temperature-dependent positive surface tension coefficient with a and b being constants. Calculate the temperature change in the process where a spherical liquid drop of radius r is divided into two identical smaller liquid drops. The process is assumed to be adiabatic and reversible, and the specific heat per unit volume c_V at constant volume of the liquid is assumed to be constant.

Solution

The differential of the internal energy is in this case given by

$$dU = TdS + \gamma dA \Rightarrow \frac{\partial(T,S)}{\partial(\gamma,A)} = -1. \tag{14.152}$$

Using the Jacobian identity in (14.152), we further have

$$\left(\frac{\partial S}{\partial A}\right)_T = \frac{\partial(S,T)}{\partial(A,T)} = -\frac{\partial(A,\gamma)}{\partial(A,T)} = -\left(\frac{\partial \gamma}{\partial T}\right)_A = -a. \tag{14.153}$$

Integrating the equation (14.153), we obtain

$$S(A, T) = -aA + f(T). \tag{14.154}$$

The definition of the specific heat at constant volume yields

$$T\left(\frac{\partial S}{\partial T}\right)_V = Tf'(T) = c_V V. \tag{14.155}$$

Integrating (14.155), we obtain $f(T) = c_V V \ln T + \text{Constant}$, and the entropy of the liquid becomes

$$S(A, T) = -aA + c_V V \ln T + \text{Constant}. \tag{14.156}$$

The process is adiabatic, and we have $\Delta S = S_2 - S_1 = 0$, such that

$$-a\Delta A + c_V V \ln \frac{T_2}{T_1} = 0 \Rightarrow T_2 = T_1 \exp\left(\frac{a\Delta A}{c_V V}\right). \tag{14.157}$$

Since the volume of the two resulting smaller drops is $V_1 = 1/2V$, the radius of these two drops is $r_1 = (1/2)^{1/3}r$. Thus the area change in the process is equal to $\Delta A = 2 \times 4\pi r_1^2 - 4\pi r^2$, or

$$\Delta A = 4\pi\left(2r_1^2 - r^2\right) = 4\pi r^2\left[2 \times \left(\frac{1}{2}\right)^{2/3} - 1\right]. \tag{14.158}$$

Substituting the expressions $\Delta A = 4\pi r^2(2^{1/3} - 1)$ and $V = \frac{4}{3}\pi r^3$, into (14.157), we obtain the final temperature T_2 as follows:

$$T_2 = T_1 \exp\left(\frac{3a\left(2^{1/3} - 1\right)}{c_V r}\right). \tag{14.159}$$

Part Four

Quantum Statistical Physics

15 Quantum Distribution Functions

15.1 Entropy Maximization in Quantum Statistics

The last part of this book is devoted to quantum-statistical physics of ideal gases. We will incorporate the quantum effects ignored in our previous discussions of classical ideal gases. Indeed, already in Chapter 6, "Number of Accessible States and Entropy," we saw that the classical count of the number of microstates, W, is not exact, and, in the same chapter, we performed the exact count of microstates for bosons and fermions. As in the case of classical systems discussed in Chapter 7, "Equilibrium States of Systems," the equilibrium state of a quantum ideal gas maximizes the entropy $S = k \ln(W)$ at a fixed number of the constituent particles ($N = $ Constant) and a fixed internal energy ($U = $ Constant) of the system. Much like in our discussions in Chapter 7, the entropy maximization gives the condition,

$$dS = k \cdot d(\ln W) = 0 \Rightarrow d(\ln W) = 0, \tag{15.1}$$

to be satisfied by small variations of the occupation numbers n_i of energy levels around their equilibrium values. In order to analyze the condition (15.1) for quantum gases, we need to calculate the quantity $d(\ln W)$ for bosons and fermions, using the results for the numbers of accessible states (6.18) and (6.21), respectively.

15.1.1 The Case of Bosons

Let us first consider the case of bosons. For the case $n_i \gg 1$ and $g_i \gg 1$, by the equation (6.18), i.e.,

$$W = \prod_i \frac{(g_i + n_i - 1)!}{(g_i - 1)! n_i!}, \tag{15.2}$$

we can write

$$\ln W = \sum_i \ln(g_i + n_i - 1)! - \ln n_i! - \ln(g_i - 1)!$$

$$\approx \sum_i \ln(g_i + n_i)! - \ln n_i! - \ln g_i!. \tag{15.3}$$

This equation, in combination with the Stirling approximation $\ln(n!) \approx n \ln n - n$, given by the equation (7.8), yields

$$\ln W = \sum_i (g_i + n_i) \ln (g_i + n_i) - (g_i + n_i) - n_i \ln n_i + n_i - g_i \ln g_i + g_i, \quad (15.4)$$

or

$$\ln W = \sum_i (g_i + n_i) \ln (g_i + n_i) - n_i \ln n_i - g_i \ln g_i. \quad (15.5)$$

Recalling that g_i here are constants, the differential $d \ln W$ is now calculated as follows:

$$d \ln W = \sum_i d n_i \ln (g_i + n_i) + (g_i + n_i) \frac{d n_i}{g_i + n_i} - d n_i \ln n_i - n_i \frac{d n_i}{n_i}, \quad (15.6)$$

or finally

$$d \ln W = \sum_i d n_i \ln \frac{g_i + n_i}{n_i}. \quad (15.7)$$

Thus, the differential of entropy for bosons is given by

$$dS = k \cdot d(\ln W) = k \sum_i d n_i \ln \frac{g_i + n_i}{n_i}. \quad (15.8)$$

15.1.2 The Case of Fermions

Next, we consider the case of fermions. For the case $g_i \gg 1$, $n_i \gg 1$, and $g_i - n_i \gg 1$, the equation (6.21), i.e.,

$$W = \prod_i \frac{g_i!}{n_i! (g_i - n_i)!}, \quad (15.9)$$

or

$$\ln W = \sum_i \ln g_i! - \ln n_i! - \ln (g_i - n_i)!, \quad (15.10)$$

combined with the Stirling approximation $\ln(n!) \approx n \ln n - n$, given by the equation (7.8), yields

$$\ln W = \sum_i g_i \ln g_i - g_i - n_i \ln n_i + n_i - (g_i - n_i) \ln (g_i + n_i) + (g_i - n_i), \quad (15.11)$$

or

$$\ln W = \sum_i g_i \ln g_i - n_i \ln n_i - (g_i - n_i) \ln (g_i - n_i). \qquad (15.12)$$

Recalling that g_i here are constants, the differential $d\ln W$ is now calculated as follows:

$$d\ln W = \sum_i -dn_i \ln n_i - n_i \frac{dn_i}{n_i}$$

$$+ dn_i \ln (g_i - n_i) + (g_i - n_i) \frac{dn_i}{g_i - n_i}, \qquad (15.13)$$

or finally

$$d\ln W = \sum_i dn_i \ln \frac{g_i - n_i}{n_i}. \qquad (15.14)$$

Thus, the differential of entropy for fermions is given by

$$dS = k \cdot d(\ln W) = k \sum_i dn_i \ln \frac{g_i - n_i}{n_i}. \qquad (15.15)$$

15.2   Quantum Equilibrium Distribution

The results (15.8) and (15.15) for the differential of entropy for bosons and fermions, respectively, can both be written by the single formula,

$$dS = k \cdot d(\ln W) = k \sum_i dn_i \ln \frac{g_i \pm n_i}{n_i}. \qquad (15.16)$$

Above (and in the following), the upper sign applies to bosons and the lower sign applies to fermions. The three differential equilibrium conditions (7.4–7.6), stated for the classical distribution in Chapter 7, are also valid in the quantum statistics. The only difference is that we now use the exact quantum-mechanical result for the differential of entropy (15.16). Thus, we have

$$dU = \sum_i E_i dn_i = 0 \qquad / \cdot \frac{1}{T}, \qquad (15.17)$$

$$dN = \sum_i dn_i = 0 \qquad / \cdot \left(-\frac{\mu}{T}\right), \qquad (15.18)$$

$$dS = k \sum_i dn_i \ln \frac{g_i \pm n_i}{n_i} = 0 \qquad / \cdot (-1). \qquad (15.19)$$

Using now the method of Lagrange multipliers, as indicated in the equations (15.17–15.19), we get

$$\sum_i \left(\frac{E_i - \mu}{T} - k \ln \frac{g_i \pm n_i}{n_i} \right) dn_i = 0. \tag{15.20}$$

The Lagrange multipliers in the equations (15.17–15.19) are now chosen in advance to secure the classical limit, given by the equation (8.8). From the result (15.20), we obtain

$$\ln \frac{g_i \pm n_i}{n_i} = \frac{E_i - \mu}{kT}, \tag{15.21}$$

or

$$\frac{g_i \pm n_i}{n_i} = \exp\left(\frac{E_i - \mu}{kT} \right). \tag{15.22}$$

Solving the equation (15.22) for the occupation number n_i, we obtain

$$g_i = n_i \left[\exp\left(\frac{E_i - \mu}{kT} \right) \mp 1 \right], \tag{15.23}$$

or

$$n_i = \frac{g_i}{\exp\left(\frac{E_i - \mu}{kT} \right) \mp 1}, \tag{15.24}$$

where, as before, the upper sign applies to bosons and the lower sign applies to fermions. The occupation number n_i is the equilibrium number of quantum particles distributed into g_i sublevels of the energy level E_i. Thus, the mean occupation number of a sublevel within the energy level E_i is given by

$$\tilde{n}_i = \frac{n_i}{g_i} = \frac{1}{\exp\left(\frac{E_i - \mu}{kT} \right) \mp 1}. \tag{15.25}$$

The result (15.25) is, for the case of bosons (− sign), known as *Bose–Einstein formula*, whereas for the case of fermions (+ sign), as *Fermi–Dirac formula*, giving the mean (average) number of identical particles occupying a given sublevel, i.e., a one-particle state that is specified by a full set of quantum numbers, such as, for example, the four numbers $(\vec{p}, \sigma_z)$ telling the three components of the particle momentum vector and the z-component of the particle spin. In the limit, when

$$\exp\left(\frac{E_i - \mu}{kT} \right) \gg 1, \tag{15.26}$$

the equation (15.25) reduces to the classical Maxwell–Boltzmann approximation (8.8) for the mean (average) number of particles occupying a sublevel with energy E_i, i.e.,

$$\tilde{n}_i = \frac{n_i}{g_i} = \exp\left(\frac{\mu - E_i}{kT}\right). \tag{15.27}$$

Using here the result (8.5), we have

$$\exp\left(\frac{\mu}{kT}\right) = \frac{N}{Z}. \tag{15.28}$$

Substituting (15.28) into (15.27), we obtain the classical result for the mean occupation number as follows:

$$\tilde{n}_i = \frac{n_i}{g_i} = \frac{N}{Z}\exp\left(-\frac{E_i}{kT}\right). \tag{15.29}$$

The classical limit (15.29) is identical to the classical result (8.20) obtained earlier. Let us now discuss the limit (15.26). It is fulfilled when

$$\frac{E_i - \mu}{kT} \gg 1. \tag{15.30}$$

In the classical limit, when the condition (15.26) is fulfilled, we can use (15.28) in the form

$$-\frac{\mu}{kT} = \ln\left(\frac{Z}{N}\right). \tag{15.31}$$

Substituting (15.31) into (15.30), we obtain

$$\frac{E_i}{kT} + \ln\left(\frac{Z}{N}\right) \gg 1. \tag{15.32}$$

Using now the classical result (11.26) for the partition function Z of an ideal gas, we obtain

$$\frac{E_i}{kT} + \ln\left[\frac{V}{N}\left(\frac{2\pi mkT}{h^2}\right)^{3/2}\right] \gg 1. \tag{15.33}$$

Since the energy levels of the system can be small and even approach zero, the condition (15.33) becomes

$$\ln\left[\frac{V}{N}\left(\frac{2\pi mkT}{h^2}\right)^{3/2}\right] \gg 1, \tag{15.34}$$

or

$$\frac{V}{N}\left(\frac{2\pi mkT}{h^2}\right)^{3/2} \gg 1. \tag{15.35}$$

Thus, we finally obtain the condition for validity of the classical approximation in the form

$$\frac{N}{V} \ll \left(\frac{2\pi mkT}{h^2}\right)^{3/2}$$

$$\Rightarrow T \gg T_{\text{quantum}} = \frac{h^2}{2\pi mk}\left(\frac{N}{V}\right)^{2/3}. \tag{15.36}$$

From the condition (15.36), we see that the classical approximation is valid for systems of relatively low density N/V and relatively high temperature T that needs to be much higher than a characteristic quantum temperature T_{quantum}. By the equation (15.36), T_{quantum} increases with decreasing mass of constituent particles, and for light particles such as electrons in metals (with $N/V \approx 10^{28}\text{m}^{-3}$), this characteristic temperature is very high, $T_{\text{quantum}} \approx 10^4 - 10^5$ K. Thus, under normal conditions ($T \approx$ room temperature), classical statistical physics is inapplicable to the electron gas in metals, which is dominated by quantum effects discussed in Chapter 16, "Electron Gases in Metals." On the other hand, for heavy constituent particles such as atoms in molecules, the T_{quantum} in the equation (15.36) is very small under normal conditions (densities) realized in ordinary gases. Thus, the quantum distribution functions (15.25) for bosons or fermions are not needed for these gases, and they are quantitatively well described by the classical distribution function (15.29). The quantum-temperature scale T_{quantum} for these gases is not accessible because it is well in the sub-Kelvin range. Well before this range is reached, typical gases undergo condensation into liquids due to inter-molecular forces.

Let us now elucidate the physical meaning of the classicality condition (15.36). Recall that the quantum de Broglie wavelength associated with a particle moving with velocity v is given by

$$\lambda = \frac{h}{mv}. \tag{15.37}$$

As particles move with different velocities (see Section 11.4, "Kinetic Theory of Ideal Monatomic Gases"), the relevant quantity is

$$<\lambda> = \frac{h}{m}<\frac{1}{v}>. \tag{15.38}$$

The average $<1/v>$ in the above equation was calculated in Problem 5 of Chapter 11, "Ideal Monatomic Gases." By using this result, we find,

$$<\lambda> = \text{Constant} \times \left(\frac{h^2}{mkT}\right)^{1/2} \tag{15.39}$$

so that the condition (15.36) for the applicability of classical statistical mechanics can be written simply as,

$$<\lambda> \ll d = \left(\frac{V}{N}\right)^{1/3}. \tag{15.40}$$

As V/N is the volume per particle, the length-scale d in the above relation is essentially the distance between the nearest neighbor particles in the gas. For the classical physics to work well, the above condition, $<\lambda> \ll d$, has to be required. It implies that a given particle is seen by *other* particles of the gas as a classical object. Hence, the classical physics is applicable to such ideal gases. On the other hand, in the opposite case (which occurs for $T < T_{quantum}$), a particle is seen by other particles as a genuine quantum-mechanical object. The results of this are nontrivial quantum correlations between identical particles occurring in ideal gases, even without any interparticle interaction. Their origin is in the symmetry requirements imposed on N-particle wave functions, discussed in Section 6.2, "Calculation of the Number of Accessible States." Due to these requirements, a particle can affect other particles even without any interparticle interaction. These effects become strong only if $<\lambda>$ exceeds d, which occurs when the gas temperature T is decreased below $T_{quantum}$. Thus, for $T < T_{quantum}$, the above quantum correlation effects dominate and give rise to interesting phenomena present both in fermionic systems (such as the so-called *linear specific heat*, $c_v \sim T$, of electron gases in metals) and in bosonic systems (such as the Bose–Einstein condensation). These phenomena are discussed in the following chapters.

15.3 Helmholtz Thermodynamic Potential

Let us now substitute the result (15.24) into (7.2). Thus, we obtain the total number of particles of the quantum system as follows:

$$N = \sum_i n_i = \sum_i \frac{g_i}{\exp\left(\frac{E_i - \mu}{kT}\right) \mp 1}. \tag{15.41}$$

The upper sign applies to bosons and the lower sign applies to fermions. On the other hand, using the result (10.42), we can express the total number of particles of a quantum system in terms of the Helmholtz thermodynamic potential Ω as follows:

$$N = -\left(\frac{\partial \Omega}{\partial \mu}\right)_T. \tag{15.42}$$

By examination of the result (15.41), we conclude that the Helmholtz thermodynamic potential Ω is given by

$$\Omega = \pm kT \sum_i g_i \ln\left[1 \mp \exp\left(\frac{\mu - E_i}{kT}\right)\right]. \tag{15.43}$$

In order to verify that the above result for the Helmholtz thermodynamic potential Ω is correct, we substitute (15.43) into (15.42) to obtain

$$N = -\left(\frac{\partial \Omega}{\partial \mu}\right)_T = \mp kT \sum_i g_i \frac{\partial}{\partial \mu} \ln\left[1 \mp \exp\left(\frac{\mu - E_i}{kT}\right)\right]$$

$$= \mp kT \sum_i g_i \frac{\mp \exp\left(\frac{\mu - E_i}{kT}\right)}{1 \mp \exp\left(\frac{\mu - E_i}{kT}\right)} \frac{1}{kT}$$

$$= \sum_i \frac{g_i}{\exp\left(\frac{E_i - \mu}{kT}\right) \mp 1}. \tag{15.44}$$

Comparing the results (15.44) and (15.42), we see that the expression (15.43) for the Helmholtz thermodynamic potential Ω indeed reproduces the correct result for the total number of particles in the quantum system (15.42).

15.4 Thermodynamics of Quantum Systems

As discussed in Section 15.2, "Quantum Equilibrium Distribution," the ordinary gases, with atoms or molecules as constituent particles, are, under normal conditions, well described by classical thermodynamics. Thus, the quantum thermodynamics is of interest primarily for the gases consisting of lighter elementary particles. The boson quantum statistics is, therefore, mostly used for photons (quanta of electromagnetic radiation) with spin $\sigma = 1$, while the fermion quantum statistics is mostly used for electrons with spin $\sigma = 1/2$.

Therefore, in this section, we study the thermodynamics of general gases, consisting of elementary particles. We will employ the general formulae valid for both bosons and fermions, such as the equation (15.25), where the upper sign always applies to bosons and the lower sign always applies to fermions. The elementary particles have only the translational degrees of freedom and spin. As for the classical case, it is possible to use the continuum limit for the distribution of the energy levels in the fictitious n-space and replace the thermodynamic sums by integrals. Thus, we can use the classical kinetic energy of translations of an elementary constituent particle, i.e.,

$$E(p) = \frac{p^2}{2m} = \frac{1}{2m}\left(p_x^2 + p_y^2 + p_z^2\right), \tag{15.45}$$

where again p is the magnitude of the linear momentum vector of a constituent particle, which should not be confused with the pressure p of a thermodynamic system. The two conflicting notations will not be used in the same context, and there should be no risk for confusion here. The number of the continuously distributed momentum states $dg(p)$, inside a thin spherical shell $(p, p + dp)$, is obtained by multiplying the classical result (11.8) with the number of possible spin states $g(\sigma) = 2\sigma + 1$. Thus, we can write

$$dg(p) = g(\sigma)\frac{V}{h^3}4\pi p^2 dp. \tag{15.46}$$

Using (15.45), we can write

$$p = \sqrt{2mE} \Rightarrow dp = \sqrt{\frac{m}{2E}}dE. \tag{15.47}$$

Substituting (15.47) into (15.46), we obtain

$$dg(E) = g(\sigma)\frac{V}{8\pi^3\hbar^3}4\pi 2mE\sqrt{\frac{m}{2E}}dE, \tag{15.48}$$

or

$$dg(E) = g(\sigma)\frac{m^{3/2}V}{\sqrt{2}\pi^2\hbar^3}\sqrt{E}dE. \tag{15.49}$$

The continuous limit of the result (15.24) with (15.49) gives the continuous energy distribution function

$$dn_E = \frac{dg(E)}{\exp\left(\frac{E-\mu}{kT}\right)\mp 1}, \tag{15.50}$$

or

$$dN_E = g(\sigma)\frac{m^{3/2}V}{\sqrt{2}\pi^2\hbar^3}\frac{\sqrt{E}dE}{\exp\left(\frac{E-\mu}{kT}\right)\mp 1}. \tag{15.51}$$

The total number of particles in the system is now given by

$$N = \int dN_E = g(\sigma)\frac{m^{3/2}V}{\sqrt{2}\pi^2\hbar^3}\int_0^\infty\frac{\sqrt{E}dE}{\exp\left(\frac{E-\mu}{kT}\right)\mp 1}. \tag{15.52}$$

The internal energy of the quantum gas is then calculated as follows:

$$U = \int_0^\infty EdN_E = \int Ev(E)dE, \quad v(E) = \frac{dN_E}{dE}. \tag{15.53}$$

Thus, we obtain the caloric-state equation for the quantum gas, consisting of elementary particles, in the following form

$$U(V,T) = g(\sigma)\frac{m^{3/2}V}{\sqrt{2}\pi^2\hbar^3}\int_0^\infty\frac{E^{3/2}dE}{\exp\left(\frac{E-\mu}{kT}\right)\mp 1}. \tag{15.54}$$

The upper sign in the above results applies to bosons and the lower sign applies to fermions. The Helmholtz thermodynamic potential (15.43) in the continuous limit is given by

$$\Omega = \pm g(\sigma) \frac{m^{3/2} V k T}{\sqrt{2}\pi^2 \hbar^3} \int_0^\infty \sqrt{E} \ln\left[1 \mp \exp\left(\frac{\mu - E}{kT}\right)\right] dE. \tag{15.55}$$

The integral on the right-hand side of the result (15.55) can be integrated by parts as follows:

$$I = \int_0^\infty \sqrt{E} \ln\left[1 \mp \exp\left(\frac{\mu - E}{kT}\right)\right] dE$$

$$= \left[\frac{\frac{2}{3}E^{3/2}}{\exp\left(\frac{E-\mu}{kT}\right) \mp 1}\right]_0^\infty - \int_0^\infty \frac{\mp \exp\left(\frac{\mu-E}{kT}\right)\left(-\frac{1}{kT}\right)\frac{2}{3}E^{3/2} dE}{1 \mp \exp\left(\frac{\mu-E}{kT}\right)}$$

$$= \mp \frac{1}{kT} \int_0^\infty \frac{\frac{2}{3}E^{3/2} dE}{\exp\left(\frac{E-\mu}{kT}\right) \mp 1}. \tag{15.56}$$

Substituting (15.56) into (15.55), we obtain

$$\Omega = -\frac{2}{3} g(\sigma) \frac{m^{3/2} V}{\sqrt{2}\pi^2 \hbar^3} \int_0^\infty \frac{E^{3/2} dE}{\exp\left(\frac{E-\mu}{kT}\right) \mp 1}. \tag{15.57}$$

Comparing the results (15.57) and (15.54), we obtain

$$\Omega = -pV = -\frac{2}{3} U. \tag{15.58}$$

Thus, we obtain the thermal-state equation for the quantum gas, consisting of elementary particles, in the following form

$$pV = \frac{2}{3} U. \tag{15.59}$$

This is an exact result, which remains valid in the classical limit. Indeed, if we use the classical result for the internal energy of an ideal gas $U = 3/2 NkT$, the equation (15.59) reduces to the classical thermal-state equation for the ideal gas, i.e.,

$$pV = NkT. \tag{15.60}$$

The enthalpy (9.19) for the quantum gas, consisting of elementary particles, is now calculated as follows:

$$H = U + pV = U + \frac{2}{3}U = \frac{5}{3}U. \tag{15.61}$$

This is also an exact result. If we substitute here the classical result for the internal energy of an ideal gas $U = 3/2NkT$, we obtain the classical result for enthalpy of an ideal gas, i.e.,

$$H = \frac{5}{2}NkT. \tag{15.62}$$

From the above results, we see that if we want to calculate the thermodynamic variables of the quantum gas, we need to calculate the integral

$$I_E = \int\limits_0^\infty \frac{E^{3/2}dE}{\exp\left(\frac{E-\mu}{kT}\right) \mp 1}. \tag{15.63}$$

When the chemical potential μ is not equal to zero, the integral (15.63) cannot be solved in the closed form and we can only find approximate solutions. Some physically interesting solutions will be discussed in the following chapter.

15.5 Evaluation of Integrals *)

In the analysis of quantum gases, we will often need the solutions of the integrals involving the quantum distribution functions. It is, therefore, of interest to outline the solution for one important class of integrals, i.e.,

$$I(x) = \int\limits_0^\infty \frac{w^{x-1}dw}{e^w \mp 1}, \tag{15.64}$$

where $x > 1$ is a real number. In order to solve the integral (15.64), we first use the geometric series

$$\frac{1}{1-a} = \sum_{n=0}^\infty a^n (a < 1) \Rightarrow \frac{1}{1 \mp e^{-w}} = \sum_{n=0}^\infty (\pm 1)^n e^{-nw} \tag{15.65}$$

to calculate

$$\frac{1}{e^w \mp 1} = \frac{e^{-w}}{1 \mp e^{-w}} = \sum_{n=0}^\infty (\pm 1)^n e^{-(n+1)w}. \tag{15.66}$$

Substituting (15.66) into (15.64), we obtain

$$I(x) = \sum_{n=0}^{\infty} (\pm 1)^n \int_0^{\infty} w^{x-1} e^{-(n+1)w} dw. \tag{15.67}$$

For integer j, we have

$$\int_0^{\infty} w^{j-1} e^{-\alpha w} dw = \frac{(j-1)!}{\alpha^j} = \frac{\Gamma(j)}{\alpha^j}, \tag{15.68}$$

where $\Gamma(j) = (j-1)!$ is the *gamma function* of an integer argument. The result (15.68) can easily be verified for integer j ($j = 1, 2, 3 \ldots$), by elementary partial integration. Furthermore, the result (15.68) can be generalized to the case of noninteger $j = x$ as follows:

$$\int_0^{\infty} w^{x-1} e^{-\alpha w} dw = \frac{\Gamma(x)}{\alpha^x}. \tag{15.69}$$

Substituting (15.69) into (15.67), we obtain

$$I(x) = \sum_{n=0}^{\infty} (\pm 1)^n \frac{\Gamma(x)}{(n+1)^x} = \Gamma(x) \sum_{n=0}^{\infty} \frac{(\pm 1)^n}{(n+1)^x}. \tag{15.70}$$

If we now make a change of the argument in the sum, i.e., $n + 1 \to n$, we obtain

$$I(x) = \Gamma(x) \sum_{n=1}^{\infty} \frac{(\pm 1)^{n-1}}{n^x} = \Gamma(x) \sum_{n=1}^{\infty} \frac{(\pm 1)^{n+1}}{n^x}. \tag{15.71}$$

Thus, we obtain for bosons

$$I(x) = \int_0^{\infty} \frac{w^{x-1} dw}{e^w - 1} = \Gamma(x) \sum_{n=1}^{\infty} \frac{1}{n^x} = \Gamma(x) \zeta(x), \tag{15.72}$$

where

$$\zeta(x) = \sum_{n=1}^{\infty} \frac{1}{n^x}, \quad (x > 1), \tag{15.73}$$

is the *Riemann zeta function*. For fermions, we have

$$I(x) = \int_0^{\infty} \frac{w^{x-1} dw}{e^w + 1} = \Gamma(x) \sum_{n=1}^{\infty} \frac{(-1)^{n+1}}{n^x}. \tag{15.74}$$

On the other hand, the difference of the two series for bosons and fermions only includes the even terms ($n = 2j$), and we have

$$\sum_{n=1}^{\infty} \frac{1}{n^x} - \sum_{n=1}^{\infty} \frac{(-1)^{n+1}}{n^x} = \sum_{n=1}^{\infty} \frac{1+(-1)^n}{n^x} = 2\sum_{j=1}^{\infty} \frac{1}{(2j)^x}, \tag{15.75}$$

or

$$\sum_{n=1}^{\infty} \frac{1}{n^x} - \sum_{n=1}^{\infty} \frac{(-1)^{n+1}}{n^x} = 2^{1-x}\sum_{j=1}^{\infty} \frac{1}{j^x}. \tag{15.76}$$

Using (15.73), we obtain

$$\zeta(x) - \sum_{n=1}^{\infty} \frac{(-1)^{n+1}}{n^x} = 2^{1-x}\zeta(x), \tag{15.77}$$

or

$$\sum_{n=1}^{\infty} \frac{(-1)^{n+1}}{n^x} = (1 - 2^{1-x})\zeta(x). \tag{15.78}$$

Substituting (15.78) into (15.74), we finally obtain the result for fermions

$$I(x) = \int_0^{\infty} \frac{w^{x-1}dw}{e^w + 1} = \left(1 - 2^{1-x}\right)\Gamma(x)\zeta(x). \tag{15.79}$$

In the special case, when $x = 2j$ is an even integer, the Riemann zeta functions can be expressed in terms of Bernoulli numbers B_{2j}, and we obtain for bosons

$$I(2j) = \int_0^{\infty} \frac{w^{2j-1}dw}{e^w - 1} = \Gamma(2j)\zeta(2j) = \frac{(2\pi)^{2j}B_{2j}}{4j}, \tag{15.80}$$

and for the fermions

$$I(2j) = \int_0^{\infty} \frac{w^{2j-1}dw}{e^w + 1} = \left(1 - 2^{1-2j}\right)\Gamma(2j)\zeta(2j)$$

$$= \left(1 - 2^{1-2j}\right)\frac{(2\pi)^{2j}B_{2j}}{4j} = \left(2^{2j-1} - 1\right)\frac{\pi^{2j}B_{2j}}{2j}. \tag{15.81}$$

The first few Bernoulli numbers are listed below

$$B_2 = \frac{1}{6}, B_4 = \frac{1}{30}, B_6 = \frac{1}{42}, B_8 = \frac{1}{30}. \tag{15.82}$$

As an example, for $j = 2$, we have for bosons

$$I(2j = 4) = \int_0^\infty \frac{w^3 dw}{e^w - 1} = \frac{(2\pi)^4 B_4}{8} = \frac{\pi^4}{15}, \tag{15.83}$$

and for fermions

$$I(2j = 4) = \int_0^\infty \frac{w^3 dw}{e^w + 1} = \frac{7}{8} \times \frac{\pi^4}{15} = \frac{7\pi^4}{120}. \tag{15.84}$$

The above integrals are important for calculation of thermodynamic variables for all quantum gases.

15.6   Problems with Solutions

Problem 1

(a) Can entropy be negative?
(b) Show that the classical statistical physics predicts negative entropy for ideal gases at temperatures below a certain characteristic temperature. Find this temperature and explain its nature in the context of this chapter.

Solution

(a) Recall that $S = k \ln W$. The number of microstates W satisfies the inequalities,

$$W \geq 1 \Rightarrow \ln W \geq 0, \tag{15.85}$$

hence

$$S = k \ln W \geq 0. \tag{15.86}$$

Thus, the entropy cannot be negative. At best, entropy can approach zero and this happens only if $W \rightarrow 1$. For example, in thermodynamic equilibrium of quantum systems, the limit $W \rightarrow 1$, i.e., $S = k \ln W \rightarrow 0$, is reached only in the limit of zero temperature $T \rightarrow 0$ (see Section 9.2.4, "Third Law of Thermodynamics"). On the other hand, for any nonzero temperature, $W > 1$, i.e., $S = k \ln W > 0$. Thus, in thermodynamic equilibrium, entropy is positive at any finite temperature, and it approaches zero value only in the limit $T \rightarrow 0$.

(b) Consider the classical statistical mechanics result for the entropy of an ideal gas, given by the equation (11.81). It can be rewritten as follows:

$$S = \frac{3}{2}Nk\left[1 + \frac{2}{3}\ln\frac{V}{N} + \ln\left(\frac{2\pi mkT}{h^2}\right)\right] + Nk$$

$$= \frac{3}{2}Nk\left[\frac{5}{3} + \frac{2}{3}\ln\frac{V}{N} + \ln\left(\frac{2\pi mkT}{h^2}\right)\right]$$

$$= \frac{3}{2}Nk\left[\ln e^{5/3} + \frac{2}{3}\ln\frac{V}{N} + \ln\left(\frac{2\pi mkT}{h^2}\right)\right]$$

$$= \frac{3}{2}Nk\ln\left[e^{5/3}\left(\frac{V}{N}\right)^{2/3}\left(\frac{2\pi mkT}{h^2}\right)\right]. \tag{15.87}$$

By the above result, the classical result for the entropy of an ideal gas, can be written shortly as,

$$S = \frac{3}{2}Nk\ln\left(\frac{T}{T^*}\right), \tag{15.88}$$

where we introduced a characteristic temperature T^* defined by the expression,

$$T^* = e^{-5/3}\frac{h^2}{2\pi mk}\left(\frac{N}{V}\right)^{2/3}. \tag{15.89}$$

By the above, we see that the classical statistical physics predicts entropy that is positive for $T > T^*$, vanishes at a nonzero temperature $T = T^*$, and then it becomes negative for the temperatures $T < T^*$ (and, moreover, it approaches negative infinity as the temperature goes to zero). In view of our discussions in the item (a) above, the exact expression for an entropy cannot be negative. Thus, it is obvious that the classical physics dramatically fails at low temperatures, below the above-defined characteristic temperature T^*. It is easy to see that this pathological behavior of classical entropy at low temperatures is because of the quantum effects discussed in Section 15.2, "Quantum Equilibrium Distribution." Indeed, by the equation (15.36), one can see that T^* is (up to inessential numerical constant $e^{-5/3}$) the same as T_{quantum}.

Problem 2

Consider a general quantum-mechanical system in equilibrium at temperature T. Its partition function is, as usual, given by the sum over the energy states

$$Z(T) = \sum_\alpha \exp\left(-\frac{E_\alpha}{kT}\right), \tag{15.90}$$

whereas the corresponding free energy is

$$F(T) = -kT \ln Z(T). \tag{15.91}$$

(a) Use the thermodynamic relation

$$U(T) = kT^2 \frac{\partial}{\partial T} \ln Z(T) \tag{15.92}$$

to show that the internal energy $U(T)$ is given by the sum

$$U(T) = \sum_\alpha E_\alpha P_\alpha, \tag{15.93}$$

with

$$P_\alpha(T) = \frac{1}{Z(T)} \exp\left(-\frac{E_\alpha}{kT}\right). \tag{15.94}$$

What is the value of the sum $\sum_\alpha P_\alpha$? What is the physical meaning of $P_\alpha(T)$?

(b) Show that the system entropy,

$$S = \frac{U(T) - F(T)}{T} = -\frac{\partial}{\partial T} F(T), \tag{15.95}$$

can be written as

$$S(T) = -k \sum_\alpha P_\alpha \ln P_\alpha. \tag{15.96}$$

Use this general expression for the entropy to

(i) prove that the entropy S cannot be negative;
(ii) find the condition under which the entropy S vanishes.

(c) Show that the heat capacity of this system,

$$C(T) = \frac{\partial}{\partial T} U(T), \tag{15.97}$$

satisfies the so-called *fluctuation–dissipation theorem*

$$C(T) = \frac{<(E_\alpha)^2> - <E_\alpha>^2}{kT^2}. \tag{15.98}$$

In the above, we employ the notation $<Q_\alpha> = \sum_\alpha Q_\alpha P_\alpha$, so,

$$<(E_\alpha)^2> = \sum_\alpha (E_\alpha)^2 P_\alpha, \quad <E_\alpha> = \sum_\alpha E_\alpha P_\alpha = U(T). \tag{15.99}$$

(d) Prove the identity

$$< [E_\alpha - U(T)]^2 > = < (E_\alpha)^2 > - < E_\alpha >^2 . \tag{15.100}$$

(e) Next, use the fluctuation–dissipation theorem to,

    (i) prove that heat capacity $C(T)$ cannot be negative;
    (ii) prove that the entropy $S(T)$ cannot decrease with increasing temperature T, i.e., $S(T)$ is, in general, an increasing function of temperature T.

(f) Show that the heat capacity of a quantum-mechanical system must vanish in zero-temperature limit.

Solution

(a) By using the chain rule for the derivative, we have

$$U(T) = kT^2 \frac{\partial}{\partial T} \ln Z(T) = kT^2 \frac{1}{Z(T)} \frac{\partial}{\partial T} Z(T)$$

$$= kT^2 \frac{1}{Z(T)} \frac{\partial}{\partial T} \sum_\alpha \exp\left(-\frac{E_\alpha}{kT}\right)$$

$$= kT^2 \frac{1}{Z(T)} \sum_\alpha \exp\left(-\frac{E_\alpha}{kT}\right) \frac{E_\alpha}{kT^2}$$

$$= \sum_\alpha E_\alpha \frac{1}{Z(T)} \exp\left(-\frac{E_\alpha}{kT}\right) = \sum_\alpha E_\alpha P_\alpha. \tag{15.101}$$

This proves the equation (15.93). Its form is telling us that the quantity P_α is the probability to find this quantum-mechanical system in the state with energy E_α. In accord with the quantities P_α being a set of probabilities, are the facts that these quantities are obviously positive [see the equation (15.94)], and that their sum is

$$\sum_\alpha P_\alpha = \sum_\alpha \frac{1}{Z(T)} \exp\left(-\frac{E_\alpha}{kT}\right) = \frac{Z(T)}{Z(T)} = 1. \tag{15.102}$$

(b) We have

$$S = \frac{U(T) - F(T)}{T} = \frac{\sum_\alpha E_\alpha P_\alpha + kT \ln Z(T)}{T}. \tag{15.103}$$

By equation (15.90), we have

$$E_\alpha = -kT[\ln P_\alpha + \ln Z(T)]. \tag{15.104}$$

Inserting (15.104) into (15.103), we obtain

$$S = \frac{1}{T}\left\{\sum_\alpha -kT[\ln P_\alpha + \ln Z(T)]P_\alpha + kT\ln Z(T)\right\}$$

$$= -k\sum_\alpha P_\alpha \ln P_\alpha - k\ln Z(T)\sum_\alpha P_\alpha + k\ln Z(T)$$

$$= -k\sum_\alpha P_\alpha \ln P_\alpha - k\ln Z(T) + k\ln Z(T), \tag{15.105}$$

or

$$S = -k\sum_\alpha P_\alpha \ln P_\alpha. \tag{15.106}$$

This proves the equation (15.96) for the entropy of a general quantum-mechanical system. From this result, it is easy to see that the entropy S cannot be negative. Indeed, all state probabilities satisfy the condition $0 \le P_\alpha \le 1$ or $\ln P_\alpha \le 0$, hence, for all states,

$$-P_\alpha \ln P_\alpha \ge 0 \Rightarrow S = -k\sum_\alpha P_\alpha \ln P_\alpha \ge 0. \tag{15.107}$$

This proves that entropy S cannot be negative, for any quantum-mechanical system. By the equation (15.107), the entropy can be either positive or zero. For the entropy to be zero, a very special condition needs to be satisfied. Indeed, by (15.107), the entropy is the sum of nonnegative terms $-P_\alpha \ln P_\alpha$, so S can vanish if and only if each of the terms in the sum vanishes. That is, S vanishes only if $-P_\alpha \ln P_\alpha = 0$ for all states. Recall now that the only solutions to the equation $x\ln x = 0$ are $x = 0$ and $x = 1$. So, if entropy is zero, the probabilities P_α are bound to be either 0 or 1. Recall, however, that these probabilities add up to 1, i.e., $\sum_\alpha P_\alpha = 1$. It is then easy to see that, if entropy is zero, then only one of these probabilities can be 1, whereas all remaining probabilities are 0. Thus, zero entropy corresponds to essentially the non-random situation in which the quantum-mechanical system occupies with certainty (probability $= 1$) one of the energy states, whereas the probabilities to find the system in any of the other states are zero. Such a situation, however, can never happen at a nonzero temperature, as can be easily seen from the form of the probabilities in (15.94). By this equation, for any finite T, all $P_\alpha(T)$ are nonzero; hence, the entropy cannot be zero for a nonzero temperature. The condition for the entropy to vanish is, however, satisfied in the limit $T \to 0$, in which the probability to find the system in the ground state (minimum energy state, with energy E_0) indeed approaches unity, whereas the probabilities of all other excited states (with energies $E_\alpha > E_0$) approach 0. Indeed, by equation (15.94), we have

$$\frac{P_\alpha(T)}{P_0(T)} = \exp\left(-\frac{E_\alpha - E_0}{kT}\right) \to 0 \quad \text{for} \quad T \to 0. \tag{15.108}$$

This, combined with the condition,

$$1 = \sum_\alpha P_\alpha = P_0 \left[1 + \sum_{\alpha \neq 0} \frac{P_\alpha(T)}{P_0(T)} \right], \tag{15.109}$$

implies that, for $T \to 0$,

$$P_0 = \frac{1}{1 + \sum_{\alpha \neq 0} \exp\left(-\frac{E_\alpha - E_0}{kT}\right)} \to 1, \tag{15.110}$$

whereas $P_\alpha(T) \to 0$ for all excited states. So, in thermodynamic equilibrium, the condition to have zero entropy is satisfied only in the limit $T \to 0$.

(c) For the heat capacity, equation (15.97) with (15.93) gives

$$C(T) = \frac{\partial U}{\partial T} = \frac{\partial}{\partial T} \sum_\alpha E_\alpha P_\alpha = \sum_\alpha E_\alpha \frac{\partial}{\partial T} P_\alpha(T). \tag{15.111}$$

Here,

$$\frac{\partial}{\partial T} P_\alpha(T) = \frac{\partial}{\partial T} \left[\frac{1}{Z(T)} \exp\left(-\frac{E_\alpha}{kT}\right) \right]$$

$$= \frac{1}{[Z(T)]^2} \left\{ \left[\frac{\partial}{\partial T} \exp\left(-\frac{E_\alpha}{kT}\right) \right] Z(T) - \exp\left(-\frac{E_\alpha}{kT}\right) \frac{\partial}{\partial T} Z(T) \right\}$$

$$= \frac{1}{[Z(T)]^2} \left\{ \exp\left(-\frac{E_\alpha}{kT}\right) \frac{E_\alpha}{kT^2} Z(T) - \exp\left(-\frac{E_\alpha}{kT}\right) \frac{\partial}{\partial T} Z(T) \right\}$$

$$= \frac{1}{kT^2} \frac{1}{Z(T)} \exp\left(-\frac{E_\alpha}{kT}\right) \left[E_\alpha - kT^2 \frac{1}{Z(T)} \frac{\partial}{\partial T} Z(T) \right]. \tag{15.112}$$

Using equations (15.92) and (15.94), we can rewrite the equation (15.112) in a simple form,

$$\frac{\partial}{\partial T} P_\alpha(T) = \frac{P_\alpha}{kT^2} [E_\alpha - U(T)] = \frac{P_\alpha}{kT^2} [E_\alpha - <E_\alpha>]. \tag{15.113}$$

Inserting (15.113) into (15.111) yields

$$C(T) = \frac{1}{kT^2} \sum_\alpha E_\alpha P_\alpha [E_\alpha - <E_\alpha>]$$

$$= \frac{1}{kT^2} \sum_\alpha P_\alpha \left[E_\alpha^2 - <E_\alpha> E_\alpha \right]$$

$$= \frac{1}{kT^2} \left\{ \sum_\alpha E_\alpha^2 P_\alpha - <E_\alpha> \sum_\alpha E_\alpha P_\alpha \right\}. \tag{15.114}$$

Using the equations (15.99), we finally obtain

$$C(T) = \frac{<(E_\alpha)^2> - <E_\alpha>^2}{kT^2}. \tag{15.115}$$

Thus, we have proved the fluctuation–dissipation theorem (15.98).

(d) Let us now prove the identity (15.100). We have,

$$< [E_\alpha - U(T)]^2 > = < (E_\alpha)^2 - 2U(T)E_\alpha + U^2(T) >$$

$$= \sum_\alpha P_\alpha [(E_\alpha)^2 - 2U(T)E_\alpha + U^2(T)]$$

$$= \sum_\alpha P_\alpha (E_\alpha)^2 - 2U(T) \sum_\alpha P_\alpha E_\alpha + \sum_\alpha P_\alpha U^2(T)$$

$$= \sum_\alpha P_\alpha (E_\alpha)^2 - 2U(T)U(T) + U^2(T)$$

$$= \sum_\alpha P_\alpha (E_\alpha)^2 - [U(T)]^2 = < (E_\alpha)^2 > - < E_\alpha >^2. \tag{15.116}$$

Thus, we have proved the identity (15.100).

(e) Using the identity (15.100), the fluctuation–dissipation theorem (15.98) can be written as

$$C(T) = \frac{<(E_\alpha)^2> - <E_\alpha>^2}{kT^2} = \frac{<[E_\alpha - U(T)]^2>}{kT^2}$$

$$= \frac{1}{kT^2} \sum_\alpha P_\alpha [E_\alpha - U(T)]^2 \geq 0, \tag{15.117}$$

and from this form of the fluctuation–dissipation theorem, it is obvious that $C(T)$ cannot be negative $C(T) \geq 0$, and we expect it to be, in general, a positive quantity. This has an interesting implication on the behavior of entropy with changing temperature. Indeed, recalling the equation (15.95), i.e.,

$$U(T) = F + TS = F - T\frac{\partial F}{\partial T}, \tag{15.118}$$

gives

$$C(T) = \frac{\partial U}{\partial T} = \frac{\partial}{\partial T}\left(F - T\frac{\partial F}{\partial T}\right)$$

$$= \frac{\partial F}{\partial T} - \frac{\partial T}{\partial T}\frac{\partial F}{\partial T} - T\left(\frac{\partial}{\partial T}\right)^2 F = -T\left(\frac{\partial}{\partial T}\right)^2 F$$

$$= T\frac{\partial}{\partial T}\left(-\frac{\partial F}{\partial T}\right) = T\frac{\partial}{\partial T}S. \tag{15.119}$$

By the equation (15.117), we then have

$$T\frac{\partial}{\partial T}S = C(T) \geq 0. \tag{15.120}$$

From the equation (15.120), we see that the entropy $S(T)$ cannot decrease with increasing temperature T. That is, $S(T)$ is, in general, an increasing function of temperature T.

(f) By the equation (15.120), it is obvious that for quantum-mechanical systems, $C(T) = T\partial S/\partial T \rightarrow 0$, as $T \rightarrow 0$. Recall that, for example, the classical monatomic ideal gas violates this property. For it, $C(T) = 3/2Nk = $ Constant, i.e., $C(T)$ is a constant. This behavior with constant $C(T)$ will be present only in the high-temperature classical regime, for T above the characteristic quantum temperature $T_{quantum}$ defined in (15.36). On the other hand, for the temperatures below $T_{quantum}$, by the above results, we expect a genuinely quantum-statistical behavior in which $C(T) \rightarrow 0$ as $T \rightarrow 0$.

Problem 3

A two-dimensional ideal gas is confined to a two-dimensional square box with area $A = a^2$, and its quantum energy levels are given by the two-dimensional analog of the equation (3.42), i.e.,

$$E = \frac{h^2}{8ma^2}\left(n_x^2 + n_y^2\right) = \frac{h^2}{8mA}\left(n_x^2 + n_y^2\right), \tag{15.121}$$

where $n_x, n_y = 1, 2, 3, 4\ldots$ are the quantum numbers labeling the energy states. Determine the chemical potential μ of the gas if the particles are fermions, bosons, and classical particles. Verify that the quantum results obtained for fermions and bosons have the classical result as a high-temperature limit.

Hint: The following integral is useful for the solution of the present problem

$$\int_0^\infty \frac{dw}{be^w \pm 1} = \pm\ln(1 \pm 1/b). \tag{15.122}$$

Solution

In order to calculate the chemical potential, we need to find the density of states for the two-dimensional gas. We use the two-dimensional analog of the equation (15.46) that reads

$$dg(p) = g(\sigma)\frac{A}{h^2}2\pi p\,dp, \tag{15.123}$$

where $g(\sigma) = 2\sigma + 1$ is the spin factor. Using the definition of the kinetic energy $E = p^2/2m$, we can write

$$p = \sqrt{2mE} \Rightarrow dp = \sqrt{\frac{m}{2E}}\,dE. \tag{15.124}$$

Substituting (15.124) into (15.123), we obtain

$$dg(E) = g(\sigma)\frac{2\pi mA}{h^2}\,dE, \tag{15.125}$$

whereby we show that the density of states of a two-dimensional gas is a constant. The number of particles in a fermion gas is then

$$N = g(\sigma)\frac{2\pi mA}{h^2}\int_0^\infty \frac{dE}{\exp\left(\frac{E-\mu}{kT}\right)+1}, \tag{15.126}$$

or

$$N = g(\sigma)\frac{2\pi mA}{h^2}\int_0^\infty \frac{dE}{\exp\left(\frac{-\mu}{kT}\right)\exp\left(\frac{E}{kT}\right)+1}. \tag{15.127}$$

Introducing here a new variable $w = E/kT$, we obtain

$$N = g(\sigma)\frac{2\pi mAkT}{h^2}\int_0^\infty \frac{dw}{\exp\left(\frac{-\mu}{kT}\right)e^w+1}, \tag{15.128}$$

or

$$\frac{Nh^2}{2\pi g(\sigma)mAkT} = \int_0^\infty \frac{dw}{\exp\left(\frac{-\mu}{kT}\right)e^w+1}. \tag{15.129}$$

Let us now define a temperature T_0 by means of the equation

$$T_0 = \frac{Nh^2}{2\pi g(\sigma)mAk}. \tag{15.130}$$

Substituting (15.130) into (15.129), we obtain

$$\frac{T_0}{T} = \int_0^\infty \frac{dw}{\exp\left(\frac{-\mu}{kT}\right)e^w+1}. \tag{15.131}$$

Using now the integral (15.122), yields

$$\frac{T_0}{T} = \ln\left[1 + \exp\left(\frac{\mu}{kT}\right)\right].\tag{15.132}$$

The final result for the chemical potential μ for fermions is

$$\mu = kT\ln\left[\exp\left(\frac{T_0}{T}\right) - 1\right].\tag{15.133}$$

The equation (15.131) in the case of bosons reads

$$\frac{T_0}{T} = \int\limits_0^\infty \frac{dw}{\exp\left(\frac{-\mu}{kT}\right)e^w - 1}.\tag{15.134}$$

Using again the integral (15.122) with minus sign yields

$$\frac{T_0}{T} = -\ln\left[1 - \exp\left(\frac{\mu}{kT}\right)\right].\tag{15.135}$$

The final result for the chemical potential μ for bosons is

$$\mu = kT\ln\left[1 - \exp\left(-\frac{T_0}{T}\right)\right].\tag{15.136}$$

In case of the classical statistics, the integral in (15.131) becomes

$$\frac{T_0}{T} = \exp\left(\frac{\mu}{kT}\right)\int\limits_0^\infty e^{-w}dw = \exp\left(\frac{\mu}{kT}\right).\tag{15.137}$$

The final result for the chemical potential μ for classical particles is

$$\mu = kT\ln\left(\frac{T_0}{T}\right).\tag{15.138}$$

Let us now check the limit of the expressions (15.133) and (15.136) when $T \gg T_0$. The expression (15.133) yields

$$\mu \approx kT\ln\left[\left(1 + \frac{T_0}{T}\right) - 1\right] = kT\ln\left(\frac{T_0}{T}\right),\tag{15.139}$$

and the expression (15.136) yields

$$\mu \approx kT\ln\left[1 - \left(1 - \frac{T_0}{T}\right)\right] = kT\ln\left(\frac{T_0}{T}\right).\tag{15.140}$$

Thus, the results for both fermions and bosons have the correct classical limit.

Problem 4

The equation (15.25) actually gives the *average* number of identical particles (bosons or fermions) occupying a given sublevel with the energy E_i,

$$\bar{n}_i = <\tilde{n}_i> = \frac{1}{\exp\left(\frac{E_i-\mu}{kT}\right) \mp 1}, \tag{15.141}$$

with upper (lower) sign applying for bosons (fermions). In thermodynamic equilibrium, the actual number of these particles fluctuates around the average given by the equation (15.141). The probability to find $\tilde{n}_i$ particles occupying the sublevel with the energy E_i is given by the expression

$$P_i(\tilde{n}_i) = \frac{1}{\Xi_i} \exp\left(\frac{\mu - E_i}{kT}\tilde{n}_i\right). \tag{15.142}$$

The equation (15.142) applies to both bosons and fermions. Here, $\tilde{n}_i = 0, 1, 2, \ldots,$ $\tilde{n}_{max}$, where $\tilde{n}_{max} = \infty$ for bosons, whereas, because of the Pauli exclusion principle, $\tilde{n}_{max} = 1$ for fermions.

(a) Calculate the value of the normalization factor Ξ_i in the equation (15.142).
(b) Show that

$$<\tilde{n}_i> = -\frac{\partial}{\partial \mu}\Omega_i, \tag{15.143}$$

where

$$\Omega_i = -kT \ln \Xi_i. \tag{15.144}$$

(c) Use the equations (15.142) and (15.144) to rederive the equation (15.141) for both bosons and fermions.
(d) Consider the average of the total number of particles

$$<N> = \sum_i g_i\bar{n}_i = \sum_i g_i <\tilde{n}_i>. \tag{15.145}$$

Show that this average is given by

$$<N> = -\frac{\partial}{\partial \mu}\Omega, \tag{15.146}$$

with

$$\Omega = \sum_i g_i\Omega_i. \tag{15.147}$$

Show also that the above Ω coincides with the Helmholtz potential (15.43).

Solution

(a) Since $P_i(\tilde{n}_i)$ must satisfy the condition

$$\sum_{\tilde{n}_i=0}^{\tilde{n}_{\max}} P_i(\tilde{n}_i) = 1, \tag{15.148}$$

by the equation (15.142), we have

$$1 = \sum_{\tilde{n}_i=0}^{\tilde{n}_{\max}} P_i(\tilde{n}_i) = \frac{1}{\Xi_i} \sum_{\tilde{n}_i=0}^{\tilde{n}_{\max}} \exp\left(\frac{\mu - E_i}{kT} \tilde{n}_i\right)$$

$$\Rightarrow \Xi_i = \sum_{\tilde{n}_i=0}^{\tilde{n}_{\max}} \exp\left(\frac{\mu - E_i}{kT} \tilde{n}_i\right). \tag{15.149}$$

For bosons, with $\tilde{n}_{\max} = \infty$, the equation (15.149) is a standard geometric series of the form

$$\Xi_i = \sum_{n=0}^{\infty} a^n = \frac{1}{1-a}, \quad |a| = \left|\exp\left(\frac{\mu - E_i}{kT}\right)\right| < 1. \tag{15.150}$$

The convergence of the series requires that $E_i > \mu$ for all states of bosonic systems. This will be obviously assured if this condition is satisfied for the minimum energy state (ground state) with energy, $E_{\min} > \mu$. Thus, for bosons,

$$\Xi_i = \left[1 - \exp\left(\frac{\mu - E_i}{kT}\right)\right]^{-1}. \tag{15.151}$$

For fermions, with $\tilde{n}_{\max} = 1$, the sum (15.149) contains only two terms, $\tilde{n}_i = 0$ and $\tilde{n}_i = 1$, so

$$\Xi_i = 1 + \exp\left(\frac{\mu - E_i}{kT}\right) = \left[1 + \exp\left(\frac{\mu - E_i}{kT}\right)\right]^{+1}. \tag{15.152}$$

The results (15.151) and (15.152) can be combined into one formula

$$\Xi_i = \left[1 \mp \exp\left(\frac{\mu - E_i}{kT}\right)\right]^{\mp 1}, \tag{15.153}$$

with upper (lower) sign applying for bosons (fermions).

(b) By equations (15.142), (15.144), and (15.149),

$$-\frac{\partial}{\partial\mu}\Omega_i = kT\frac{\partial}{\partial\mu}\ln\Xi_i = kT\frac{1}{\Xi_i}\frac{\partial}{\partial\mu}\Xi_i$$

$$= kT\frac{1}{\Xi_i}\frac{\partial}{\partial\mu}\sum_{\tilde{n}_i=0}^{\tilde{n}_{\max}}\exp\left(\frac{\mu-E_i}{kT}\tilde{n}_i\right)$$

$$= kT\frac{1}{\Xi_i}\sum_{\tilde{n}_i=0}^{\tilde{n}_{\max}}\exp\left(\frac{\mu-E_i}{kT}\tilde{n}_i\right)\frac{\tilde{n}_i}{kT}$$

$$= \sum_{\tilde{n}_i=0}^{\tilde{n}_{\max}}\frac{1}{\Xi_i}\exp\left(\frac{\mu-E_i}{kT}\tilde{n}_i\right)\tilde{n}_i$$

$$= \sum_{\tilde{n}_i=0}^{\tilde{n}_{\max}}P_i(\tilde{n}_i)\tilde{n}_i = <\tilde{n}_i>. \tag{15.154}$$

This proves the equation (15.143), i.e., that

$$<\tilde{n}_i> = -\frac{\partial}{\partial\mu}\Omega_i = kT\frac{\partial}{\partial\mu}\ln\Xi_i. \tag{15.155}$$

(c) By equations (15.153) and (15.155), we have

$$<\tilde{n}_i> = kT\frac{\partial}{\partial\mu}\ln\Xi_i = kT\frac{\partial}{\partial\mu}\ln\left[1\mp\exp\left(\frac{\mu-E_i}{kT}\right)\right]^{\mp1}$$

$$= \mp kT\frac{\partial}{\partial\mu}\ln\left[1\mp\exp\left(\frac{\mu-E_i}{kT}\right)\right]$$

$$= \mp kT\left[1\mp\exp\left(\frac{\mu-E_i}{kT}\right)\right]^{-1}\frac{\partial}{\partial\mu}\left[1\mp\exp\left(\frac{\mu-E_i}{kT}\right)\right]$$

$$= \mp kT\left[1\mp\exp\left(\frac{\mu-E_i}{kT}\right)\right]^{-1}\left[\mp\frac{1}{kT}\exp\left(\frac{\mu-E_i}{kT}\right)\right]$$

$$= \left[1\mp\exp\left(\frac{\mu-E_i}{kT}\right)\right]^{-1}\exp\left(\frac{\mu-E_i}{kT}\right)$$

$$= \frac{1}{\exp\left(\frac{E_i-\mu}{kT}\right)\mp1}. \tag{15.156}$$

This proves the equation (15.141).

(d) We have, by the equation (15.143),

$$<N> = \sum_i g_i \bar{n}_i = \sum_i g_i < \tilde{n}_i >$$

$$= \sum_i g_i \left(-\frac{\partial}{\partial \mu} \Omega_i\right) = -\frac{\partial}{\partial \mu} \sum_i g_i \Omega_i, \tag{15.157}$$

that is,

$$<N> = -\frac{\partial}{\partial \mu}\Omega \quad \text{with} \quad \Omega = \sum_i g_i \Omega_i. \tag{15.158}$$

This proves the equation (15.146). Let us finally show that the above Ω coincides with the Helmholtz potential (15.43). Indeed, by the equations (15.144) and (15.153), we have

$$\Omega = \sum_i g_i \Omega_i = -kT \sum_i g_i \ln \Xi_i$$

$$= -kT \sum_i g_i \ln \left[1 \mp \exp\left(\frac{\mu - E_i}{kT}\right)\right]^{\mp 1}$$

$$= \pm kT \sum_i g_i \ln \left[1 \mp \exp\left(\frac{\mu - E_i}{kT}\right)\right], \tag{15.159}$$

coinciding with the expression in (15.43).

Problem 5

For fermions,

(a) Show that, in the notations of Problem 4,

$$P_i(\tilde{n}_i = 1) = 1 - P_i(\tilde{n}_i = 0) = <\tilde{n}_i > = \frac{1}{\exp\left(\frac{E_i - \mu}{kT}\right) + 1}. \tag{15.160}$$

(b) Calculate the average $<(\tilde{n}_i)^{137}>$.

Solution

(a) For fermions, i.e., $\tilde{n}_{max} = 1$, we can have only $\tilde{n}_i = 0$ and $\tilde{n}_i = 1$, so

$$<\tilde{n}_i> = \sum_{\tilde{n}_i=0}^{\tilde{n}_i=1} P_i(\tilde{n}_i)\tilde{n}_i = P_i(0)\cdot 0 + P_i(1)\cdot 1 = P_i(1). \tag{15.161}$$

Thus, for fermions (but *not* for bosons), we have

$$P_i(\tilde{n}_i = 1) = <\tilde{n}_i> = \frac{1}{\exp\left(\frac{E_i - \mu}{kT}\right) + 1}, \tag{15.162}$$

or

$$P_i(\tilde{n}_i = 1) = \frac{\exp\left(\frac{\mu - E_i}{kT} \cdot 1\right)}{\exp\left(\frac{\mu - E_i}{kT}\right) + 1}. \tag{15.163}$$

On the other hand, we have

$$1 = <1> = \sum_{\tilde{n}_i = 0}^{\tilde{n}_i = 1} P_i(\tilde{n}_i) \cdot 1 = P_i(0) + P_i(1) \tag{15.164}$$

so that

$$P_i(\tilde{n}_i = 0) = 1 - P_i(\tilde{n}_i = 1) = 1 - \tilde{n}_i$$

$$= 1 - \frac{1}{\exp\left(\frac{E_i - \mu}{kT}\right) + 1} = \frac{\exp\left(\frac{E_i - \mu}{kT}\right)}{\exp\left(\frac{E_i - \mu}{kT}\right) + 1}, \tag{15.165}$$

or

$$P_i(\tilde{n}_i = 0) = \frac{\exp\left(\frac{\mu - E_i}{kT} \cdot 0\right)}{\exp\left(\frac{\mu - E_i}{kT}\right) + 1}. \tag{15.166}$$

The final forms of $P_i(1)$ and $P_i(0)$ written in equations (15.163) and (15.166), respectively, are results of simple algebra. The results in equations (15.163) and (15.166) are written in such a way to emphasize that they coincide with the general result in equation (15.142), for the special case of fermions.

(b) We have

$$<(\tilde{n}_i)^{137}> = \sum_{\tilde{n}_i = 0}^{\tilde{n}_i = 1} P_i(\tilde{n}_i) \cdot (\tilde{n}_i)^{137} = P_i(0) \cdot 0^{137} + P_i(1) \cdot 1^{137}$$

$$= P_i(1) = <\tilde{n}_i> = \frac{1}{\exp\left(\frac{E_i - \mu}{kT}\right) + 1}. \tag{15.167}$$

By a similar reasoning, $<(\tilde{n}_i)^m> = <\tilde{n}_i>$ for any positive m.

16 Electron Gases in Metals

Metals are well known to be very good electric current conductors. Conductivity of metals comes from a fraction of all electrons in metals, the so-called *conducting electrons*, that are nearly free to move in the background of the crystalline lattice of positively charged ions. The system consisting of a large number of these free electrons in a metal specimen can be described as an *electron gas*. The electrons are elementary particles with half-integer spin ($\sigma = 1/2$) and they obey the quantum statistics for fermions. In this analysis, the electron–electron interactions and the influence of the periodic potential of the ions in the crystal lattice on the motion of free electrons are neglected. In such an approximation, the electron gas can be treated as an ideal quantum gas. If the metal specimen is surrounded by nonconducting environment (e.g., air), the electrons cannot escape from the metal. Assuming that the metal specimen is a cube with the side a, we can describe the electron gas as a system consisting of N noninteracting point-like particles, enclosed in the cubic box with impenetrable walls, having the volume $V = a^3$. Apart from their spin, the electrons have only the translational degrees of freedom. The quantum-mechanical problem is thereby the same as the one studied in Chapter 3, "Kinetic Energy of Translational Motion," and the translational energy of a single electron is given by the result (3.41), i.e.,

$$E = \frac{\hbar^2 \kappa^2}{2m} = \frac{\hbar^2 \pi^2}{2ma^2} \left(n_x^2 + n_y^2 + n_z^2 \right),$$

(16.1)

where

$$\vec{\kappa} = \frac{2\pi}{a} \vec{n}, \quad \vec{n} = \left(n_x, n_y, n_z \right)$$

(16.2)

is the wave vector of the electron. From the above result for the energy E of free electrons in the electron gas, we see that the electron gas is a regular nonrelativistic quantum gas. Thus, the general results for such quantum gases, obtained in the previous chapter, are applicable to the electron gas. In particular, the number of electrons with energies within the interval $(E, E + dE)$ is given by the result (15.49) with $g(\sigma) = 2$, i.e.,

$$dg(E) = 2 \frac{m^{3/2} V}{\sqrt{2} \pi^2 \hbar^3} \sqrt{E} dE = \frac{1}{2\pi^2} \left(\frac{2m}{\hbar^2} \right)^{3/2} V E^{1/2} dE.$$

(16.3)

Introductory Statistical Thermodynamics

16.1 Ground State of Electron Gases in Metals

In the ground state of an N-electron gas ($T = 0$), according to the Pauli exclusion principle, the lowest $N/2$ energy states are occupied, each by two electrons with opposite spins. All the other energy states above these $N/2$ states are empty. The highest occupied energy level has the energy E_F, which is called the *Fermi energy*. In the ground state, all N electrons have energies lower than or equal to the Fermi energy. The Fermi energy can thus be obtained from

$$N = \int_{E<E_F} dg(E) = \frac{1}{2\pi^2}\left(\frac{2m}{\hbar^2}\right)^{3/2} V \int_0^{E_F} E^{1/2} dE, \tag{16.4}$$

or

$$N = \frac{1}{3\pi^2}\left(\frac{2mE_F}{\hbar^2}\right)^{3/2} V. \tag{16.5}$$

The density of electrons in the electron gas is then obtained as follows:

$$n = \frac{N}{V} = \frac{1}{3\pi^2}\left(\frac{2mE_F}{\hbar^2}\right)^{3/2}. \tag{16.6}$$

The Fermi energy, in terms of the density of the electron gas, is given by

$$E_F = \frac{\hbar^2}{2m}\left(3\pi^2 n\right)^{2/3}. \tag{16.7}$$

The so-called *Fermi temperature* T_F is obtained from $E_F = kT_F$, and we have

$$T_F = \frac{E_F}{k} = \frac{\hbar^2}{2mk}\left(3\pi^2 n\right)^{2/3}. \tag{16.8}$$

By the equation (16.8), we see that T_F is (up to an inessential numerical constant) the same as the characteristic quantum temperature T_{quantum} in the equation (15.36). Thus, the classical statistical thermodynamics should be applicable for $T \gg T_F$, whereas quantum effects discussed in Chapter 15, "Quantum Distribution Functions," become dominant at temperatures below T_F. In order to calculate the density of the electron gas, we first look at the crystal lattice of a metal with atomic mass A (kg/mole) and mass density ρ_m (kg/m^3). The crystal lattice of a metal consists of a large number of unit cells with volume V_C (m^3). In this analysis, we consider only metals with one atom per unit cell. In such a case, the mass of a single atom can be obtained either as $\rho_m V_C$ or A/N_A, where N_A (atoms/mole) is the Avogadro number given by (11.85). Thus, we can write

$$m_A = \rho_m V_C = \frac{A}{N_A} \quad \Rightarrow \quad V_C = \frac{A}{\rho_m N_A}. \tag{16.9}$$

Each metal atom has Z valence electrons, i.e., free electrons that constitute the electron gas. The density of the electron gas in the metal specimen is then given by

$$n = \frac{Z}{V_C} = ZN_A \frac{\rho_m}{A} \sim 10^{29} \frac{\text{electrons}}{\text{m}^3}.$$ (16.10)

From the order-of-magnitude estimate given in (16.10), we see that the density of the electron gas is very high compared with the densities of the ordinary gases. Substituting the estimate (16.10) into (16.8), we obtain $T_F \sim 90,000$ K. The condition for the classical statistics to be applicable to the case of an electron gas is $T \gg T_F$. Obviously, for realistic metals, $T \ll T_F$. Indeed at $T \sim T_F$, all known metals would evaporate. Thus, the electron gas in realistic metals is a strongly degenerate quantum gas, and the quantum statistics must be applied. The ground-state internal energy of the electron gas is given by

$$U_0 = \int_{E<E_F} E dg(E) = \frac{1}{2\pi^2} \left(\frac{2m}{\hbar^2} \right)^{3/2} V \int_0^{E_F} E^{3/2} dE,$$ (16.11)

or

$$U_0 = \frac{1}{5\pi^2} \left(\frac{2m}{\hbar^2} \right)^{3/2} V E_F^{5/2} = \frac{3}{5} \frac{1}{3\pi^2} \left(\frac{2mE_F}{\hbar^2} \right)^{3/2} V E_F.$$ (16.12)

Substituting (16.5) into (16.12), we obtain

$$U_0 = \frac{3}{5} N E_F.$$ (16.13)

16.2   Electron Gases in Metals at Finite Temperatures

At realistic temperatures for metals, the free electrons are nonrelativistic, so the equation (16.13) is applicable, in combination with the equation (15.24). Thus, the number of electrons with energies within the interval $(E, E + dE)$ is given by

$$dn_E = \frac{dg(E)}{\exp\left(\frac{E-\mu}{kT} \right) + 1}.$$ (16.14)

In the ground state for $T = 0$, we have used the energy distribution function

$$dn_E = dg(E) \times \begin{Bmatrix} 1 & E < E_F \\ 0 & E > E_F \end{Bmatrix}$$ (16.15)

to calculate the total number of particles (16.4) in the ground state. On the other hand, in the limit $T \to 0$, the equation (16.14) becomes

$$dn_E = dg(E) \times \begin{Bmatrix} 1 & E < \mu \\ 0 & E > \mu \end{Bmatrix}. \tag{16.16}$$

Comparing the results (16.15) and (16.16), we see that in the ground state, we have $\mu(T = 0) = E_F$. Furthermore, even at relatively low finite temperatures, the distribution function (16.14) is close to the ground-state limit (16.16). Thus, we can often use the approximation $\mu \approx E_F$ at low temperatures $T \ll T_F$.

Using the results (15.52) and (15.54) with $g(\sigma) = 2$ and the correct sign for fermion gas, we obtain the total number of electrons

$$N = \frac{1}{2\pi^2} \left(\frac{2m}{\hbar^2} \right)^{3/2} V \int_0^\infty \frac{E^{1/2} dE}{\exp\left(\frac{E-\mu}{kT} \right) + 1}, \tag{16.17}$$

and the internal energy of the electron gas

$$U = \frac{1}{2\pi^2} \left(\frac{2m}{\hbar^2} \right)^{3/2} V \int_0^\infty \frac{E^{3/2} dE}{\exp\left(\frac{E-\mu}{kT} \right) + 1}. \tag{16.18}$$

We proceed to calculate the integrals on the right-hand sides of the equations (16.17) and (16.18), respectively. Let us first introduce a new variable w as follows:

$$E = \mu + kTw \quad \Rightarrow \quad dE = kT dw. \tag{16.19}$$

Substituting (16.19) into (16.17), we obtain the total number of electrons

$$N = \frac{1}{2\pi^2} \left(\frac{2m}{\hbar^2} \right)^{3/2} V \int_{-\mu/kT}^\infty \frac{(\mu + kTw)^{1/2} kT dw}{e^w + 1}, \tag{16.20}$$

or

$$N = \frac{1}{2\pi^2} \left(\frac{2m}{\hbar^2} \right)^{3/2} V(kT)^{3/2} \int_{-\mu/kT}^\infty \frac{(w + \mu/kT)^{1/2} dw}{e^w + 1}. \tag{16.21}$$

Substituting (16.19) into (16.18), we obtain the internal energy of the electron gas

$$U = \frac{1}{2\pi^2} \left(\frac{2m}{\hbar^2} \right)^{3/2} V \int_{-\mu/kT}^\infty \frac{(\mu + kTw)^{3/2} kT dw}{e^w + 1}, \tag{16.22}$$

or

$$U = \frac{1}{2\pi^2} \left(\frac{2m}{\hbar^2}\right)^{3/2} V(kT)^{5/2} \int\limits_{-\mu/kT}^{\infty} \frac{(w+\mu/kT)^{3/2}dw}{e^w + 1}. \tag{16.23}$$

The integrals on the right-hand side of the equations (16.21) and (16.23) are of the type

$$I = \int\limits_{-\mu/kT}^{\infty} \frac{f(w+\mu/kT)dw}{e^w + 1}, \tag{16.24}$$

where in our case

$$f(w+\mu/kT) = (w+\mu/kT)^{j/2}, \quad j = 1, 3. \tag{16.25}$$

Thus, we have

$$I = \int\limits_{-\mu/kT}^{0} \frac{f(w+\mu/kT)dw}{e^w + 1} + \int\limits_{0}^{\infty} \frac{f(w+\mu/kT)dw}{e^w + 1}. \tag{16.26}$$

Changing the sign of the variable $w \rightarrow -w$ in the first integral on the right-hand side of (16.26), we obtain

$$I = \int\limits_{0}^{\mu/kT} \frac{f(-w+\mu/kT)dw}{e^{-w} + 1} + \int\limits_{0}^{\infty} \frac{f(w+\mu/kT)dw}{e^w + 1}. \tag{16.27}$$

In the first integral on the right-hand side of (16.27), we use the identity

$$\frac{1}{e^{-w} + 1} = 1 - \frac{1}{e^w + 1} \tag{16.28}$$

to obtain

$$I = \int\limits_{0}^{\mu/kT} f(-w+\mu/kT)dw - \int\limits_{0}^{\mu/kT} \frac{f(-w+\mu/kT)dw}{e^w + 1} + \int\limits_{0}^{\infty} \frac{f(w+\mu/kT)dw}{e^w + 1}. \tag{16.29}$$

In the first integral on the right-hand side of (16.29), we can introduce a new variable $\lambda = -w + \mu/kT$. Thus, we obtain

$$\int\limits_{0}^{\mu/kT} f(-w+\mu/kT)dw = \int\limits_{\mu/kT}^{0} f(\lambda)d(-\lambda) = \int\limits_{0}^{\mu/kT} f(\lambda)d\lambda. \tag{16.30}$$

In the strongly degenerate quantum limit, it is justified to assume $\mu/kT \approx T_F/T \gg 1$. In such a case, the second integral on the right-hand side of (16.29) is rapidly convergent and we can extend the upper integration limit to infinity. Using (16.30), the equation (16.29) gives

$$I = \int_0^{\mu/kT} f(\lambda)\mathrm{d}\lambda + \int_0^{\infty} \frac{[f(w+\mu/kT) - f(-w+\mu/kT)]\mathrm{d}w}{e^w + 1}. \tag{16.31}$$

Let us now expand the functions $f(w+\mu/kT)$ and $f(-w+\mu/kT)$ as the Taylor series in powers of w as follows:

$$f\left(w+\frac{\mu}{kT}\right) = f\left(\frac{\mu}{kT}\right) + f^{(1)}\left(\frac{\mu}{kT}\right)w$$
$$+ \frac{1}{2!}f^{(2)}\left(\frac{\mu}{kT}\right)w^2 + \frac{1}{3!}f^{(3)}\left(\frac{\mu}{kT}\right)w^3 + \cdots \tag{16.32}$$

and

$$f\left(-w+\frac{\mu}{kT}\right) = f\left(\frac{\mu}{kT}\right) - f^{(1)}\left(\frac{\mu}{kT}\right)w$$
$$+ \frac{1}{2!}f^{(2)}\left(\frac{\mu}{kT}\right)w^2 - \frac{1}{3!}f^{(3)}\left(\frac{\mu}{kT}\right)w^3 + \cdots. \tag{16.33}$$

Thus, we can write

$$f\left(w+\frac{\mu}{kT}\right) - f\left(-w+\frac{\mu}{kT}\right) = 2f^{(1)}\left(\frac{\mu}{kT}\right)w + 2\frac{1}{3!}f^{(3)}\left(\frac{\mu}{kT}\right)w^3 + \cdots. \tag{16.34}$$

Substituting (16.34) into (16.31), we obtain

$$I = + \int_0^{\mu/kT} f(\lambda)\mathrm{d}\lambda + 2f^{(1)}\left(\frac{\mu}{kT}\right)\int_0^{\infty} \frac{w\mathrm{d}w}{e^w + 1} + \frac{1}{3}f^{(3)}\left(\frac{\mu}{kT}\right)\int_0^{\infty} \frac{w^3\mathrm{d}w}{e^w + 1}. \tag{16.35}$$

The solutions of the two integrals on the right-hand side of the equation (16.35) are obtained using the significant integral (15.81). Thus, we have

$$\int_0^{\infty} \frac{w\mathrm{d}w}{e^w + 1} = \frac{\pi^2}{12}, \quad \int_0^{\infty} \frac{w^3\mathrm{d}w}{e^w + 1} = \frac{7\pi^4}{120}. \tag{16.36}$$

Substituting (16.36) into (16.35), we get

$$I = \int_0^{\mu/kT} f(\lambda)\mathrm{d}\lambda + \frac{\pi^2}{6}f^{(1)}\left(\frac{\mu}{kT}\right) + \frac{7\pi^4}{360}f^{(3)}\left(\frac{\mu}{kT}\right). \tag{16.37}$$

Let us now consider the functions (16.25). For $j = 1$, we have

$$f^{(1)}\left(\frac{\mu}{kT}\right) = \frac{1}{2}\left(\frac{\mu}{kT}\right)^{-1/2}, \quad f^{(3)}\left(\frac{\mu}{kT}\right) = \frac{3}{8}\left(\frac{\mu}{kT}\right)^{-5/2}, \tag{16.38}$$

and

$$\int_0^{\mu/kT} f(\lambda)d\lambda = \int_0^{\mu/kT} \lambda^{1/2}d\lambda = \frac{2}{3}\left(\frac{\mu}{kT}\right)^{3/2}. \tag{16.39}$$

Substituting (16.38) and (16.39) into (16.37), we obtain

$$I_N = \frac{2}{3}\left(\frac{\mu}{kT}\right)^{3/2} + \frac{\pi^2}{12}\left(\frac{\mu}{kT}\right)^{-1/2} + \frac{7\pi^4}{960}\left(\frac{\mu}{kT}\right)^{-5/2}. \tag{16.40}$$

Using the result (16.21) with (16.40), we obtain the number of electrons in the electron gas as follows:

$$N = \frac{1}{2\pi^2}\left(\frac{2m}{\hbar^2}\right)^{3/2} V(kT)^{3/2}I_N, \tag{16.41}$$

or

$$N = \frac{1}{2\pi^2}\left(\frac{2m}{\hbar^2}\right)^{3/2} V\left(\frac{2}{3}\mu^{3/2} + \frac{\pi^2}{12}\frac{k^2T^2}{\mu^{1/2}} + \frac{7\pi^4}{960}\frac{k^4T^4}{\mu^{5/2}}\right). \tag{16.42}$$

On the other hand, for $j = 3$, we have

$$f^{(1)}\left(\frac{\mu}{kT}\right) = \frac{3}{2}\left(\frac{\mu}{kT}\right)^{1/2}, \quad f^{(3)}\left(\frac{\mu}{kT}\right) = -\frac{3}{8}\left(\frac{\mu}{kT}\right)^{-3/2}, \tag{16.43}$$

and

$$\int_0^{\mu/kT} f(\lambda)d\lambda = \int_0^{\mu/kT} \lambda^{3/2}d\lambda = \frac{2}{5}\left(\frac{\mu}{kT}\right)^{5/2}. \tag{16.44}$$

Substituting (16.43) and (16.44) into (16.37), we obtain

$$I_U = \frac{2}{5}\left(\frac{\mu}{kT}\right)^{5/2} + \frac{\pi^2}{4}\left(\frac{\mu}{kT}\right)^{1/2} - \frac{7\pi^4}{960}\left(\frac{\mu}{kT}\right)^{-3/2}. \tag{16.45}$$

Using the result (16.23) with (16.45), we obtain the internal energy of the electron gas as follows:

$$U = \frac{1}{2\pi^2}\left(\frac{2m}{\hbar^2}\right)^{3/2} V(kT)^{5/2}I_U, \tag{16.46}$$

or

$$U = \frac{1}{2\pi^2} \left(\frac{2m}{\hbar^2} \right)^{3/2} V \left(\frac{2}{5} \mu^{5/2} + \frac{\pi^2}{4} \frac{k^2 T^2}{\mu^{-1/2}} - \frac{7\pi^4}{960} \frac{k^4 T^4}{\mu^{3/2}} \right). \tag{16.47}$$

Thus, we have obtained the approximate results for the number of constituent particles and internal energy of the electron gas. These results will be used in the next two sections to calculate some important thermodynamic quantities of the electron gas.

16.3 Chemical Potential at Finite Temperatures

In order to calculate the chemical potential of the electron gas as a function of temperature, we can use the result (16.42), where we drop the fourth-order term in temperature. Thus, we obtain the density of the electron gas $n = N/V$ as follows:

$$n = \frac{N}{V} = \frac{1}{2\pi^2} \left(\frac{2m}{\hbar^2} \right)^{3/2} \left(\int\limits_0^\mu \lambda^{1/2} d\lambda + \frac{\pi^2}{12} \frac{k^2 T^2}{\mu^{1/2}} \right). \tag{16.48}$$

Let us now calculate the integral

$$\int\limits_0^\mu \lambda^{1/2} d\lambda = \int\limits_0^{E_F} \lambda^{1/2} d\lambda + \int\limits_{E_F}^\mu \lambda^{1/2} d\lambda. \tag{16.49}$$

Using here the approximation $\mu \approx E_F$, we get

$$\int\limits_0^\mu \lambda^{1/2} d\lambda \approx \frac{2E_F^{3/2}}{3} + E_F^{1/2} \int\limits_{E_F}^\mu d\lambda = \frac{2E_F^{3/2}}{3} + E_F^{1/2}(\mu - E_F). \tag{16.50}$$

Substituting (16.50) into (16.48), we obtain

$$n = \frac{1}{2\pi^2} \left(\frac{2m}{\hbar^2} \right)^{3/2} \left(\frac{2E_F^{3/2}}{3} + E_F^{1/2}(\mu - E_F) \right)$$
$$+ \frac{1}{2\pi^2} \left(\frac{2m}{\hbar^2} \right)^{3/2} \frac{\pi^2}{12} \frac{k^2 T^2}{E_F^{1/2}}, \tag{16.51}$$

where we used the approximation $\mu \approx E_F$ in the last term of the equation (16.51). After some algebra, we obtain

$$n = \frac{1}{3\pi^2} \left(\frac{2mE_F}{\hbar^2} \right)^{3/2} + \frac{1}{2\pi^2} \left(\frac{2mE_F}{\hbar^2} \right)^{3/2} \left(\frac{\mu}{E_F} - 1 + \frac{\pi^2}{12} \frac{k^2 T^2}{E_F^2} \right). \tag{16.52}$$

If we now recall the result for the density n of the electron gas (16.16), we see that the first term on the right-hand side of the equation (16.52) cancels the left-hand side of that equation. Thus, we may write

$$0 = \frac{1}{2\pi^2} \left(\frac{2mE_F}{\hbar^2} \right)^{3/2} \left(\frac{\mu}{E_F} - 1 + \frac{\pi^2}{12} \frac{k^2 T^2}{E_F^2} \right). \tag{16.53}$$

By the equation (16.53), we obtain the approximate result for the chemical potential of the degenerate electron gas in the following form

$$\mu = E_F \left(1 - \frac{\pi^2}{12} \frac{k^2 T^2}{E_F^2} \right). \tag{16.54}$$

16.4 Thermodynamics of Electron Gases

In order to derive the approximate caloric-state equation for the degenerate electron gas, we can use the result for the internal energy (16.47), where we drop the fourth-order term in temperature. Thus, we obtain

$$U = \frac{1}{2\pi^2} \left(\frac{2m}{\hbar^2} \right)^{3/2} V \left(\int_0^{\mu/kT} \lambda^{3/2} d\lambda + \frac{\pi^2}{4} \frac{k^2 T^2}{E_F^{-1/2}} \right), \tag{16.55}$$

where we used the approximation $\mu \approx E_F$ in the last term of the equation (16.55). Let us now calculate the integral

$$\int_0^{\mu} \lambda^{3/2} d\lambda = \int_0^{E_F} \lambda^{3/2} d\lambda + \int_{E_F}^{\mu} \lambda^{3/2} d\lambda. \tag{16.56}$$

Using here the approximation $\mu \approx E_F$, we get

$$\int_0^{\mu} \lambda^{3/2} d\lambda \approx \frac{2E_F^{5/2}}{5} + E_F^{3/2} \int_{E_F}^{\mu} d\lambda = \frac{2E_F^{5/2}}{5} + E_F^{3/2}(\mu - E_F). \tag{16.57}$$

From the result for the chemical potential (16.54), we have

$$\mu - E_F = -\frac{\pi^2}{12} \frac{k^2 T^2}{E_F}. \tag{16.58}$$

Substituting (16.58) into (16.57), we obtain

$$\int_0^{\mu} \lambda^{3/2} d\lambda = \frac{2E_F^{5/2}}{5} - \frac{\pi^2}{12} \frac{k^2 T^2}{E_F^{-1/2}}. \tag{16.59}$$

Further, substituting (16.59) into (16.55), we obtain

$$U = \frac{1}{2\pi^2} \left(\frac{2m}{\hbar^2}\right)^{3/2} V \left(\frac{2E_F^{5/2}}{5} - \frac{\pi^2}{12} \frac{k^2 T^2}{E_F^{-1/2}} + \frac{\pi^2}{4} \frac{k^2 T^2}{E_F^{-1/2}}\right), \tag{16.60}$$

or

$$U = \frac{1}{2\pi^2} \left(\frac{2m}{\hbar^2}\right)^{3/2} V \left(\frac{2E_F^{5/2}}{5} + \frac{\pi^2}{6} \frac{k^2 T^2}{E_F^{-1/2}}\right). \tag{16.61}$$

After some algebra, the result (16.61) becomes

$$U = \frac{1}{5\pi^2} \left(\frac{2mE_F}{\hbar^2}\right)^{3/2} V E_F \left(1 + \frac{5\pi^2}{12} \frac{k^2 T^2}{E_F^2}\right). \tag{16.62}$$

If we now recall the result for the ground-state internal energy U_0 of the electron gas (16.12), we finally obtain

$$U = U_0 \left(1 + \frac{5\pi^2}{12} \frac{k^2 T^2}{E_F^2}\right), \quad U_0 = \frac{3}{5} N E_F. \tag{16.63}$$

The heat capacity at constant volume is given by

$$m_{\text{sample}} \, c_V = \left(\frac{\partial U}{\partial T}\right)_V = \frac{\pi^2}{2} N k \frac{kT}{E_F}. \tag{16.64}$$

On the other hand, the specific heat at constant volume for the classical ideal gas is given by $m_{\text{sample}} \, c_{VC} = 3/2 Nk$. Thus, the ratio between the quantum and classical specific heats at realistic temperatures ($T \sim 1000\,\text{K}$) is

$$\frac{c_V}{c_{VC}} = \frac{\pi^2}{3} \frac{kT}{E_F} = \frac{\pi^2}{3} \frac{T}{T_F} \sim 10^{-2}. \tag{16.65}$$

Using now the significant integral (15.59), we obtain the approximate thermal-state equation for the degenerate electron gas in the following form

$$pV = \frac{2}{3} U = \frac{2}{5} N E_F \left(1 + \frac{5\pi^2}{12} \frac{k^2 T^2}{E_F^2}\right). \tag{16.66}$$

In terms of electron density $n = N/V$, the thermal-state equation becomes

$$pV = \frac{2}{5}nVE_F\left(1 + \frac{5\pi^2}{12}\frac{k^2T^2}{E_F^2}\right).$$

(16.67)

The pressure of the electron gas is then given by

$$p = \frac{2}{5}nE_F\left(1 + \frac{5\pi^2}{12}\frac{k^2T^2}{E_F^2}\right).$$

(16.68)

The enthalpy (15.61) for the electron gas is given by

$$H = \frac{5}{3}U = NE_F\left(1 + \frac{5\pi^2}{12}\frac{k^2T^2}{E_F^2}\right).$$

(16.69)

The Helmholtz thermodynamic potential is given by

$$\Omega = -pV = -\frac{2}{5}nVE_F\left(1 + \frac{5\pi^2}{12}\frac{k^2T^2}{E_F^2}\right).$$

(16.70)

Using the same order of approximation, it is also possible to calculate a number of other thermodynamic variables for a degenerate electron gas.

16.5 Problems with Solutions

Problem 1

The density of the electron gas in a metal is approximately equal to $n = 2.6 \times 10^{28}\,\text{m}^{-3}$. Calculate the Fermi energy E_F and the molar heat capacity c_V at $T = 300\,\text{K}$. This metal has one free electron per each atom.

Solution

The Fermi energy, in terms of the density of the electron gas, is calculated as follows:

$$E_F = \frac{\hbar^2}{2m}\left(3\pi^2 n\right)^{2/3} = 5.13 \times 10^{-19}\,\text{J}.$$

(16.71)

The molar specific heat at constant volume is given by

$$c_V = \frac{\pi^2 N_{AV} k^2 T}{2E_F} = 0.33\,\frac{\text{J}}{\text{K mol}}.$$

(16.72)

Problem 2

Calculate the isothermal compressibility coefficient γ for a free electron gas at $T = 0$ in terms of the density n of the gas.

Solution

The isothermal compressibility coefficient is defined by

$$\gamma = -\frac{1}{V}\left(\frac{\partial V}{\partial p}\right)_T \quad \Rightarrow \quad \gamma^{-1} = -V\left(\frac{\partial p}{\partial V}\right)_T. \tag{16.73}$$

Using now the result for pressure p in terms of the free energy F, i.e.,

$$p = -\left(\frac{\partial F}{\partial V}\right)_T, \tag{16.74}$$

we obtain

$$\gamma^{-1} = V\left(\frac{\partial^2 F}{\partial V^2}\right)_T. \tag{16.75}$$

On the other hand at $T = 0$, the free energy $F = U + TS = U$ is equal to the internal energy U of the electron gas. Thus, we have

$$\gamma^{-1} = V\left(\frac{\partial^2 U}{\partial V^2}\right)_T. \tag{16.76}$$

The internal energy of a free electron gas at $T = 0$ is given by

$$U = \frac{3}{5}NE_F = \frac{3N\hbar^2}{10m}\left(3\pi^2\frac{N}{V}\right)^{2/3} = AV^{-2/3}, \tag{16.77}$$

where we define a parameter A as follows:

$$A = \frac{3N\hbar^2}{10m}\left(3\pi^2 N\right)^{2/3}. \tag{16.78}$$

Thus, we can calculate

$$\gamma^{-1} = V\frac{\partial^2}{\partial V^2}\left(AV^{-2/3}\right) = \frac{10}{9}AV^{-5/3}, \tag{16.79}$$

or finally

$$\gamma^{-1} = \frac{10U}{9V} = \frac{2NE_F}{3V} = \frac{2}{3}nE_F. \tag{16.80}$$

Problem 3

An ideal gas of spin-$\frac{1}{2}$ fermions, of density $n = N/V$ at a temperature T has the zero kinetic energy state half-filled (for *each* of the two spin orientations). Determine the chemical potential μ and the temperature T of the gas exactly.

Solution

The probability of finding a fermion in the zero kinetic energy state, i.e., the mean occupation number of the zero kinetic energy state (for *each* of the two spin orientations), is given by the equation (15.25) with the positive sign in the denominator and with $E = 0$

$$\bar{n}_0 = \frac{1}{\exp\left(\frac{0-\mu}{kT}\right) + 1} = \frac{1}{2}, \tag{16.81}$$

and in this case is equal to $1/2$ since this state is said to be exactly half-filled. From the equation (16.81), we see that at this particular temperature, the chemical potential of the fermion gas must be equal to zero $\mu = 0$. The temperature T is now determined using the definition of the number of spin-$\frac{1}{2}$ fermions (16.17), which for $\mu = 0$ reads

$$N = \frac{1}{2\pi^2}\left(\frac{2m}{\hbar^2}\right)^{3/2} V \int_0^\infty \frac{E^{1/2}dE}{\exp\left(\frac{E}{kT}\right) + 1}. \tag{16.82}$$

Introducing a new variable $w = E/kT$, we obtain

$$N = \frac{1}{2\pi^2}\left(\frac{2mkT}{\hbar^2}\right)^{3/2} V \int_0^\infty \frac{w^{1/2}dw}{e^w + 1}. \tag{16.83}$$

The integral in (16.83) is solved using (15.79) for $x = 3/2$, and we have

$$N = \frac{1}{2\pi^2}\left(\frac{2mkT}{\hbar^2}\right)^{3/2} V(1 - 2^{-1/2})\Gamma\left(\frac{3}{2}\right)\zeta\left(\frac{3}{2}\right), \tag{16.84}$$

where $\Gamma(3/2) = \sqrt{\pi}/2$ and $\zeta(3/2) \approx 2.612$. Introducing all dimensionless numerical constants and dividing by V, we obtain

$$n = \frac{N}{V} = 0.09715 \times \left(\frac{mkT}{\hbar^2}\right)^{3/2}. \tag{16.85}$$

Finally, the expression for the temperature of the fermion gas reads

$$T = 4.732 \times \frac{\hbar^2 n^{2/3}}{mk}. \tag{16.86}$$

17 Photon Gas in Equilibrium

17.1 Planck Distribution

One of the most important applications of the quantum statistics for bosons is the description of the electromagnetic radiation, i.e., a photon gas in thermal equilibrium. This radiation is usually called the *black-body radiation*. Maxwell's equations describing electromagnetic field dynamics equations are linear equations, which imply that there are no interactions between photons in media that can be modeled by linear electric and magnetic susceptibilities, e.g., in vacuum. Thus, a photon gas typically fulfills the requirements for an ideal gas. Photons are bosons with integral spin $\sigma = 1$. From the usual rule for spin angular momentum, we would thus expect $g(\sigma) = 2\sigma + 1 = 3$ different spin states for photons. However, due to a symmetry of the electromagnetic field equations (related to the fact that photon has zero rest mass), the state with zero-spin projection onto the direction of photon motion turns out to be forbidden. Thus, there are only two allowed spin states, i.e., we have $g(\sigma) = 2$.

Unlike the ordinary gases, the photon gas does not have a constant number of constituent particles ($dN \neq 0$). The equilibrium of a photon gas is maintained by constant absorption and emission of photons by the surrounding matter, e.g., by the walls of a closed container (cavity). The total number of photons inside the cavity is thus not constant in time. Because of this, the condition (15.18) does not apply to the photon gas and has to be omitted. Using the method of Lagrange multipliers, the two remaining equilibrium conditions (15.17), with the $+$ sign, as appropriate for bosons and (15.19) are, i.e.,

$$dU = \sum_i E_i dn_i = 0 \quad / \cdot \frac{1}{T}, \tag{17.1}$$

$$dS = k \sum_i dn_i \ln \frac{g_i + n_i}{n_i} = 0 \quad / \cdot (-1), \tag{17.2}$$

given the following equation

$$\sum_i \left(\frac{E_i}{T} - k \ln \frac{g_i + n_i}{n_i} \right) dn_i = 0. \tag{17.3}$$

Eq. (17.3) has to be valid for any choice of dn_i, so the prefactor of each dn_i therein has to vanish. This yields the result,

$$n_i = \frac{g_i}{\exp\left(\frac{E_i}{kT}\right) - 1}. \tag{17.4}$$

Comparing the result (17.4) with the general result (15.24), we conclude that the photon gas is a boson gas with zero chemical potential ($\mu = 0$). Assuming that the volume of the photon gas is large, we can use the continuum from the result (17.4) to obtain the number of photons within the energy interval $(E, E + dE)$,

$$dn_E = \frac{dg(E)}{\exp\left(\frac{E}{kT}\right) - 1}. \tag{17.5}$$

In order to calculate $dg(E)$, we recall the result (15.46), i.e.,

$$dg(p) = g(\sigma)\frac{V}{h^3}4\pi p^2 dp. \tag{17.6}$$

It should be noted here that photons are massless ($m = 0$) relativistic particles and that the relation between energy E and momentum $\vec{p}$ is not given by the result (15.45) used in the general analysis of quantum gases in Chapter 15, "Quantum Distribution Functions." The relation between energy E and momentum $\vec{p}$ of the gas particles has an impact on the general thermodynamic behavior of the quantum gas. It is, therefore, not possible to apply the limit $\mu \to 0$ of the general results obtained in Chapter 15, to the photon gas. In other words, for the extremely relativistic photon gas, we need to redo some of the analysis made in Section 15.4, "Thermodynamics of Quantum Systems." The relativistic relation between energy E and momentum $\vec{p}$ of a massive particle is given by

$$E = \sqrt{p^2 c^2 + m^2 c^4}. \tag{17.7}$$

For massless particles, we obtain a very simple relation $E = pc$. If we now apply this relation to the general quantum–mechanical result (2.4), we can write

$$\vec{p} = \hbar\vec{\kappa} = \frac{h}{2\pi}\vec{\kappa}, \quad E = \hbar\omega = \hbar|\vec{\kappa}|c = \hbar\kappa c. \tag{17.8}$$

Using $g(\sigma) = 2$ and the equations (17.6) and (17.8), we obtain

$$dg(\kappa) = 2\frac{V}{h^3}4\pi\left(\frac{h}{2\pi}\right)^3\kappa^2 d\kappa, \tag{17.9}$$

or

$$dg(\omega) = 2\frac{V}{(2\pi)^3}4\pi\frac{\omega^2 d\omega}{c^3} = \frac{V}{\pi^2 c^3}\omega^2 d\omega. \tag{17.10}$$

Substituting $E = \hbar\omega$ and (17.10) into (17.5), we obtain the number of photons in the interval of frequencies $(\omega, \omega + d\omega)$ corresponding to the energy interval $(E, E + dE)$ as follows:

$$dn_\omega = \frac{dg(\omega)}{\exp\left(\frac{\hbar\omega}{kT}\right) - 1} = \frac{V}{\pi^2 c^3} \frac{\omega^2 d\omega}{\exp\left(\frac{\hbar\omega}{kT}\right) - 1}. \tag{17.11}$$

The contribution to the internal energy of the photon gas dU_ω from photons in the frequency range $(\omega, \omega + d\omega)$ is then obtained as a product of the number of photons dn_ω in this range and the energy of an individual photon in this range $E = \hbar\omega$. Thus, we obtain the *Planck formula* for the spectral energy distribution of black-body radiation

$$dU_\omega = \frac{V\hbar}{\pi^2 c^3} \frac{\omega^3 d\omega}{\exp\left(\frac{\hbar\omega}{kT}\right) - 1}. \tag{17.12}$$

In the low-frequency limit $(\omega \to 0)$, we can use the approximation $\exp(x) \approx 1 + x$, valid for $x \ll 1$, to obtain

$$\frac{1}{\exp\left(\frac{\hbar\omega}{kT}\right) - 1} \approx \frac{kT}{\hbar\omega}. \tag{17.13}$$

Substituting (17.13) into (17.12), we obtain the classical *Rayleigh–Jeans formula* in the following form

$$dU_\omega = \frac{V\hbar}{\pi^2 c^3} \frac{kT}{\hbar\omega} \omega^3 d\omega = \frac{VkT}{\pi^2 c^3} \omega^2 d\omega. \tag{17.14}$$

If we now recall the result (17.10), we see that the classical formula (17.14) can be rewritten as follows:

$$dU_\omega = \left[\frac{V}{\pi^2 c^3} \omega^2 d\omega\right] kT = dg(\omega)kT. \tag{17.15}$$

The physical meaning of this result is that the contribution to the internal energy of the photon gas dU_ω from photons in the frequency range $(\omega, \omega + d\omega)$ is obtained as a product of the number of radiation modes (i.e., the "vibrational degrees of freedom") in this frequency range and the average classical internal energy per mode kT. This result can be obtained from the classical equipartition theorem, proviso each mode is interpreted as a classical harmonic oscillator with internal energy $U_{osc} = kT$ (see Section 11.4.6, "Equipartition Theorem"). On the other hand, we already know that the correct internal energy of a single quantum-mechanical oscillator is given by the equation (12.43), i.e.,

$$U_{osc} = \frac{\hbar\omega}{2} + \frac{1}{\exp\left(\frac{\hbar\omega}{kT}\right) - 1}. \tag{17.16}$$

Here, the first term, the so-called *vacuum-energy*, is temperature independent. Such term would not affect, for example, the constant volume specific heat. If one leaves the vacuum energy term in the above quantum–mechanical expression for U_{osc} and then uses it to replace the kT term (the classical U_{osc}) figuring in the equation (17.15), one regains the exact Planck formula as written in the equation (17.12). In fact, historically, this was the line of thinking pursued by Planck himself, who considered the above formula for U_{osc} *without* the vacuum energy term. From our previous derivation of the Planck formula based on the concept of photon gas (rather than on the oscillator picture), one can see that the vacuum energy term simply did not appear in the course of this theory. It should be stressed though that in various physical problems, the vacuum energy of this sort may play a certain role, as evidenced, for example, by modern cosmological studies of inflatory universe.

Going back to our discussions, we finally note that in the high-frequency limit $(\omega \to \infty)$, we can use the approximation $\exp(x) \gg 1$ to obtain the so-called *Wien formula* in the form

$$dU_\omega = \frac{V\hbar}{\pi^2 c^3} \exp\left(-\frac{\hbar\omega}{kT}\right) \omega^3 d\omega. \tag{17.17}$$

The formulae (17.14) and (17.17) had been suggested before the development of the quantum theory. They were applicable in their respective frequency limits, but they failed to provide an acceptable explanation of the entire frequency spectrum. The excellent agreement of the Planck formula (17.12) with the experimental data over the entire frequency spectrum was one of the early proofs of the validity of the quantum hypothesis.

17.2 Thermodynamics of Photon Gas in Equilibrium

The objective of the present section is to calculate various thermodynamic quantities for the photon gas. In order to analyze the thermodynamic properties of the photon gas, let us first recall the general relation between the free energy F and the Helmholtz thermodynamic potential Ω, given by the equation (10.40) with (10.16), i.e.,

$$\Omega = F - \Phi = F - N\mu \Rightarrow F = \Omega. \tag{17.18}$$

As the chemical potential for the photon gas is equal to zero ($\mu = 0$), we see that the free energy F of the photon gas is equal to its Helmholtz thermodynamic potential Ω. The total number of particles in the photon gas is equal to

$$N = \int dn_\omega = \frac{V}{\pi^2 c^3} \int_0^\infty \frac{\omega^2 d\omega}{\exp\left(\frac{\hbar\omega}{kT}\right) - 1}. \tag{17.19}$$

On the other hand, using the result (10.42), we can express the total number of particles of the photon gas in terms of the Helmholtz thermodynamic potential Ω as follows:

$$N = -\left(\frac{\partial \Omega}{\partial \mu}\right)_T . \tag{17.20}$$

Analogously to the result (15.43), we conclude that the Helmholtz thermodynamic potential Ω is given by

$$F = \Omega = \frac{VkT}{\pi^2 c^3} \int\limits_0^\infty \ln\left[1 - \exp\left(-\frac{\hbar\omega}{kT}\right)\right] \omega^2 d\omega. \tag{17.21}$$

Let us now introduce a new integration variable w as follows:

$$\frac{\hbar\omega}{kT} = w \Rightarrow \omega = \frac{kT}{\hbar} w \Rightarrow d\omega = \frac{kT}{\hbar} dw. \tag{17.22}$$

Substituting (17.22) into (17.21), we obtain

$$F = \Omega = \frac{VkT}{\pi^2 c^3}\left(\frac{kT}{\hbar}\right)^3 \int\limits_0^\infty \ln\left(1 - e^{-w}\right) w^2 dw, \tag{17.23}$$

or

$$F = \Omega = \frac{Vk^4 T^4}{\pi^2 c^3 \hbar^3} \int\limits_0^\infty \ln\left(1 - e^{-w}\right) w^2 dw. \tag{17.24}$$

The integral on the right-hand side of the equation (17.24) can be integrated by parts as follows:

$$I = \int\limits_0^\infty \ln\left(1 - e^{-w}\right) w^2 dw = \left[\frac{w^3}{3} \ln\left(1 - e^{-w}\right)\right]_0^\infty$$
$$- \frac{1}{3}\int\limits_0^\infty \frac{w^3 e^{-w} dw}{1 - e^{-w}} = -\frac{1}{3}\int\limits_0^\infty \frac{w^3 dw}{e^w - 1}. \tag{17.25}$$

Substituting (17.25) into (17.24), we obtain

$$F = \Omega = -\frac{Vk^4 T^4}{3\pi^2 c^3 \hbar^3} \int\limits_0^\infty \frac{w^3 dw}{e^w - 1}. \tag{17.26}$$

On the other hand, from the Planck formula, we obtain the following result for the total internal energy of the photon gas

$$U = \frac{V\hbar}{\pi^2 c^3} \int_0^\infty \frac{\omega^3 d\omega}{\exp\left(\frac{\hbar\omega}{kT}\right) - 1}.$$ (17.27)

Introducing again a new integration variable w, defined by (17.22), we obtain

$$U = \frac{V\hbar}{\pi^2 c^3} \left(\frac{kT}{\hbar}\right)^4 \int_0^\infty \frac{w^3 dw}{e^w - 1},$$ (17.28)

or

$$U(T, V) = \frac{Vk^4 T^4}{\pi^2 c^3 \hbar^3} \int_0^\infty \frac{w^3 dw}{e^w - 1}.$$ (17.29)

The equation (17.29) is the caloric-state equation for the photon gas. Comparing the results (17.26) and (17.29), we see that

$$\Omega = -pV = -\frac{U}{3} \Rightarrow pV = \frac{1}{3}U.$$ (17.30)

Thus, for an extremely relativistic gas like the photon gas, the thermal-state equation (15.59) is replaced by the equation (17.30). The value of the constant integral on the right-hand side of the equation (17.29) is given by (15.82), i.e.,

$$I(4) = \int_0^\infty \frac{w^3 dw}{e^w - 1} = \frac{\pi^4}{15}.$$ (17.31)

Substituting (17.31) into (17.29), we obtain the caloric-state equation for the photon gas in the form

$$U(T, V) = \frac{\pi^2 Vk^4 T^4}{15 c^3 \hbar^3}.$$ (17.32)

It is customary to introduce here the *Stefan–Boltzmann constant*, defined by

$$\sigma_{SB} = \frac{\pi^2 k^4}{60 \hbar^3 c^2} = \frac{2\pi^5 k^4}{15 h^3 c^2} = 5.667 \times 10^{-8} \frac{kg}{s^3 K^4}.$$ (17.33)

Substituting (17.33) into (17.32), we obtain

$$U(T, V) = \frac{4\sigma_{SB} VT^4}{c}.$$ (17.34)

The specific heat at constant volume is given by

$$c_V = \frac{1}{m}\left(\frac{\partial U}{\partial T}\right)_V = \frac{16\sigma_{SB}VT^3}{mc}. \tag{17.35}$$

From the results (17.30) and (17.34), we obtain the thermal-state equation for the photon gas

$$pV = \frac{4\sigma_{SB}VT^4}{3c}. \tag{17.36}$$

The free energy F of the photon gas is obtained in the form

$$F = \Omega = -pV = -\frac{4\sigma_{SB}VT^4}{3c}. \tag{17.37}$$

The entropy S of the photon gas is obtained using the general result $U = F + TS$ as follows:

$$S = \frac{U-F}{T} = \frac{-3F-F}{T} = -\frac{4F}{T} = \frac{16\sigma_{SB}VT^3}{3c}. \tag{17.38}$$

Thus, in an adiabatic process ($S =$ Constant), we have the condition $VT^3 =$ Constant. Using the thermal-state equation (17.36), we see that we also have $pV^{4/3} =$ Constant.

Using the caloric-state equation for the photon gas (17.34), we can calculate the rate of the radiation energy emerging from a perfect absorber in thermal equilibrium ("black body"), dQ/dt. Let us consider an area A, which is irradiated by the photons propagating at the angle θ relative to the normal to the area. During the time dt, the photons travel the distance cdt and sweep the volume $dV = Acdt\cos\theta$. The contribution to the internal energy of the photon gas from the photons within the volume dV, with the direction of motion within the solid angle element $dO = \sin\theta d\theta d\phi$ and within the frequency range $(\omega, \omega + d\omega)$, is given by

$$du(\omega,\theta,\phi) = dU_\omega \frac{dV}{V}\frac{dO}{4\pi}, \tag{17.39}$$

or

$$du(\omega,\theta,\phi) = dU_\omega \frac{Acdt\cos\theta}{V}\frac{\sin\theta d\theta d\phi}{4\pi}. \tag{17.40}$$

Thus, we have

$$\frac{du(\omega,\theta,\phi)}{dt} = dU_\omega \frac{Ac\cos\theta}{V}\frac{\sin\theta d\theta d\phi}{4\pi}. \tag{17.41}$$

The total rate of the radiation energy emerging from the black body is obtained by integrating the result (17.41) over the entire range of frequencies and space angles, i.e.,

$$\frac{dQ}{dt} = \frac{Ac}{4\pi V} \int_0^\infty dU_\omega \int_0^{\pi/2} \cos\theta \sin\theta d\theta \int_0^{2\pi} d\phi, \tag{17.42}$$

or

$$\frac{dQ}{dt} = \frac{Ac}{4V} U(V, T). \tag{17.43}$$

Substituting the caloric-state equation for the photon gas (17.34) into (17.43), we obtain the *Stefan–Boltzmann law* in the form

$$\frac{dQ}{dt} = \frac{Ac}{4V} \frac{4\sigma_{SB} V T^4}{c} = A\sigma_{SB} T^4. \tag{17.44}$$

Thus, the total radiant energy per second of a black body is obtained using the result (17.44).

17.3   Problems with Solutions

Problem 1

The energy density per unit volume emitted by a black body in the wavelength range $(\lambda, \lambda + d\lambda)$ is given by

$$dU_\lambda = u(\lambda)d\lambda = \frac{8\pi hc}{\lambda^5} \frac{d\lambda}{\exp\left(\frac{hc}{\lambda kT}\right) - 1}. \tag{17.45}$$

(a) Find the value of the wavelength λ_m, where the function $u(\lambda)$ has its maximum value.

(b) If we treat the Sun as a black body with $\lambda_m = 480$ nm, calculate the temperature of the Sun.

Solution

(a) The maximum of the function $u(\lambda)$ is obtained from the condition

$$\frac{du}{d\lambda} = \frac{8\pi h^2 c^2}{\lambda^7 kT} \frac{\exp\left(\frac{hc}{\lambda kT}\right)}{\left[\exp\left(\frac{hc}{\lambda kT}\right) - 1\right]^2} - \frac{8\pi hc}{\lambda^6} \frac{5}{\exp\left(\frac{hc}{\lambda kT}\right) - 1} = 0, \tag{17.46}$$

or

$$\frac{8\pi hc}{\lambda^6} \frac{1}{\exp\left(\frac{hc}{\lambda kT}\right) - 1} \left[\frac{hc}{\lambda kT} \frac{\exp\left(\frac{hc}{\lambda kT}\right)}{\exp\left(\frac{hc}{\lambda kT}\right) - 1} - 5\right] = 0, \tag{17.47}$$

or

$$\frac{hc}{\lambda kT} \frac{\exp\left(\frac{hc}{\lambda kT}\right)}{\exp\left(\frac{hc}{\lambda kT}\right) - 1} = 5, \tag{17.48}$$

or

$$\frac{hc}{\lambda kT} = 5\left[1 - \exp\left(-\frac{hc}{\lambda kT}\right)\right]. \tag{17.49}$$

Introducing here a new variable $w = \frac{hc}{\lambda kT}$, we obtain the equation

$$w = 5\left(1 - e^{-w}\right). \tag{17.50}$$

In order to find a numerical solution of this equation, we use the following iteration formula

$$w_{j+1} = 5\left[1 - \exp(-w_j)\right], \quad w_0 = 5. \tag{17.51}$$

After a few iteration steps, we obtain

$$w_m = \frac{hc}{\lambda_m kT} = 4.965. \tag{17.52}$$

Thus, we finally obtain the formula for the wavelength λ_m

$$\lambda_m = \frac{hc}{4.965kT}. \tag{17.53}$$

(b) For a given $\lambda_m = 480$ nm, the temperature of the Sun can be calculated as follows:

$$T = \frac{hc}{4.965\lambda_m k} = 6037\,\text{K}. \tag{17.54}$$

Problem 2

The Sun can be treated as a black body with an approximate surface temperature $T = 5800\,\text{K}$. The radius of the Sun is $R_S \approx 7 \times 10^8$ m and the distance from Sun–Earth is $R_{SE} \approx 1.5 \times 10^{11}$ m.

(a) Calculate the radiant energy per second (power) emitted by the Sun.
(b) Calculate the energy rate (power) per unit area that reaches the Earth's atmosphere.

Solution

(a) Using the Stefan–Boltzmann Law, we have

$$\frac{dQ}{dt} = A\sigma_{SB}T^4 = 4\pi R_S^2 \sigma_{SB}T^4 = 3.95 \times 10^{26}\,\text{W}. \tag{17.55}$$

(b) The energy rate per unit area that reaches the Earth's atmosphere is obtained as

$$\frac{dQ_E}{dt} = \frac{1}{4\pi R_{SE}^2}\frac{dQ}{dt} = 1.40\,\frac{\text{kW}}{\text{m}^2}. \tag{17.56}$$

Problem 3

The light from an ordinary lamp can approximately be treated as a black-body radiation. If the light from the lamp has a maximum intensity at a wavelength of $\lambda_m = 1200\,\text{nm}$, calculate the temperature of the lamp.

Solution

Using the formula (17.53) for the wavelength λ_m, i.e.,

$$\lambda_m = \frac{hc}{4.965kT}, \tag{17.57}$$

we obtain

$$T = \frac{hc}{4.965\lambda_m k} = 2415\,\text{K}. \tag{17.58}$$

Problem 4

Just below the surface of the Sun, photons are in thermal equilibrium at a temperature $T = 6000\,\text{K}$. The radiation into a certain space angle Ω can be described as a radiation through a surface $A = r^2\Omega$, where at the surface of the Sun, we have $r = R_S = 7 \times 10^8\,\text{m}$, being the radius of the Sun. Calculate the linear momentum of the photons radiated into the solid angle Ω per unit time. Assume that $\Omega \ll 4\pi$. If the radiation is totally reflected from an object covering this solid angle, calculate the force exerted by photons on the object. Assume that the object is a perfect reflector and that it is flat. Also assume that the photons strike the object perpendicularly. If a spaceship at the distance $r = R_{SE} = 1.5 \times 10^{11}\,\text{m}$ from the Sun wants to use such an object (solar sail) for acceleration, how large of an area of the solar sail needs to be in order to generate the acceleration $a = 0.01$ g? Total mass of the solar sail and the spaceship is $M = 1000$ kg.

Solution

The total radiant energy per second of the Sun is obtained using the result (17.44), i.e.,

$$\frac{dQ}{dt} = A_S \sigma_{SB} T^4 = 4\pi R_S^2 \sigma_{SB} T^4. \tag{17.59}$$

The energy radiated into a space angle Ω is then given by

$$\frac{dE_\Omega}{dt} = 4\pi R_S^2 \sigma_{SB} T^4 \frac{\Omega}{4\pi} = R_S^2 \Omega \sigma_{SB} T^4. \tag{17.60}$$

The linear momentum of a single photon is related to its energy by $p = E/c$. Thus, the momentum of radiation into a solid angle Ω per unit time is

$$\frac{dp_\Omega}{dt} = \frac{R_S^2 \Omega \sigma_{SB} T^4}{c}. \tag{17.61}$$

If the photons are totally reflected from an object, the linear momentum transferred from the photons to the object per unit time is twice the momentum given by (17.61). The force on the object caused by the total reflection of the photons is equal to this momentum transfer per unit time, i.e.,

$$F = Ma = \frac{2R_S^2 \Omega \sigma_{SB} T^4}{c}. \tag{17.62}$$

At a distance $r = R_{SE}$, the solid angle of such a solar sail with irradiated surface A is $\Omega = A/R_{SE}^2$. The equation (17.62) then gives

$$Ma = \frac{2R_S^2 \sigma_{SB} T^4}{c} \frac{A}{R_{SE}^2}. \tag{17.63}$$

The area of the solar sail needed to generate the given acceleration is thus

$$A = \frac{MacR_{SE}^2}{2R_S^2 \sigma_{SB} T^4} = 9200014 \, \text{m}^2 \sim 3 \, \text{km} \times 3 \, \text{km}. \tag{17.64}$$

18 Other Examples of Boson Systems

18.1 Lattice Vibrations and Phonons

The atoms or molecules in a solid-state specimen are frequently arranged in a periodic structure called the *crystal lattice*. In conducting materials, e.g., metals, the conducting electrons are free to move. They are delocalized from the lattice sites and they constitute an electron gas in the metal. The properties of the free electron gas have been studied in Chapter 16, "Electron Gases in Metals," using Fermi–Dirac statistics. The remaining positively charged ions are localized at the lattice sites. At zero temperature ($T = 0$), the ions are frozen in their equilibrium positions in the crystal lattice. However, at finite temperatures ($T > 0$), the ions are not at rest and they vibrate about their equilibrium positions.

The magnitude of displacements of ions from their equilibrium position is typically small compared with the lattice constant (the size of the unit cell of the lattice) so that the ion vibrations are approximately harmonic. In quantum theory, lattice vibrations can be described in terms of the particle-like objects (quasi-particles) called *phonons*. Conceptual relation between phonons and classical lattice vibrations is the same as the one between photons and classical electromagnetic fields. In both cases, the classical field can be represented as set of modes each modeled by a simple harmonic oscillator. The quantization of this field reduces thus to the quantization of these oscillators (see Chapter 4, "Energy of Vibrations,"). The n-th energy state of each of these oscillators, with energy $E = E_0 + n\hbar\omega$, can be depicted as a state occupied by n bosons each with energy $\hbar\omega$. In the case of lattice vibrations, these bosonic quasi-particles are called phonons, whereas in the case of electromagnetic fields, we call them photons. By being bosonic, both photons and phonons have to obey Bose–Einstein statistics discussed in Chapter 15, "Quantum Distribution Functions." Unlike the ordinary gases and similarly to the photon gas, the phonon gas does not have a constant number of constituent particles ($dN \neq 0$). Thus, similarly to the case of the photon gas, the chemical potential for phonons is equal to zero ($\mu = 0$).

Like a photon, a single phonon has the energy $E = \hbar\omega$, so the number of phonons in the interval of frequencies ($\omega, \omega + d\omega$), corresponding to the energy interval ($E, E + dE$), can be obtained using the result (17.11), obtained in the previous chapter for photons as follows:

$$dn_\omega = \frac{dg(\omega)}{\exp\left(\frac{\hbar\omega}{kT}\right) - 1}, \tag{18.1}$$

where $dg(\omega)$ is the number of states in the energy interval $(E, E + dE)$ corresponding to the interval of frequencies $(\omega, \omega + d\omega)$. In order to determine the number of states $dg(\omega)$ for phonons, we need to determine the dispersion relation $\omega = \omega(\vec{\kappa})$ for the lattice vibrations, where $\vec{\kappa} = \vec{p}/\hbar$ is the wave vector (related to the momentum) of phonons.

18.1.1 Vibration Modes

In order to discuss the dispersion relation $\omega = \omega(\vec{\kappa})$ for the lattice vibrations, let us consider a simple one-dimensional lattice with nearest-neighbor coupling only and with two ions per unit cell, as shown in Fig. 18.1. The two ions within a unit cell have different masses M_A and M_B, as shown in Fig. 18.1. The nearest neighbor interactions between ions can be modeled by harmonic springs with the spring constant g. The small displacements from the equilibrium position, due to the lattice vibrations, are denoted by u_A and u_B for ions with masses M_A and M_B, respectively. The classical Newtonian equations of motion for lattice vibrations are in this case given by

$$
\begin{aligned}
M_A \ddot{u}_{Aj} &= -g\left(u_{Aj} - u_{Bj}\right) - g\left(u_{Aj} - u_{Bj-1}\right), \\
M_B \ddot{u}_{Bj} &= -g\left(u_{Bj} - u_{Aj}\right) - g\left(u_{Bj} - u_{Aj+1}\right),
\end{aligned}
\tag{18.2}
$$

where j is the label of a particular unit cell. Here, $1 \le j \le N$, where N is the total number of unit cells in the one-dimensional lattice.

Much like for photon states, where wave functions are plane-wave solutions to Maxwell's equations, the wave functions of phonon states are plane wave solutions of the equations (18.2) that can be written as

$$
u_{Aj}(t) = u_A(\kappa)e^{-i\kappa R_j + i\omega t}, \tag{18.3}
$$

$$
u_{Bj}(t) = u_B(\kappa)e^{-i\kappa R_j + i\omega t}, \tag{18.4}
$$

where $R_j = j a$ is the position of the j-th unit cell, and κ and ω are the wave vector and (angular) frequency of the mode, respectively. By the above equations, it is easy to see that

$$
u_{Aj+1}(t) = e^{-i\kappa a}u_{Aj}(t), \quad u_{Bj-1}(t) = e^{+i\kappa a}u_{Bj}(t), \tag{18.5}
$$

$$
\ddot{u}_{Aj}(t) = -\omega^2 u_{Aj}(t), \quad \ddot{u}_{Bj}(t) = -\omega^2 u_{Bj}(t). \tag{18.6}
$$

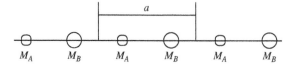

Figure 18.1 One-dimensional lattice with two ions per unit cell.

Inserting the above results into the equations (18.2) yields [after canceling the common factor $\exp(-i\kappa R_j + i\omega t)$]

$$\left(-M_A\omega^2 + 2g\right) u_A(\kappa) - g\left(1 + e^{i\kappa a}\right) u_B(\kappa) = 0, \tag{18.7}$$

$$- g\left(1 + e^{-i\kappa a}\right) u_A(\kappa) + \left(-M_B\omega^2 + 2g\right) u_B(\kappa) = 0, \tag{18.8}$$

or in the matrix form

$$\begin{bmatrix} -M_A\omega^2 + 2g & -g\left(1 + e^{i\kappa a}\right) \\ -g\left(1 + e^{-i\kappa a}\right) & -M_B\omega^2 + 2g \end{bmatrix} \begin{bmatrix} u_A(\kappa) \\ u_B(\kappa) \end{bmatrix} = 0. \tag{18.9}$$

This matrix equation has nontrivial solutions, if

$$\det \begin{bmatrix} -M_A\omega^2 + 2g & -g\left(1 + e^{i\kappa a}\right) \\ -g\left(1 + e^{-i\kappa a}\right) & -M_B\omega^2 + 2g \end{bmatrix} = 0, \tag{18.10}$$

or

$$\left(\omega^2 - \frac{2g}{M_A}\right)\left(\omega^2 - \frac{2g}{M_B}\right)$$
$$\frac{g^2}{M_A M_B}\left(1 + e^{i\kappa a}\right)\left(1 + e^{-i\kappa a}\right) = 0, \tag{18.11}$$

or

$$\omega^4 - \frac{2g}{M}\omega^2 + \frac{4g^2}{M_A M_B} - \frac{g^2}{M_A M_B}\left(2 + 2\cos\kappa a\right) = 0, \tag{18.12}$$

where

$$M = \frac{M_A M_B}{M_A + M_B}. \tag{18.13}$$

From (18.12), we further get

$$\omega^4 - \frac{2g}{M}\omega^2 + \frac{2g^2}{M_A M_B}\left(1 - \cos\kappa a\right) = 0, \tag{18.14}$$

or finally

$$\omega^4 - \frac{2g}{M}\omega^2 + \frac{4g^2}{M_A M_B}\sin^2\left(\frac{\kappa a}{2}\right) = 0. \tag{18.15}$$

The two solutions of this equation represent the dispersion relations for the two vibration modes $\omega_\pm^2 = \omega_\pm^2(\kappa)$, and they are given by

$$\omega_\pm^2(\kappa) = \frac{g}{M}\left\{1 \pm \left[1 - \frac{4M^2}{M_A M_B}\sin^2\left(\frac{\kappa a}{2}\right)\right]^{1/2}\right\}. \tag{18.16}$$

In the limit of small κ, we can use $\sqrt{1-x} \approx 1 - x/2$ as well as $\sin x \approx x$ to obtain

$$\omega_-^2(\kappa) \approx \frac{g}{M}\left\{1 - \left[1 - \frac{1}{2}\frac{4M^2}{M_A M_B}\left(\frac{\kappa a}{2}\right)^2\right]\right\},$$
(18.17)

or

$$\omega_-^2(\kappa) \approx \frac{g}{2(M_A + M_B)}a^2\kappa^2 \Rightarrow \omega_-(\kappa) \approx c_S\kappa,$$
(18.18)

where c_S is the speed of sound in this simple crystal lattice given by

$$c_S = \sqrt{\frac{g}{2(M_A + M_B)}}a.$$
(18.19)

On the other hand in the limit of small κ, we also have

$$\omega_+^2(\kappa) \approx \frac{2g}{M} \Rightarrow \omega_+(\kappa) \approx \sqrt{\frac{2g}{M}} = \text{Constant}.$$
(18.20)

The mode with the dispersion relation $\omega_-(\kappa) \approx c_S\kappa$ is called the *acoustic mode*, while the mode with the (nearly) constant dispersion relation $\omega_+(\kappa) \approx \text{Constant}$ is called the *optical mode*.

From the above analysis, it may appear that the choice of the wave vector κ is arbitrary. However, it is not so. To see this, let us look at the wave functions in the equations (18.3 and 18.4). They are of the form,

$$u_j(t, \kappa) = Ce^{-i\kappa R_j + i\omega(\kappa)t} = Ce^{-i\kappa aj + i\omega(\kappa)t}.$$
(18.21)

Using here the Euler formula ($e^{ix} = e^{ix+i2\pi}$), and the fact that, by equations (18.16) and (18.17),

$$\omega(\kappa) = \omega\left(\kappa + \frac{2\pi}{a}\right),$$
(18.22)

it can be easily seen that

$$u_j(t, \kappa) = u_j\left(t, \kappa + \frac{2\pi}{a}\right),$$
(18.23)

meaning that $u_j(t, \kappa)$ is periodic in κ, with the period $2\pi/a$. Thus, the modes with wave vectors κ and $\kappa + 2\pi/a$ are identical to each other. The modes, i.e., the phonon states that are different from each other, are to be confined within one period of $u_j(t, \kappa)$ as a function of κ, e.g., the range

$$-\frac{\pi}{a} \leq \kappa \leq +\frac{\pi}{a}.$$
(18.24)

This range in the wave vector space is known as the *first Brillouin zone*. Thus, the wave vectors of all physically different modes, i.e., of the phonon states, must be confined within the first Brillouin zone. Its existence is a direct consequence of the discreteness imposed by the presence of the crystal lattice. Indeed, in the continuum limit of zero lattice constant $a \to 0$, the size of the first Brillouin zone diverges, so it would cover the entire wave vector space. Such a situation is actually realized for photons, since the physical space can be thought of as a continuum limit of a three-dimensional lattice. In this limit, there are no restrictions on the magnitudes of photon wave vectors, as we tacitly assumed in Chapter 17, "Photon Gas in Equilibrium." Phonons, however, live on a crystalline lattice and the first Brillouin zone has always a finite size. Its existence must be taken into account, as done in our discussions in the following.

Another restriction on the possible values of wave vectors comes from a finite size of the crystalline sample, such as the length $L = Na$ of the one-dimensional crystal in our example above. Such restrictions are due boundary conditions imposed on the wave functions. In statistical physics, we are usually interested in large systems $(L \gg a)$, and the significant extensive quantities (e.g., internal energy) are insensitive to detailed form of boundary conditions. For the wave functions such as (18.21), the mathematically convenient form is the so-called *periodic boundary condition*,

$$u_j(t, \kappa) = u_{j+N}(t, \kappa). \tag{18.25}$$

By equations (18.21) and (18.25), one easily obtains the condition

$$e^{-i\kappa Na} = e^{-i\kappa L} = 1 \Rightarrow \kappa = \kappa_m = \frac{2\pi}{L} m, \tag{18.26}$$

where $m = 0, \pm 1, \pm 2, \pm 3, \dots$. By the equation (18.26), we see that the spectrum of allowed wave vectors is discrete, with the uniform spacing between the successive wave vectors given by

$$\kappa_{m+1} - \kappa_m = \frac{2\pi}{L}. \tag{18.27}$$

By equations (18.24) and (18.27), we can see that the total number of the allowed wave vectors within the first Brillouin zone is actually equal to the number of lattice cells,

$$\frac{\frac{2\pi}{a}}{\frac{2\pi}{L}} = \frac{L}{a} = N. \tag{18.28}$$

These wave vectors are uniformly distributed in the κ-space and within the interval $(\kappa, \kappa + d\kappa)$, there are

$$\frac{d\kappa}{\frac{2\pi}{L}} = \frac{L}{2\pi} d\kappa, \tag{18.29}$$

allowed wave vectors.

The above discussions can be easily generalized to lattices in any number of special dimensions d, in which atomic vibration displacements $\vec{u}(\vec{R}, t)$ are d-dimensional vectors. The vibration modes are again plane waves of the form,

$$\vec{u}(\vec{R}, t) = \vec{C} \exp\left(-i\vec{\kappa} \cdot \vec{R} + i\omega(\kappa)t\right). \tag{18.30}$$

In this case, the allowed wave vectors $\vec{\kappa}$ are d-dimensional vectors in a d-dimensional $\vec{\kappa}$-space. For a d-dimensional crystal shaped as a cube with side L, periodic boundary conditions demand that each component $\vec{\kappa}$ satisfies the equation (18.26). Due to this, the allowed wave vectors form, in the d-dimensional $\vec{\kappa}$-space, a cubic lattice with the lattice constant $2\pi/L$ [just as in equation (18.27)], so the d-dimensional $\vec{\kappa}$-space volume that comes per each allowed wave vector is the volume of a cube with side $2\pi/L$, i.e.,

$$\left(\frac{2\pi}{L}\right)^d = \frac{(2\pi)^d}{V}, \tag{18.31}$$

where $V = L^d$ is the volume of the crystal. Thus, the number of allowed wave vectors in the $\vec{\kappa}$-space volume element $d^d\kappa$ is

$$\frac{d^d\kappa}{\frac{(2\pi)^d}{V}} = \frac{V}{(2\pi)^d} d^d\kappa. \tag{18.32}$$

Due to this, for example, for $d = 3$, the number of allowed wave vectors within the spherical shell with radii κ and $\kappa + d\kappa$ is

$$dg(\kappa) = \frac{\text{shell volume}}{\frac{(2\pi)^3}{V}} = \frac{V}{(2\pi)^3} 4\pi\kappa^2 d\kappa. \tag{18.33}$$

For the so-called cubic d-dimensional lattices, the lattice cells are cubes with side a (= the lattice constant). For these lattices, the first Brillouin zone in wave vector space is a d-dimensional cube with the side $2\pi/a$ [just as in equation (18.24)], and its $\vec{\kappa}$-space volume is thus

$$V_{BZ} = \left(\frac{2\pi}{a}\right)^d = \frac{(2\pi)^d}{a^d} = \frac{(2\pi)^d}{v_{\text{cell}}}, \tag{18.34}$$

with $v_{\text{cell}} = a^d$, the real-space lattice cell volume. Thus, by equations (18.30) and (18.34), the total number of allowed wave vectors is

$$\frac{V_{BZ}}{\frac{(2\pi)^d}{V}} = \frac{\frac{(2\pi)^d}{v_{\text{cell}}}}{\frac{(2\pi)^d}{V}} = \frac{V}{v_{\text{cell}}} = N. \tag{18.35}$$

So, the number of allowed wave vectors is equal to the number of the crystal lattice cells N.

For noncubic lattices, the shape of the Brillouin zone is more complicated; however, the number of allowed wave vectors is still equal to the number of the crystal lattice cells N. Thus, the equation (18.35) applies to whatever lattice and in general,

$$V_{BZ} = \frac{(2\pi)^d}{v_{\text{cell}}} = (2\pi)^d \frac{N}{V}. \tag{18.36}$$

The Brillouin zone is never spherical for realistic crystals; however, as done for the first time by the Dutch physicist Peter Debye, it is sometimes convenient to approximate it by a spherical shape. Such a spherical Brillouin zone is called the *Debye sphere*, and its radius is chosen to respect the general rule stated in equation (18.36). For example, for $d = 3$, the radius κ_D of the Debye sphere is determined from the equation

$$\int_0^{\kappa_D} 4\pi \kappa^2 d\kappa = \frac{4\pi}{3} (\kappa_D)^3 = V_{BZ} = (2\pi)^3 \frac{N}{V}. \tag{18.37}$$

We will employ the above relation in the next section.

Thus, for any crystal, the number of allowed wave vectors is equal to the number of lattice cells N. For each of these wave vectors, there may be several phonon states (vibration modes) present. Indeed, for example, for the $d = 1$ lattice with two atoms per each cell, for each wave vector κ, we found two such modes (acoustic and optical modes). Thus, the total number of vibration modes is $2N$. For a $d = 1$ lattice, with v atoms in each cell, we would likewise find vN modes. In particular, with one atom per cell ($v = 1$), there are only acoustic modes, so there are N modes for such $d = 1$ lattices. For a general d-dimensional lattice with v atoms per each of the N cells (with the total of $vN = N_{\text{atoms}}$ atoms), the number of modes is $vNd = N_{\text{atoms}}d$, so for $d = 3$ lattices, there are $3vN = 3N_{\text{atoms}}$ modes. This rule reflects the following mathematical facts: in a d-dimensional lattice, atomic displacements are d-dimensional vectors. Any atomic configuration is representable as a superposition of a complete set of phonon modes with suitably chosen amplitudes. Thus, specifying all atomic displacement vectors (by giving their $vNd = N_{\text{atoms}}d$ co-ordinates) is in one-to-one correspondence with specifying the amplitudes of all phonon modes. This is mathematically possible only if the number of modes is equal to $vNd = N_{\text{atoms}}d$.

This result can be easily used to derive the classical prediction for the crystal internal energy due to its thermal vibrations. Each phonon mode can be associated with a simple harmonic oscillator with internal energy kT, according to the classical equipartition theorem (see Sections 11.4.6, "Equipartition Theorem" and 11.6.2 "Harmonic and Anharmonic Oscillators"). Thus, the classical internal energy is simply the number of modes times kT, that is,

$$U = d \cdot v \cdot NkT = d \cdot N_{\text{atoms}}kT, \tag{18.38}$$

for a d-dimensional crystal of N lattice cells with v atoms per cell. The corresponding heat capacity is

$$C_V = \left(\frac{\partial U}{\partial T}\right)_V = d \cdot v \cdot Nk = d \cdot N_{\text{atoms}}k.$$ (18.39)

For $d = 3$, this reduces to the equation

$$C_V = 3v \cdot Nk = 3N_{\text{atoms}}k$$ (18.40)

known as Dulong–Petit law. Experimentally, many crystals do seem to approach such a heat capacity at high temperatures. However, with decreasing temperature, it has been observed that heat capacity also decreases. This is a signature of quantum effects. Indeed, by the general result of Chapter 15 (see the Problem 4 therein), the heat capacity must approach zero in the zero temperature limit of a quantum-mechanical system.

Before proceeding we note that, for a general dimension d, the results (18.33) and (18.37) can be directly generalized. For example, for any d, the number of allowed wave vectors within the spherical shell with radii κ and $\kappa + d\kappa$ is

$$dg(\kappa) = \frac{\text{shell volume}}{\frac{(2\pi)^d}{V}} = \frac{V}{(2\pi)^d} S(d)\kappa^{d-1}d\kappa$$ (18.41)

with V, the volume of the d-dimensional system (e.g., $V = L^d$, for a cubic sample with the side L). In the equation (18.41), the $S(d)$ labels the surface area of the d-dimensional sphere with unit radius, given by the formula

$$S(d) = \frac{(2\pi)^{d/2}2^{1-d/2}}{\Gamma(d/2)},$$ (18.42)

reproducing the familiar results $S(3) = 4\pi$, $S(2) = 2\pi$, and $S(1) = 2$. The equation (18.42) is, however, meaningful even for noninteger d! By using it, one can, for example, generalize the equation (18.37) for the volume of the Debye sphere to any, even noninteger d,

$$\int_0^{k_D} S(d)\kappa^{d-1}d\kappa = \frac{S(d)}{d}(\kappa_D)^d = V_{\text{BZ}} = (2\pi)^d\frac{N}{V}.$$ (18.43)

There is a lot of contemporary interest, both experimental and theoretical, in low-dimensional systems with $d = 2$ and $d = 1$. In view of this, in the next section, we will discuss the quantum-statistical physics for systems in any number of dimensions. In addition, the extension to a noninteger d is also interesting in its own right, as exemplified by the renormalization group theories of critical points, as well as by the studies of low-dimensional systems. Some examples for this are provided through the problems accompanying this chapter.

18.1.2 Internal Energy of Lattice Vibrations

We proceed to discuss the quantum theory of lattice vibrations. By Section 18.1.1, "**Vibration Modes**," for each d-dimensional wave vector $\vec{\kappa}$, there are $d \cdot v$ modes (phonon states), for a lattice with v atoms per each of its N cells. Each of these modes has a different dispersion relation $\omega_\alpha(\vec{\kappa})$, where the subscripts $\alpha = 1, 2, \ldots, d \cdot v$ are used to label different modes. Each of these modes can be treated as quantum-mechanical harmonic oscillator with angular frequency $\omega_\alpha(\vec{\kappa})$ and energy levels given in equation (4.21). By equations (12.46) and (12.47), in thermal equilibrium, the oscillator contribution to the internal energy is

$$U_\alpha(\vec{\kappa}) = \frac{\hbar\omega_\alpha(\vec{\kappa})}{2} + <n_\alpha(\vec{\kappa})> \hbar\omega_\alpha(\vec{\kappa}), \tag{18.44}$$

or

$$U_\alpha(\vec{\kappa}) = \frac{\hbar\omega_\alpha(\vec{\kappa})}{2} + \frac{\hbar\omega_\alpha(\vec{\kappa})}{\exp\left(\frac{\hbar\omega_\alpha(\vec{\kappa})}{kT}\right) - 1}. \tag{18.45}$$

Here, the first term is the ground-state contribution, the so-called *zero-point ("vacuum") energy*. The second term in the equation (18.45) can be interpreted as the contribution of the phonons occupying the state labeled by $\vec{\kappa}$ and α. Each of them carries the energy $\hbar\omega_\alpha(\vec{\kappa})$, and their average occupation number for this state is

$$<n_\alpha(\vec{\kappa})> = \frac{1}{\exp\left(\frac{\hbar\omega_\alpha(\vec{\kappa})}{kT}\right) - 1}. \tag{18.46}$$

The equation (18.46) coincides with the Bose–Einstein statistics expression in the equation (15.25), with zero chemical potential. Phonons can be thus interpreted as bosons, which actual total number is fixed by temperature, much like we found for photons in Chapter 17.

The total vibrational internal energy is obtained by summing the contributions in the equation (18.45) over all modes. This involves summing the equation (18.45) over all N allowed values of $\vec{\kappa}$ within the first Brillouin zone, as well as summing over all values of $\alpha = 1, 2, \ldots, d \cdot v$. The number of allowed states within the d-dimensional $\vec{\kappa}$-space element is given by (18.32). Using this, we can sum the single-mode contributions in the equation (18.45) over all modes, to find the internal vibration energy of the crystal in the form,

$$U_{\text{VIB}}(T) = \sum_{\alpha=1}^{d \cdot v} V \int_{\text{BZ}} \frac{d^d\kappa}{(2\pi)^d} U_\alpha(\vec{\kappa}) = U_0 + U_{\text{phonon}}(T). \tag{18.47}$$

Above, the integrals are done over the first Brillouin zone. The vibration internal energy in the equation (18.47) is written as a sum of the term

$$U_0 = \sum_{\alpha=1}^{d \cdot v} V \int_{BZ} \frac{d^d \kappa}{(2\pi)^d} \frac{\hbar \omega_\alpha(\vec{\kappa})}{2} \tag{18.48}$$

representing the zero-point energy contribution, and the term

$$U_{phonon}(T) = \sum_{\alpha=1}^{d \cdot v} V \int_{BZ} \frac{d^d \kappa}{(2\pi)^d} < n_\alpha(\vec{\kappa}) > \hbar \omega_\alpha(\vec{\kappa})$$

$$= \sum_{\alpha=1}^{d \cdot v} V \int_{BZ} \frac{d^d \kappa}{(2\pi)^d} \frac{\hbar \omega_\alpha(\vec{\kappa})}{\exp\left(\frac{\hbar \omega_\alpha(\vec{\kappa})}{kT}\right) - 1}, \tag{18.49}$$

representing the contribution coming from thermally excited phonons. By the equation (18.49), this contribution vanishes for $T \to 0$, so the net vibration energy $U_{VIB}(T) \to U_0$ for $T \to 0$. The zero-point energy is thus a vibrational contribution to the systems ground-state energy. The zero-point energy $U_0 \sim \hbar$ is of purely quantum-mechanical origin, yet it may be significant for a crystal comprised of lighter atoms such as helium or argon. Being temperature independent, U_0 does not affect the experimentally significant constant volume heat capacity, which is governed by the phonon contribution given in the equation (18.49), i.e.,

$$C_{VIB} = \left(\frac{\partial U_{VIB}(T)}{\partial T}\right)_V = \left(\frac{\partial U_{phonon}(T)}{\partial T}\right)_V. \tag{18.50}$$

In the high-temperature limit, by using $\exp(x) \approx 1 + x$ valid for $x \ll 1$, we can see that the integrands in the equation (18.49) are all approximated by kT (the classical expression for the harmonic oscillator internal energy). Thus, for high enough temperatures, we have

$$U_{VIB}(T) \approx U_{phonon}(T) \approx \sum_{\alpha=1}^{d \cdot v} V \int_{BZ} \frac{d^d \kappa}{(2\pi)^d} kT$$

$$= d \cdot v \frac{V}{(2\pi)^d} V_{BZ} kT = d \cdot v \cdot N \cdot kT, \tag{18.51}$$

where we used the equation (18.36), for the first Brillouin zone volume, to calculate

$$\int_{BZ} d^d \kappa = V_{BZ} = (2\pi)^d \frac{N}{V}. \tag{18.52}$$

The corresponding high-temperature heat capacity is thus

$$C_{\text{VIB}} = \left(\frac{\partial U_{\text{VIB}}(T)}{\partial T} \right)_V = \left(\frac{\partial U_{\text{phonon}}(T)}{\partial T} \right)_V = d \cdot v \cdot N \cdot k, \tag{18.53}$$

coinciding with the classical Dulong–Pettit law in the equation (18.40), obtained from the equipartition theorem. Thus, the high-temperature limit of our quantum theory is the classical limit. Notably from the above derivation, detailed forms of the dispersion relations $\omega_\alpha(\vec{\kappa})$ simply disappear in the classical limit. It is thus clear that the dispersion relations $\omega_\alpha(\vec{\kappa})$ will play a significant role as the temperature decreases and quantum effects become prominent. In the following, we will show that the difference between the dispersion relations for the acoustic ($\omega_\alpha(\vec{\kappa}) \to 0$ as $\kappa = |\vec{\kappa}| \to 0$) and the optical modes ($\omega_\alpha(\vec{\kappa} = 0) \neq 0$) plays a significant role at low temperatures. For d-dimensional crystals with v atoms per lattice cell, for each $\vec{\kappa}$, there are d acoustic modes and $d \cdot (v - 1)$ optical modes, as exemplified in Section 18.1.1, "**Vibration Modes**," for the case $d = 1$ and $v = 2$. Thus, in crystals with only one atom per cell ($v = 1$), all d modes are acoustic. These are essentially modes of sound, with the velocities c_α, i.e.,

$$\omega_\alpha(\vec{\kappa}) = c_\alpha |\vec{\kappa}| = c_\alpha \kappa, \quad \alpha = 1, 2, \dots, d, \tag{18.54}$$

for small $|\vec{\kappa}| = \kappa \ll \pi/a$, i.e., close to the center of the first Brillouin zone. On the other hand, the remaining $d \cdot (v - 1)$ optical modes have

$$\omega_\alpha(\vec{\kappa}) = \omega_\alpha(\vec{\kappa} = 0) \neq 0, \quad \alpha = d+1, d+2, \dots, d \cdot v, \tag{18.55}$$

for small $|\vec{\kappa}| = \kappa \ll \pi/a$. By the equation (18.49), one can separate the internal energy contributions coming from acoustic and optical modes,

$$U_{\text{ac}}(T) = U_{\text{ac}}(T) + U_{\text{opt}}(T), \tag{18.56}$$

with the acoustic mode contribution,

$$U_{\text{ac}}(T) = \sum_{\alpha=1}^{d} V \int_{\text{BZ}} \frac{d^d \kappa}{(2\pi)^d} \frac{\hbar \omega_\alpha(\vec{\kappa})}{\exp\left(\frac{\hbar \omega_\alpha(\vec{\kappa})}{kT} \right) - 1}, \tag{18.57}$$

and the optical mode contribution,

$$U_{\text{opt}} = \sum_{\alpha=d+1}^{d \cdot v} V \int_{\text{BZ}} \frac{d^d \kappa}{(2\pi)^d} \frac{\hbar \omega_\alpha(\vec{\kappa})}{\exp\left(\frac{\hbar \omega_\alpha(\vec{\kappa})}{kT} \right) - 1}. \tag{18.58}$$

In the so-called *Einstein model*, motivated by the equation (18.55), the dispersion of optical modes is completely ignored, so in the equation (18.58), one sets $\omega_\alpha(\vec{\kappa}) = \omega_\alpha(\vec{\kappa} = 0)$ throughout the first Brillouin zone. With this, integrands in the equation (18.58) become $\vec{\kappa}$-independent and the integrals can be calculated exactly as follows:

$$
\begin{aligned}
U_{\text{opt}} &= \sum_{\alpha=d+1}^{d \cdot v} V \int_{\text{BZ}} \frac{d^d \kappa}{(2\pi)^d} \frac{\hbar \omega_\alpha(0)}{\exp\left(\frac{\hbar \omega_\alpha(0)}{kT}\right) - 1} \\
&= \sum_{\alpha=d+1}^{d \cdot v} \frac{\hbar \omega_\alpha(0)}{\exp\left(\frac{\hbar \omega_\alpha(0)}{kT}\right) - 1} \frac{V}{(2\pi)^d} \int_{\text{BZ}} d^d \kappa \\
&= \sum_{\alpha=d+1}^{d \cdot v} \frac{\hbar \omega_\alpha(0)}{\exp\left(\frac{\hbar \omega_\alpha(0)}{kT}\right) - 1} \frac{V}{(2\pi)^d} V_{\text{BZ}} \\
&= N \sum_{\alpha=d+1}^{d \cdot v} \frac{\hbar \omega_\alpha(0)}{\exp\left(\frac{\hbar \omega_\alpha(0)}{kT}\right) - 1},
\end{aligned}
\tag{18.59}
$$

where we used the equation (18.36), for the first Brillouin zone volume, to calculate

$$
\int_{\text{BZ}} d^d \kappa = V_{\text{BZ}} = (2\pi)^d \frac{N}{V}.
\tag{18.60}
$$

By the equation (18.59), at low temperatures ($T \to 0$),

$$
U_{\text{opt}} \approx N \sum_{\alpha=d+1}^{d \cdot v} \hbar \omega_\alpha(0) \exp\left(-\frac{\hbar \omega_\alpha(0)}{kT}\right).
\tag{18.61}
$$

Thus, as temperature goes to zero, the optical mode contributions go to zero rapidly, in an exponential fashion. By the equation (18.61), the associated heat capacity also rapidly goes zero , as

$$
C_{\text{opt}} = \left(\frac{\partial U_{\text{opt}}(T)}{\partial T}\right)_V \approx Nk \sum_{\alpha=d+1}^{d \cdot v} \left(\frac{\hbar \omega_\alpha(0)}{kT}\right)^2 \exp\left(-\frac{\hbar \omega_\alpha(0)}{kT}\right).
\tag{18.62}
$$

On the other hand, as it will be shown in the following, the low-temperature contribution coming from acoustic modes yields a contribution of the form

$$
U_{\text{ac}} \sim T^{d+1} \Rightarrow C_{\text{ac}} = \left(\frac{\partial U_{\text{ac}}(T)}{\partial T}\right)_V \sim T^d,
\tag{18.63}
$$

which approaches zero only as a power law of temperature and thus dominates over the optical mode contribution in the equations (18.61) and (18.62). Thus, at low enough temperatures,

$$
U_{\text{phonon}}(T) = U_{\text{ac}}(T) + U_{\text{opt}}(T) \approx U_{\text{ac}}(T) \sim T^{d+1},
\tag{18.64}
$$

and

$$C_{\text{phonon}}(T) = C_{\text{ac}}(T) + C_{\text{opt}}(T) \approx C_{\text{ac}}(T) \sim T^d. \tag{18.65}$$

The result in the equation (18.63) [for $d = 3$] and the special role played by acoustic phonons were revealed by Debye in early years of quantum-statistical physics. He observed that, at low temperatures, the dominant contribution to the integrals in the equation (18.57) comes from a range of short wave vectors such that $\hbar \omega_\alpha(\vec{\kappa}) = \hbar c_\alpha \kappa \sim kT$, i.e.,

$$\kappa \sim \frac{kT}{\hbar c_\alpha}. \tag{18.66}$$

Indeed the integrands in the equation (18.57) are exponentially depressed for wave vectors much larger than those in (18.66). Since the wave vector scale in the equation (18.66) becomes small at low temperatures, Debye further realized that one can obtain asymptotically correct results (for the $T \to 0$ limit), just by using in the equation (18.57), the short wave vector (long wavelength) form of the acoustic modes dispersion relation,

$$\omega_\alpha(\vec{\kappa}) = c_\alpha |\vec{\kappa}| = c_\alpha \kappa, \quad \alpha = 1, 2, \ldots, d. \tag{18.67}$$

Indeed, consider the temperatures much smaller than the characteristic temperature (the so-called *Debye temperature*), defined by

$$T_\alpha = \frac{\hbar c_\alpha}{k} \kappa_D, \tag{18.68}$$

with κ_D, the radius of Debye sphere (serving here just as a measure of the first Brillouin zone size). By the equation (18.68), we see that, for $T \ll T_\alpha$, the dominant wave-vector range in the equation (18.66) satisfies

$$\kappa \sim \frac{kT}{\hbar c_\alpha} \ll \frac{kT_\alpha}{\hbar c_\alpha} = \kappa_D \Rightarrow \kappa \ll \kappa_D. \tag{18.69}$$

In fact, by equations (18.66) and (18.68), for the dominant wave-vectors, we find

$$\frac{\kappa}{\kappa_D} \sim \frac{kT}{\hbar c_\alpha \kappa_D} = \frac{T}{\hbar c_\alpha \kappa_D / k} = \frac{T}{T_\alpha} \Rightarrow \frac{\kappa}{\kappa_D} \sim \frac{T}{T_\alpha} \ll 1. \tag{18.70}$$

Thus, for $T \ll T_\alpha$, the integrals in the equation (18.57) are dominated by a small portion of the first Brillouin zone, which is close to its center, where the approximation (18.54) is valid. Thus, in the calculation of the acoustic-phonons internal energy in (18.57), at low enough temperatures, one can employ the simple dispersion relations in the equation (18.67), as eventually Debye did. In view of the conclusion in the equation (18.70), the contributions to the integrals in (18.57) coming from the outer boundaries of the first Brillouin zone are expected to be small. So, at low enough temperatures, one can simply assume that the Brillouin zone has infinite size in the wave vector space. This corresponds to a crystalline lattice with zero lattice constant. In physical terms, it means that the asymptotic law gives the low-temperature phonon internal energy density

$$\tilde{u}_{\text{phonon}} = \frac{U_{\text{phonon}}}{V} \tag{18.71}$$

will eventually come out in a form that does not depend on the microscopic details such as the lattice constant. In other words, we expect this quantity to depend only on the temperature, but it should be independent of the lattice cell sizes, in the $T \to 0$ limit. This situation is dramatically different from the one we have in the classical high-temperature limit, where, by the equation (18.38), we have the result

$$\tilde{u}_{\text{phonon}} = \frac{d \cdot v \cdot NkT}{V} = \frac{d \cdot v \cdot kT}{V/N} = \frac{d \cdot v \cdot kT}{v_{\text{cell}}}, \tag{18.72}$$

in which the dependence on the crystal lattice cell v_{cell} volume is manifest. With decreasing temperature, dependence on microscopic details will simply disappear, in the asymptotically valid laws giving the internal energy density at low temperatures.

To show this, we fulfill the program outlined above, by importing the dispersion relations (18.67) into the equation (18.57). A nice feature of this step is that the integrands in (18.57) become functions of the magnitude of the wave vector only, so one can replace the d-dimensional integrals therein with the integrals over κ only, via

$$\int_{\text{BZ}} d^d\kappa \to \int_0^{\kappa_D} S(d)\kappa^{d-1} d\kappa. \tag{18.73}$$

In the equation (18.73), we anticipated that, in our calculation, we will replace the original Brillouin zone with a "spherical Brillouin zone," i.e., the Debye sphere discussed in Section 18.1.1, "**Vibration Modes**." We recall that this approximation does not change the volume of the Brillouin zone. Because of this, it is assured that our final result will have the correct classical (high-temperature) limit for the acoustic-phonon contributions,

$$U_{\text{ac}}(T) = d \cdot NkT, \tag{18.74}$$

which relies only on the volume of the Brillouin zone but not on its shape or on the form of dispersion relation [see, e.g., our derivations in (18.52)]. In addition, by the above discussions (18.71), we also expect that our result will be asymptotically exact also in the low-temperature limit. In this limit, the actual shape and size of the first Brillouin zone does not matter. It may thus be set to be of infinite size, corresponding to setting $\kappa_D \to \infty$ in the equation (18.73). Yet, to obtain a theory working also in the classical regime, one needs to keep κ_D finite, to assure that the equation (18.74) holds.

It should be stressed though that in the intermediary temperature range (in between the quantum and classical regimes), the quantitative validity of applying the concept of Debye sphere and simple dispersion relations (18.67) is questionable. In the intermediary regime, the detailed forms of dispersion relations and of the Brillouin zone do matter. However, here, we are primarily interested in the low-temperature limit in which the theory results are asymptotically exact. Thus, we import equations (18.67) and (18.73) into the equation (18.57) to obtain the major equation of the Debye theory,

$$U_{ac} = \sum_{\alpha=1}^{d} V \int_0^{\kappa_D} \frac{S(d)\kappa^{d-1}d\kappa}{(2\pi)^d} \frac{\hbar c_\alpha \kappa}{\exp\left(\frac{\hbar c_\alpha \kappa}{kT}\right) - 1}. \tag{18.75}$$

In the high-temperature limit, by using $e^x \approx 1 + x$ valid for $x \ll 1$, we see that the equation (18.75) reduces to

$$U_{ac} = \sum_{\alpha=1}^{d} V \int_0^{\kappa_D} \frac{S(d)\kappa^{d-1}d\kappa}{(2\pi)^d} kT$$

$$= \sum_{\alpha=1}^{d} \frac{V}{(2\pi)^d} \int_0^{\kappa_D} S(d)\kappa^{d-1}d\kappa\, kT$$

$$= d \cdot \frac{V}{(2\pi)^d} V_{BZ} kT = d \cdot N \cdot kT, \tag{18.76}$$

where we used the equation (18.60) for the last steps. Thus, as expected, the Debye theory has the correct classical, high-temperature limit given by the equation (18.74). At low temperatures, the equation (18.66) suggests the following change of variable κ into a variable w, via

$$\kappa = \frac{kT}{\hbar c_\alpha} w, \tag{18.77}$$

transforming the equation (18.75) into

$$U_{ac} = \sum_{\alpha=1}^{d} V \frac{(kT)^{d+1}}{(\hbar c_\alpha)^d} \frac{S(d)}{(2\pi)^d} \int_0^{w_\alpha} \frac{w^d dw}{e^w - 1}. \tag{18.78}$$

Here, the upper limits of the integrals are given by

$$w_\alpha = \frac{\hbar c_\alpha}{kT} \kappa_D = \frac{\hbar c_\alpha \kappa_D / k}{T} = \frac{T_\alpha}{T}, \tag{18.79}$$

where we recalled the definition of the Debye temperatures in the equation (18.68). The integrals in (18.78), all converge in the low-temperature limit, i.e., when

$$w_\alpha = \frac{T_\alpha}{T} \to \infty, \tag{18.80}$$

which is, by the equation (18.79), equivalent to the formal limit $\kappa_D \to \infty$, in accordance with our previous discussions. In this limit, any information on microscopic details, such as the size of the Debye sphere, i.e., lattice constant, simply disappear, and the equation (18.78) yields the universal form for the internal vibration energy density

$$\tilde{u}_{ac}(T) = \frac{U_{ac}(T)}{V} = K(d)\frac{(kT)^{d+1}}{\hbar^d}\sum_{\alpha=1}^{d}\frac{1}{(c_\alpha)^d}, \tag{18.81}$$

and the associated heat capacity per unit volume, $\tilde{c}_{ac} = C_{ac}/V$, i.e.,

$$\tilde{c}_{ac}(T) = \frac{\partial \tilde{u}_{ac}}{\partial T} = (d+1)\cdot K(d)k\frac{(kT)^d}{\hbar^d}\sum_{\alpha=1}^{d}\frac{1}{(c_\alpha)^d}. \tag{18.82}$$

The results in equations (18.81) and (18.82) are asymptotically valid for $T \ll T_\alpha$, for whatever shape and size of the first Brillouin zone. In equations (18.81) and (18.82),

$$K(d) = \frac{S(d)}{(2\pi)^d}\int_0^\infty \frac{w^d dw}{e^w - 1} \tag{18.83}$$

is a universal constant, which depends only on the dimensionality of the system. By equations (18.81) and (18.82),

$$\tilde{u}_{ac}(T) \sim T^{d+1}, \quad \tilde{c}_{ac}(T) \sim T^d \tag{18.84}$$

as anticipated before in the equation (18.63). So, the acoustic modes contribution indeed dominates over that optical modes, as argued in the equation (18.65). Whereas independent of microscopic (atomic) details such as lattice constants, the results (18.81) and (18.82) do depend only on macroscopic properties of the system, such as the temperature and experimentally measurable velocities c_α ($\alpha = 1, 2, \ldots, d$) of the d-long-wave length acoustic (sound) modes. We note that, for high-symmetry ("isotropic") crystals and glasses, there is one longitudinal acoustic mode, with the velocity c_L, and $d-1$ transverse acoustic modes, all with the velocity c_T, so in the equations (18.81) and (18.82), we have

$$\sum_{\alpha=1}^{d}\frac{1}{(c_\alpha)^d} = \frac{1}{(c_L)^d} + \frac{d-1}{(c_T)^d}. \tag{18.85}$$

For such system, there are thus only two Debye temperatures as defined by the equation (18.68),

$$T_L = \frac{\hbar c_L}{k} \kappa_D, \quad T_T = \frac{\hbar c_T}{k} \kappa_D, \tag{18.86}$$

whereas the equation (18.78) gives

$$\tilde{u}_{ac} = \frac{U_{ac}}{V} = \frac{(kT)^{d+1}}{\hbar^d} \frac{S(d)}{(2\pi)^d} \left[\frac{1}{(c_L)^d} \int_0^{w_L} \frac{w^d dw}{e^w - 1} + \frac{d-1}{(c_T)^d} \int_0^{w_T} \frac{w^d dw}{e^w - 1} \right]. \tag{18.87}$$

Here, by the equation (18.79),

$$w_L = \frac{\hbar c_L}{kT} = \frac{T_L}{T}, \quad w_T = \frac{\hbar c_T}{kT} = \frac{T_T}{T}. \tag{18.88}$$

Even though rarely realized in realistic systems, the above results are sometimes simplified by the assumption that $c_L \approx c_T$. Within this, "single sound velocity" approximation ($c_L = c_T = c_S$), there is only one Debye temperature,

$$T_D = \frac{\hbar c_S}{k} \kappa_D \Rightarrow w_D = \frac{\hbar c_S}{kT} = \frac{T_D}{T}. \tag{18.89}$$

Thus, the equations (18.75) and (18.78) give

$$\tilde{u}_{ac} = d \cdot \int_0^{\kappa_D} \frac{S(d)\kappa^{d-1} d\kappa}{(2\pi)^d} \frac{\hbar c_S \kappa}{\exp\left(\frac{\hbar c_S \kappa}{kT}\right) - 1} = \frac{(kT)^{d+1}}{(\hbar c_S)^d} \frac{d \cdot S(d)}{(2\pi)^d} \int_0^{w_D} \frac{w^d dw}{e^w - 1}. \tag{18.90}$$

The equation (18.90) is frequently displayed in the $d = 3$ form (see Problem 1), in which case, we use $S(d = 3) = 4\pi$. The $d = 2$ and $d = 1$ cases are also experimentally interesting. For them, $S(d = 2) = 2\pi$ and $S(d = 1) = 4\pi$ can be used to write down the equations such as the equation (18.75) and other equations in the above discussion. The theory can be extended even to noninteger dimensions d, by using the general result for $S(d)$ displayed in the equation (18.42), i.e.,

$$S(d) = \frac{(2\pi)^{d/2} 2^{1-d/2}}{\Gamma(d/2)}. \tag{18.91}$$

It can be used to evaluate the universal constant of the low-temperature behavior in the equation (18.83) as follows:

$$K(d) = \frac{S(d)}{(2\pi)^d} \int_0^\infty \frac{w^d dw}{e^w - 1} = \frac{2^{1-d/2}}{(2\pi)^{d/2}\Gamma(d/2)} \int_0^\infty \frac{w^d dw}{e^w - 1}. \tag{18.92}$$

Here, by the equation (15.72),

$$\int_0^\infty \frac{w^d dw}{e^w - 1} = \Gamma(d+1)\zeta(d+1), \tag{18.93}$$

so, in general dimensions d, the universal constant can be expressed in terms of a Gamma function and a Riemann Zeta function (15.73) as follows:

$$K(d) = \frac{2^{1-d/2}\Gamma(d+1)\zeta(d+1)}{(2\pi)^{d/2}\Gamma(d/2)}. \tag{18.94}$$

For the case $d = 3$, the integral in the equation (18.93) has already been calculated in (15.82), and by the same procedure, one can also calculate the integral (18.93) for $d = 1$, by using the equation (15.80), and the Bernoulli number in the equation (15.82). Thus, one finds

$$\int_0^\infty \frac{w^1 dw}{e^w - 1} = \Gamma(1+1)\zeta(1+1) = I(2) = \frac{(2\pi)^2}{4\cdot 1}B_2 = \frac{\pi^2}{6}, \tag{18.95}$$

which can be combined with the equation (18.92), to find (by recalling that $\Gamma(1/2) = \sqrt{\pi}$),

$$K(d=1) = \frac{2^{1-1/2}}{(2\pi)^{1/2}\Gamma(1/2)}\frac{\pi^2}{6} = \frac{\pi}{6}. \tag{18.96}$$

In a similar way, by using

$$\int_0^\infty \frac{w^3 dw}{e^w - 1} = \Gamma(3+1)\zeta(3+1) = I(4) = \frac{(2\pi)^4}{8}B_4 = \frac{\pi^4}{15} \tag{18.97}$$

in combination with the equation (18.92), we find

$$K(d=3) = \frac{2^{1-3/2}}{(2\pi)^{3/2}\Gamma(3/2)}\frac{\pi^4}{15} = \frac{\pi^2}{30}. \tag{18.98}$$

For $d = 2$, the equation (18.94) yields

$$K(d=2) = \frac{2^{1-2/2}\Gamma(2+1)\zeta(2+1)}{(2\pi)^{2/2}\Gamma(2/2)} = \frac{\zeta(3)}{\pi}. \tag{18.99}$$

Here,

$$\zeta(3) = \sum_{n=1}^\infty \frac{1}{n^3} = 1.2020569031595\ldots, \tag{18.100}$$

is the so-called *Apery's constant*, for which only the above numerical value is known. Thus, numerically,

$$K(d = 2) = \frac{\zeta(3)}{\pi} = 0.38262659603\ldots. \tag{18.101}$$

To summarize the results on the universal low-temperature behavior of phonon internal energy, let us introduce an average sound velocity, defined via

$$\frac{d}{(c_S)^d} = \sum_{\alpha=1}^{d} \frac{1}{(c_\alpha)^d}, \tag{18.102}$$

or, for the case of isotropic solids, by the equation (18.85), via

$$\frac{d}{(c_S)^d} = \frac{1}{(c_L)^d} + \frac{d-1}{(c_T)^d}. \tag{18.103}$$

The low-temperature universal results for the acoustic-phonon internal energy density $\tilde{u}_{ac}(T)$ and heat capacity per unit volume $\tilde{c}_{ac}(T) = C_{ac}/V$ in the equations (18.81) and (18.82), can thus be written in the simple forms

$$\tilde{u}_{ac}(T) = \frac{U_{ac}(T)}{V} = d \cdot K(d) \frac{(kT)^{d+1}}{(\hbar c_S)^d}, \tag{18.104}$$

and

$$\tilde{c}_{ac}(T) = \frac{\partial \tilde{u}_{ac}}{\partial T} = d \cdot (d+1) \cdot K(d) k \frac{(kT)^d}{(\hbar c_S)^d}, \tag{18.105}$$

respectively. Combining equations (18.104) and (18.105), with the above results for $K(d)$, we find the following:

For $d = 1$,

$$\tilde{u}_{ac}(T) = \frac{U_{ac}(T)}{V} = \frac{\pi}{6} \frac{(kT)^2}{\hbar c_S} \sim T^2, \tag{18.106}$$

and

$$\tilde{c}_{ac}(T) = \frac{\partial \tilde{u}_{ac}}{\partial T} = \frac{\pi}{3} k \frac{kT}{\hbar c_S} \sim T. \tag{18.107}$$

For $d = 2$,

$$\tilde{u}_{ac}(T) = \frac{U_{ac}(T)}{V} = \frac{2\zeta(3)}{\pi} \frac{(kT)^3}{(\hbar c_S)^2} \sim T^3, \tag{18.108}$$

and

$$\tilde{c}_{ac}(T) = \frac{\partial \tilde{u}_{ac}}{\partial T} = \frac{6\zeta(3)}{\pi} k \left(\frac{kT}{\hbar c_S}\right)^2 \sim T^2. \tag{18.109}$$

For $d = 3$,

$$\tilde{u}_{ac}(T) = \frac{U_{ac}(T)}{V} = \frac{\pi^2}{10} \frac{(kT)^4}{(\hbar c_S)^3} \sim T^4, \tag{18.110}$$

and

$$\tilde{c}_{ac}(T) = \frac{\partial \tilde{u}_{ac}}{\partial T} = \frac{2\pi^2}{5} k \left(\frac{kT}{\hbar c_S}\right)^3 \sim T^3. \tag{18.111}$$

It should finally be stressed that these results, as they stand, apply only if all d acoustic modes are in the quantum regime. It means, for $T \ll \min(T_\alpha)$, that is, by the equation (18.68), for

$$T \ll \min(T_\alpha) = \frac{\hbar \min(c_\alpha)}{k} \kappa_D. \tag{18.112}$$

In other words, the acoustic mode with the smallest sound velocity imposes the upper temperature limit for the applicability of the above universal results. Typically, this is the transverse mode, so $\min(c_\alpha) = c_T$.

18.2   Bose–Einstein Condensation

Energy of an ideal gas is simply the sum of the energies of individual particles, each having the kinetic energy

$$E(\kappa) = \frac{p^2}{2m} = \frac{\hbar^2 \kappa^2}{2m}. \tag{18.113}$$

Unlike the classical ideal gases, the ideal gases of fermions have a nonzero kinetic energy even at zero absolute temperature, i.e., in their N-particle ground state (see Chapter 16). The reason for this is the Pauli exclusion principle, which states that not more than one fermion can occupy the same quantum state. Because of this principle, the fermion gas ground state cannot be formed by putting all N fermions into the single state with zero kinetic energy [with $E(\kappa = 0) = 0$]. Rather, all energy states up to the Fermi energy level must be occupied. Unlike the case of fermions, any number of bosons can occupy a single quantum state. Because of this, the boson gas ground state can be formed simply by putting all N bosons into the state with zero kinetic energy [with $E(\kappa = 0) = 0$]. Thus, at $T = 0$, there is a *macroscopic fraction* $(= 1)$ of bosons occupying a *single quantum state*. In this section, we will discuss a closely related (but not really obvious) phenomenon of the so-called *Bose–Einstein condensation* that starts to occur *already* at nonzero temperatures. It turns out that a finite *macroscopic fraction* of bosons starts to occupy the $\kappa = 0$ state already below a certain *nonzero* critical temperature T_C. At T_C, this fraction is zero, and, as the temperature decreases from T_C down to 0, this fraction continuously increases from 0 to 1.

Much like in our discussion of phonons, the phenomenon of Bose–Einstein condensation can be studied in any number of spatial dimensions d, such as $d = 3$ (e.g., N bosons in a cubic box with the volume $V = L^3$), $d = 2$ (e.g., N bosons in a square with the area $A = L^2$), and $d = 1$ (e.g., N bosons confined to a line segment with the length L^1). Here, we will focus on the case $d = 3$, whereas a general d case is discussed in the Problem 2 of this chapter. Here, we note that for $d = 2$ and any $d < 2$ (in particular, for $d = 1$), the critical temperature T_C of the Bose–Einstein condensation is depressed to zero. Thus, in these dimensions, $T_C = 0$, i.e., there is no real Bose–Einstein condensation. On the other hand, for any $d > 2$ (in particular, for $d = 3$), the critical temperature T_C of the Bose–Einstein condensation is nonzero. Thus, for any $d > 2$, there is a real Bose–Einstein condensation. In order to study the phenomenon of Bose–Einstein condensation in $d = 3$, let us consider the equation (15.52) in the case of spinless bosons, i.e.,

$$N = \frac{m^{3/2}V}{\sqrt{2}\pi^2\hbar^3} \int\limits_0^\infty \frac{\sqrt{E}dE}{\exp\left(\frac{E-\mu}{kT}\right) - 1}. \tag{18.114}$$

It should be noted that the chemical potential μ for bosons is either negative or in some cases (e.g. for massless photons), it is equal to zero. A positive chemical potential cannot occur for ideal gases of bosons. It is because the state occupation numbers must be positive for all states, whereas a negative μ would mean that these numbers are negative in the range $E_i < \mu$ for (positive) single-particle kinetic energies E_i, i.e.,

$$\bar{n}_i = \frac{n_i}{g_i} = \frac{1}{\exp\left(\frac{E_i-\mu}{kT}\right) - 1} < 0, \quad E_i < \mu. \tag{18.115}$$

Let us now consider the process of lowering the temperature of a boson system at constant particle density N/V. In order to keep the particle density N/V constant, the integral on the right-hand side of the equation (18.114) must remain constant as T is decreasing. This requires that $|\mu|$ also decreases when the temperature T is decreasing. Interestingly, for the bosons in $d = 3$, the equation (18.115) predicts that μ approaches zero at a *nonzero* temperature T_C. The critical temperature T_C is determined from the condition that $\mu \to 0$ from the negative values. Using (18.114) with $T = T_C$ and $\mu = 0$, we have

$$N = \frac{m^{3/2}V}{\sqrt{2}\pi^2\hbar^3} \int\limits_0^\infty \frac{\sqrt{E}dE}{\exp\left(\frac{E}{kT_C}\right) - 1}. \tag{18.116}$$

Introducing a new variable $w = E/kT_C$, we obtain

$$N = \frac{V}{4\pi^2} \left(\frac{2mkT_C}{\hbar^2}\right)^{3/2} \int_0^\infty \frac{w^{1/2}dw}{e^w - 1},$$ (18.117)

or

$$4\pi^2 n = \left(\frac{2mkT_C}{\hbar^2}\right)^{3/2} \Gamma\left(\frac{3}{2}\right) \zeta\left(\frac{3}{2}\right),$$ (18.118)

where $n = N/V$, $\Gamma(3/2) = \sqrt{\pi}/2$, and $\zeta(3/2) \approx 2.612$. Thus, we obtain the following result for the critical temperature T_C, as a function of the density of the boson system

$$T_C = 3.3128 \times \frac{\hbar^2}{mk} n^{2/3}.$$ (18.119)

Note that, up to a numerical factor, the critical temperature T_C in the equation (18.119) is the same as the gas quantum temperature in the equation (15.136).

Let us now consider an ideal boson gas at temperatures below the critical temperature ($T < T_C$), where $\mu = 0$. As discussed in Problem 2, for $T < T_C$, the equation (18.114), with $\mu = 0$, actually gives the total number of particles in the excited states, i.e., the ones having kinetic energies $E > 0$. Their state occupation numbers are as in the equation (18.115), with, for $T < T_C$, the chemical potential set to zero. All the other particles are in the state with $E = 0$. Let us denote the number of these "condensed" particles, with $E = 0$, by N_0. So the total number of particles in the excited states, with $E > 0$, is $N - N_0$. As the equation (18.114) with $\mu = 0$ is valid only for these $N - N_0$ particles in the excited states, we can write

$$N - N_0 = \frac{m^{3/2}V}{\sqrt{2}\pi^2\hbar^3} \int_0^\infty \frac{\sqrt{E}dE}{\exp\left(\frac{E}{kT}\right) - 1}.$$ (18.120)

Introducing a new variable $w = E/kT$, we obtain

$$N - N_0 = \frac{V}{4\pi^2} \left(\frac{2mkT}{\hbar^2}\right)^{3/2} \int_0^\infty \frac{w^{1/2}dw}{e^w - 1},$$ (18.121)

or

$$N - N_0 = \frac{V}{4\pi^2} \left(\frac{2mkT_C}{\hbar^2}\right)^{3/2} \left(\frac{T}{T_C}\right)^{3/2} \int_0^\infty \frac{w^{1/2}dw}{e^w - 1}.$$ (18.122)

Comparing the equations (18.117) and (18.122) gives

$$N - N_0 = N \left(\frac{T}{T_C} \right)^{3/2}. \tag{18.123}$$

From the result (18.123), we obtain the number of condensed bosons N_0 at some temperature $T < T_C$, in the form

$$N_0 = N \left[1 - \left(\frac{T}{T_C} \right)^{3/2} \right]. \tag{18.124}$$

From the equation (18.124), we see that a finite macroscopic fraction ($= N_0/N$) of bosons starts to occupy the $E = 0$ state already below the nonzero critical temperature T_C. By (18.124), at $T = T_C$, this fraction is zero and, as the temperature decreases from T_C down to 0, this fraction continuously increases from 0 to 1. For example, at $T = T_C/2^{2/3}$, using the equation (18.124), we find that $N_0/N = 1/2$, that is, the 50 % of *all bosons* are in the zero kinetic energy state with $E = 0$. Thus, the Bose–Einstein condensation is an example of the so-called *macroscopic quantum phenomena*. Other notable examples are superfluidity and superconductivity. Historically, however, the Bose–Einstein theory outlined here, for an ideal boson gas, has served as a somewhat academic yet very important paradigm preceding the theories of all other macroscopic quantum phenomena. Typical values of T_C in the equation (18.119) are in the sub-Kelvin regime, where interparticle interactions would have to be included in the discussion. Typically, with cooling a gas, the interactions preclude the Bose–Einstein condensation, by causing an ordinary condensation of the gas into a liquid, well before the above T_C is reached.

Finally, let us now discuss the internal energy and pressure of an ideal boson gas below the critical temperature. Since the condensed N_0 particles all have $E = 0$, only the noncondensed fraction of bosons (in the states with $E > 0$) contributes to the internal energy of the system. Their state occupation numbers are as in the equation (18.115), with, for $T < T_C$, the chemical potential set to zero. Therefore, we can use the result (15.54) with $\mu = 0$ in the case of spinless bosons, to obtain the internal energy of a partly condensed boson system

$$U(V, T) = \frac{m^{3/2} V}{\sqrt{2} \pi^2 \hbar^3} \int_0^\infty \frac{E^{3/2} dE}{\exp\left(\frac{E}{kT} \right) - 1}. \tag{18.125}$$

Introducing a new variable $w = E/kT$, we obtain

$$U(V, T) = \frac{VkT}{4\pi^2} \left(\frac{2mkT}{\hbar^2} \right)^{3/2} \int_0^\infty \frac{w^{3/2} dw}{e^w - 1}, \tag{18.126}$$

or

$$U(V, T) = \frac{VkT}{4\pi^2} \left(\frac{2mkT}{\hbar^2}\right)^{3/2} \Gamma\left(\frac{5}{2}\right) \varsigma\left(\frac{5}{2}\right) \sim T^{5/2}. \tag{18.127}$$

Thus, the specific heat at constant volume for any $T < T_C$ is given by an exact result as follows:

$$c_V = \left(\frac{\partial U}{\partial T}\right)_V, \tag{18.128}$$

or

$$c_V = \frac{5Vk}{8\pi^2} \left(\frac{2mkT}{\hbar^2}\right)^{3/2} \Gamma\left(\frac{5}{2}\right) \varsigma\left(\frac{5}{2}\right) \sim T^{3/2}. \tag{18.129}$$

We see that the specific heat at constant volume at $T < T_C$ obeys a $T^{3/2}$ law. It should be noted here that the pressure of the system by the equation (15.59), i.e.,

$$p(T) = \frac{2U}{3V} = \frac{2kT}{12\pi^2} \left(\frac{2mkT}{\hbar^2}\right)^{3/2} \Gamma\left(\frac{5}{2}\right) \varsigma\left(\frac{5}{2}\right), \tag{18.130}$$

is not dependent on the density N/V of the system, and it is a function of temperature only for $T < T_C$. Thus, an isothermal compression (or expansion) of the system does not cause any change of the pressure of the system for $T < T_C$. This striking behavior of the system pressure for $T < T_C$ is in sharp contrast to the behavior for $T > T_C$ when the pressure depends on both the density and the temperature of the system. Recall, for example, that for $T \gg T_C \sim T_{\text{quantum}}$, we have the ideal gas law $p = (N/V)kT$, exemplifying the pressure dependence on the particle density.

18.3   Problems with Solutions

Problem 1

In $d = 3$, the internal energy in the simplified Debye model of lattice vibrations (with all acoustic mode velocities set to be c_S) is given by the equation (18.90) [recall that $S(d = 3) = 4\pi$ and $\omega = c_S\kappa$]

$$U = \frac{V\hbar}{2\pi c_S^3} \int_0^{\omega_D} \frac{\omega^3 d\omega}{\exp\left(\frac{\hbar\omega}{kT}\right) - 1}. \tag{18.131}$$

Determine the energy of the phonons that make the largest contribution to the internal energy of a silicon crystal at $T_1 = 30\,\text{K}$ and at the Debye temperature of silicon $T_2 = T_D = 625\,\text{K}$.

Solution

Up to a multiplicative constant, the integrand in the equation (18.131) can be written in the form

$$f(w) = \frac{\left(\frac{\hbar\omega}{kT}\right)^3}{\exp\left(\frac{\hbar\omega}{kT}\right) - 1} = \frac{w^3}{e^w - 1}, \quad w = \frac{\hbar\omega}{kT}. \tag{18.132}$$

The energy of the phonons that make the largest contribution to the internal energy of a crystal is obtained from the maximum of the function $f = f(w)$, which is obtained from the condition

$$f'(w) = \frac{3w^2}{e^w - 1} - \frac{w^3}{(e^w - 1)^2} e^w = 0, \tag{18.133}$$

or

$$f'(w) = \frac{w^2}{(e^w - 1)^2} \left[3(e^w - 1) - we^w\right] = 0, \tag{18.134}$$

or

$$3(e^w - 1) - we^w = 0 \Rightarrow w = 3(1 - e^{-w}). \tag{18.135}$$

This equation can be solved numerically using the following algorithm

$$w_{j+1} = 3[1 - \exp(-w_j)], \quad j = 1, 2, 3, \ldots, \tag{18.136}$$

and we obtain the result $w_m = 2.821439357 \approx 2.82$. At $T_1 = 30\,\text{K}$, the energy of the phonons that make the largest contribution to the internal energy of a crystal becomes

$$\hbar\omega_1 = 2.82 \times kT_1 = 1.16748 \times 10^{-21}\,\text{J}, \tag{18.137}$$

Similarly at $T_2 = 625\,\text{K}$, we have

$$\hbar\omega_2 = 2.82 \times kT_2 = 2.43225 \times 10^{-20}\,\text{J}. \tag{18.138}$$

Problem 2

Consider a d-dimensional problem of N zero-spin bosons in a d-dimensional hypercube with the side L and the hypervolume $V = L^d$. For $d = 1$, this hypercube is just a line segment of length L; for $d = 2$, it is a square of area L^2, and for $d = 3$, it is an ordinary three-dimensional cube of the volume L^3. Let $n = N/V$ denote the particle number density. The bosons are nonrelativistic with kinetic energies,

$$E(|\vec{k}|) = \frac{|\vec{p}|^2}{2m} = \frac{\hbar^2|\vec{k}|^2}{2m}. \tag{18.139}$$

The energies in (18.139) are obtained by solving the Schrödinger equation with periodic boundary conditions, in which case the solutions are plane waves, with allowed wave vectors $\vec{\kappa}$ forming a hypercubic lattice with the lattice constant $2\pi/L$, as discussed in Section 18.1.1, "Vibration Modes." The state with the wave vector $\vec{\kappa}$ is (on average) occupied with $\bar{n}(\vec{\kappa})$ bosons, given by the Bose–Einstein formula

$$\bar{n}(\vec{\kappa}) = \frac{1}{\exp\left(\frac{E(\kappa)-\mu}{kT}\right)-1} = \frac{1}{Y\exp\left(\frac{E(\kappa)}{kT}\right)-1}, \tag{18.140}$$

where $\kappa = |\vec{\kappa}|$ and $Y = \exp(-\mu/kT)$. The dimensionless quantity Y actually plays a more fundamental role than μ itself. To assure that the occupation numbers in the equation (18.140) are all positive, even for $\vec{\kappa} = 0$, whence $\bar{n}(\vec{\kappa} = 0) = (Y-1)^{-1}$, it is required that $Y > 1$ or equivalently $\mu < 0$.

(a) From the condition $Y \to 1$, show that the critical temperature for the Bose–Einstein condensation is given by the expression

$$T_C = Q(d)T_{\text{quantum}}, \quad T_{\text{quantum}} = \frac{2\pi\hbar^2}{mk}n^{2/d}. \tag{18.141}$$

Here, the T_{quantum} is the same as the characteristic quantum temperature defined in the equation (15.136), expressed there in terms of $h = 2\pi\hbar$. Show that the numerical factor $Q(d)$ in the equation (18.141) is given by a simple formula

$$Q(d) = \frac{1}{[\zeta(d/2)]^{2/d}}, \quad \zeta(x) = \sum_{n=1}^{\infty}\frac{1}{n^x}, \quad (x > 1), \tag{18.142}$$

where we note that the Zeta function in (18.142) is finite only for $x = d/2 > 1$, and it diverges for $x = d/2 \to 1$, as

$$\zeta(x) = \frac{1}{x-1} + \text{finite terms}. \tag{18.143}$$

Use the equation (18.143) and the result (15.72), i.e.,

$$I(x) = \int_0^{\infty}\frac{w^{x-1}dw}{e^w-1} = \Gamma(x)\sum_{n=1}^{\infty}\frac{1}{n^x} = \Gamma(x)\zeta(x), \tag{18.144}$$

to show that, for d slightly above 2, one has

$$Q(d) = \frac{d-2}{2} \Rightarrow T_C = \frac{d-2}{2}T_{\text{quantum}} \to 0, \tag{18.145}$$

for $d \to 2$. Thus, $T_C \to 0$ for $d \to 2$ from above. This implies the physically important conclusion that real Bose–Einstein condensation, with $T_C > 0$, occurs only for $d > 2$, for example, for $d = 3$. On the other hand, for $d = 2$ and for any $d < 2$ (such as $d = 1$), the critical temperature is depressed to zero, $T_C = 0$, so there is no real Bose–Einstein condensation at nonzero temperatures.

(b) For $d > 2$ and $T < T_C$, show that the number of the bosons occupying the lowest energy state ($\vec{\kappa} = \vec{0}$), i.e.,

$$N_0 = \bar{n}(\vec{\kappa} = \vec{0}) = \frac{1}{Y - 1}, \tag{18.146}$$

satisfies the relation

$$\frac{N_0}{N} = 1 - \left(\frac{T}{T_C}\right)^{d/2}. \tag{18.147}$$

Further, show that, for $T < T_C$, the gas internal energy density (per unit volume), as well as the corresponding heat capacity (per unit volume), do not depend on the gas density n but only on its temperature T, as

$$\tilde{u} = \frac{U}{V} \sim T^{1+d/2}, \quad \tilde{c}_V = \frac{\partial \tilde{u}}{\partial T} \sim T^{d/2}. \tag{18.148}$$

Remark: It can be shown that for any d and any temperature T (along the lines of Section 15.4, "Thermodynamics of Quantum Systems," done there for $d = 3$), the pressure p satisfies $p = \frac{2}{d}\tilde{u}$. So, by the result (18.148), for $T < T_C$, the pressure does not depend on the density n, but only on the temperature T, as

$$p \sim \tilde{u} \sim T^{1+d/2} \quad \text{for} \quad T < T_C. \tag{18.149}$$

(c) Next, consider the system for $d = 2$ and for any $d < 2$. By the above discussion, there is no finite-temperature Bose–Einstein condensation. In other words, in the limit $N = nV \to \infty$, at fixed density n and temperature T, the quantity Y [entering the equation (18.140)] must be given by an expression, which approaches 1 only for $T \to 0$, i.e., $Y \to 1$ only for $T \to 0$. By definition, $Y = \exp(-\mu/kT)$ and the fact that $\ln(x) \approx x - 1$ for $|x - 1| \ll 1$, we have

$$\mu = -kT \ln(Y) \approx -kT(Y - 1) \quad \text{for} \quad |Y - 1| \ll 1. \tag{18.150}$$

Thus, we also expect that $\mu \to 0$ only for $T \to 0$ (for $d = 2$ and for any $d < 2$).

In order to verify the above expectations
(a) Show that, for $d = 2$, the quantity Y is given by the following exact expression

$$Y = \frac{1}{1 - \exp\left(-\frac{T_{\text{quantum}}}{T}\right)}, \tag{18.151}$$

such that for $T \ll T_{\text{quantum}}$, we have

$$Y - 1 = \frac{\exp\left(-\frac{T_{\text{quantum}}}{T}\right)}{1 - \exp\left(-\frac{T_{\text{quantum}}}{T}\right)} \approx \exp\left(-\frac{T_{\text{quantum}}}{T}\right) \to 0, \tag{18.152}$$

which in view of the equation (18.150), indeed shows that $Y \to 1$ and that

$$\mu \approx -kT(Y - 1) = -kT \exp\left(-\frac{T_{\text{quantum}}}{T}\right) \to 0, \tag{18.153}$$

only in the limit $T \to 0$.

(b) Show that, for $d < 2$ and $T \ll T_{\text{quantum}}$, the quantity Y is given by the asymptotically exact expression

$$Y - 1 \approx A(d)\left(\frac{T}{T_{\text{quantum}}}\right)^{d/(2-d)}$$

$$\Rightarrow \mu \approx -kTA(d)\left(\frac{T}{T_{\text{quantum}}}\right)^{d/(2-d)}, \tag{18.154}$$

which indeed demonstrates that $Y \to 1$ and $\mu \to 0$ only in the limit $T \to 0$. In the equation (18.154), the numerical prefactor is given by

$$A(d) = \left[\frac{\pi}{\Gamma(d/2)\sin(\pi d/2)}\right]^{2/(2-d)}. \tag{18.155}$$

Solution

(a) By the equation (18.140),

$$N = \sum_{\vec{\kappa}} \bar{n}(\vec{\kappa}) = \sum_{\vec{\kappa}} \frac{1}{Y \exp\left(\frac{E(|\vec{\kappa}|)}{kT}\right) - 1}. \tag{18.156}$$

For $T \geq T_C$, the above sum is possible to handle as an integral, using the equations (18.32) and (18.41) of Section 18.1.1, "Vibration Modes." Thus, we obtain

$$N = \int_0^\infty \frac{dg(\kappa)}{Y \exp\left(\frac{E(\kappa)}{kT}\right) - 1}$$

$$= \frac{V}{(2\pi)^d} S(d) \int_0^\infty \frac{\kappa^{d-1} d\kappa}{Y \exp\left(\frac{E(\kappa)}{kT}\right) - 1}, \tag{18.157}$$

with

$$E(\kappa) = \frac{\hbar^2 \kappa^2}{2m} \Rightarrow \kappa = \frac{\sqrt{2mE}}{\hbar} \Rightarrow d\kappa = \frac{1}{\hbar} \left(\frac{m}{2E}\right)^{1/2} dE. \tag{18.158}$$

By changing the variable κ into the variable E in the equation (18.157), as suggested by (18.158), we obtain

$$N = V \cdot 2^{-1+d/2} S(d) \frac{m^{d/2}}{(2\pi\hbar)^d} \int_0^\infty \frac{E^{-1+d/2} dE}{Y \exp\left(\frac{E}{kT}\right) - 1}. \tag{18.159}$$

Changing further the variable E into the variable w, via $E = kTw$, yields

$$N = V \cdot 2^{-1+d/2} S(d) \frac{(mkT)^{d/2}}{(2\pi\hbar)^d} \int_0^\infty \frac{w^{-1+d/2} dw}{Y e^w - 1}. \tag{18.160}$$

As $T \to T_C$, one has $Y \to 1$ in (18.160). Thus,

$$N = V \cdot 2^{-1+d/2} S(d) \frac{(mkT_C)^{d/2}}{(2\pi\hbar)^d} \int_0^\infty \frac{w^{-1+d/2} dw}{e^w - 1}. \tag{18.161}$$

Using the result (18.144), we can calculate the integral in the equation (18.161) as follows:

$$\int_0^\infty \frac{w^{-1+d/2} dw}{e^w - 1} = \Gamma(d/2)\zeta(d/2), \tag{18.162}$$

whereas $S(d)$ is given by the equation (18.91), i.e.,

$$S(d) = \frac{(2\pi)^{d/2} 2^{1-d/2}}{\Gamma(d/2)}. \tag{18.163}$$

Substituting the results (18.162) and (18.163) into the equation (18.161), we obtain

$$N = V \left(\frac{mkT_C}{2\pi\hbar^2}\right)^{d/2} \zeta(d/2). \tag{18.164}$$

The equation (18.164) is easily solved for T_C, with the result

$$T_C = \frac{1}{[\zeta(d/2)]^{2/d}} \frac{2\pi\hbar^2}{mk} \left(\frac{N}{V}\right)^{2/d} = Q(d) T_{\text{quantum}}, \tag{18.165}$$

where we used the definition of T_{quantum} in the equation (18.141) and introduced the notation

$$Q(d) = \frac{1}{[\zeta(d/2)]^{2/d}}.$$ (18.166)

Thus, the result (18.165) with (18.166) proves the anticipated expressions (18.141) and (18.142). For $x = d/2$ slightly above 1, we have by the equation (18.143),

$$\zeta(d/2) = \frac{1}{d/2 - 1} + \text{finite terms} = \frac{2}{d - 2} + \text{finite terms}.$$ (18.167)

So, for d slightly above 2, we have by equations (18.166) and (18.167),

$$Q(d) \approx \frac{1}{\left[\frac{2}{d-2}\right]^{2/2}} = \frac{d - 2}{2}.$$ (18.168)

For d slightly above 2, by equations (18.165) and (18.168), we further have

$$T_C = Q(d)T_{\text{quantum}} \approx \frac{d - 2}{2} T_{\text{quantum}} \to 0 \quad \text{for} \quad d \to 2.$$ (18.169)

Thus, we have derived the equation (18.145) implying the major effect here that $T_C \to 0$ for $d \to 2$ from above.

(b) Let us consider the case $d > 2$ at $T < T_C$. The equation (18.156) can be written as

$$N = \bar{n}(0) + \sum_{\vec{\kappa} \neq 0} \bar{n}(\vec{\kappa}) = N_0 + \sum_{\vec{\kappa} \neq 0} \frac{1}{Y \exp\left(\frac{E(|\vec{\kappa}|)}{kT}\right) - 1}.$$ (18.170)

Here, the second term represents the total number of bosons in the excited states, with $\vec{\kappa} \neq 0$. For $d > 2$ and $T < T_C$, this term has the same value as the expression on the right-hand sides of equations (18.157), (18.159), and (18.160), with $Y = 1$ set therein (see the Comment at the end of this problem). Thus,

$$N = N_0 + V \cdot 2^{-1+d/2} S(d) \frac{(mkT)^{d/2}}{(2\pi \hbar)^d} \int_0^\infty \frac{w^{-1+d/2} dw}{e^w - 1},$$ (18.171)

or

$$N - N_0 = V \cdot 2^{-1+d/2} S(d) \frac{(mkT)^{d/2}}{(2\pi \hbar)^d} \int_0^\infty \frac{w^{-1+d/2} dw}{e^w - 1}.$$ (18.172)

Recalling here the equation (18.161), i.e.,

$$N = V \cdot 2^{-1+d/2} S(d) \frac{(mkT_C)^{d/2}}{(2\pi\hbar)^d} \int\limits_0^\infty \frac{w^{-1+d/2}dw}{e^w - 1} \tag{18.173}$$

and dividing the equation (18.172) by the equation (18.173), we easily see that

$$\frac{N - N_0}{N} = 1 - \frac{N_0}{N} = \left(\frac{T}{T_C}\right)^{d/2} \Rightarrow \frac{N_0}{N} = 1 - \left(\frac{T}{T_C}\right)^{d/2} \tag{18.174}$$

at $T < T_C$ for $d > 2$ whence $T_C > 0$. This proves the equation (18.147).

It is worthwhile noting that the statement $Y = 1$ is actually true in the limit $N = nV \to \infty$, at fixed density n and temperature $T < T_C$. This can be seen from the equation (18.140) with $E = 0$, by which

$$N_0 = \bar{n}(\vec{\kappa} = \vec{0}) = \frac{1}{Y-1} \Rightarrow Y = 1 + \frac{1}{N_0}. \tag{18.175}$$

Next, by the equation (18.174), we see, at a given $T < T_C$, that

$$Y - 1 = \frac{1}{N_0} \sim \frac{1}{N} \to 0 \quad \text{for} \quad N \to \infty. \tag{18.176}$$

In other words, $Y \to 1$ for $N \to \infty$, affirming the correctness of setting $Y = 1$ in our calculations above. The internal energy of the boson gas is calculated as follows:

$$U = \sum_{\vec{\kappa}} \bar{n}(\vec{\kappa}) E(|\vec{\kappa}|) = \sum_{\vec{\kappa}} \frac{E(|\vec{\kappa}|)}{Y \exp\left(\frac{E(|\vec{\kappa}|)}{kT}\right) - 1}. \tag{18.177}$$

Note that the $E(|\vec{\kappa} = \vec{0}|) = 0$, so only the bosons in the excited states contribute to the internal energy (at any temperature). For $T < T_C$, we can set $Y = 1$ in the equation (18.177) and handle the sum in (18.177) in the same way as we handled the sum in (18.156) [see the equations (18.157) through (18.160)]. The final result is

$$\tilde{u} = \frac{U}{V} = 2^{-1+d/2} S(d) kT \frac{(mkT)^{d/2}}{(2\pi\hbar)^d} \int\limits_0^\infty \frac{w^{d/2}dw}{e^w - 1}. \tag{18.178}$$

Thus, we have proved the equation (18.148), i.e.,

$$\tilde{u} = \frac{U}{V} \sim T^{1+d/2}, \quad \tilde{c}_V = \frac{\partial \tilde{u}}{\partial T} \sim T^{d/2}. \tag{18.179}$$

(c) To proceed, let us simplify the form of the result (18.160), by writing the temperature therein as $T = yT_{quantum}$. With this, and by the equation (18.141), we get

$$1 = y^{d/2} \frac{S(d)}{2\pi^{d/2}} \int_0^\infty \frac{w^{-1+d/2}dw}{Ye^w - 1},$$ (18.180)

or

$$\frac{2\pi^{d/2}}{S(d)} y^{-d/2} = \int_0^\infty \frac{w^{-1+d/2}dw}{Ye^w - 1},$$ (18.181)

Using here the form of $S(d)$ from (18.163), the equation (18.181) assumes a very simple form

$$\Gamma(d/2)y^{-d/2} = \int_0^\infty \frac{w^{-1+d/2}dw}{Ye^w - 1},$$ (18.182)

or as $y = T/T_{quantum}$,

$$\Gamma(d/2)\left(\frac{T_{quantum}}{T}\right)^{d/2} = \int_0^\infty \frac{w^{-1+d/2}dw}{Ye^w - 1}.$$ (18.183)

The equation (18.183), with $Y = 1$, is [in combination with the equation (18.144)] easily shown to yield our result for the critical temperature in the equation (18.141). The result shows that T_C vanishes for $d = 2$ and $d < 2$. Going with this are the results (18.151) and (18.154), which we will prove in the following:

(a) Case $d = 2$: The equation (18.183) reduces to

$$\frac{T_{quantum}}{T} = \int_0^\infty \frac{dw}{Ye^w - 1}.$$ (18.184)

The integral here can be calculated exactly as follows:

$$\int_0^\infty \frac{dw}{Ye^w - 1} = \int_0^\infty \frac{e^{-w}dw}{Y - e^{-w}} = \int_0^\infty \frac{d(Y - e^{-w})}{Y - e^{-w}}$$

$$= \ln(Y - e^{-w})|_0^\infty = \ln\frac{Y}{Y - 1}.$$ (18.185)

Substituting the result (18.185) into the equation (18.184), we obtain

$$\frac{T_{\text{quantum}}}{T} = \ln \frac{Y}{Y-1}. \tag{18.186}$$

The equation (18.186) is easily solved for Y yielding,

$$Y = \frac{1}{1 - \exp\left(-\frac{T_{\text{quantum}}}{T}\right)}. \tag{18.187}$$

This proves the equation (18.151), which by the equations (18.152) and (18.153) shows that $Y \to 1$ and $\mu \to 0$ only in the limit $T \to 0$.

(b) Case $d < 2$: In this case, we will evaluate the integral in the equation (18.183) in the asymptotic limit $Y \to 1$ whence $Y - 1 = \epsilon \ll 1$. This is possible because the integral is dominated by small values of $w \sim \epsilon$ (this w-range actually causes a divergence of this integral for $Y - 1 = \epsilon \to 0$). One can thus expand the e^w in the denominator of the integral in the equation (18.183) just to the first-order $e^w \approx 1 + w$ and then do the change of variables $w = \epsilon x$, as displayed below

$$\int_0^\infty \frac{w^{-1+d/2}dw}{Ye^w - 1} \approx \int_0^\infty \frac{w^{-1+d/2}dw}{Y(1+w) - 1} = \int_0^\infty \frac{w^{-1+d/2}dw}{Y - 1 + Yw}$$

$$\approx \int_0^\infty \frac{w^{-1+d/2}dw}{Y - 1 + w} = \int_0^\infty \frac{(\epsilon x)^{-1+d/2}d(\epsilon x)}{\epsilon + \epsilon x}$$

$$= \epsilon^{-1+d/2} \int_0^\infty \frac{x^{-1+d/2}dx}{1 + x}. \tag{18.188}$$

The result (18.188) is asymptotically exact for $\epsilon = Y - 1 \to 0$ and any $d < 2$. Here, it remains to calculate the integral in (18.188). It can be done, by the following change of variables $y = 1/(1 + x)$ or $x = -1 + 1/y$, so $dx = -dy/y^2$. With this change of variables, one easily obtains

$$\int_0^\infty \frac{x^{-1+d/2}dx}{1 + x} = \int_0^1 dy y^{-d/2}(1 - y)^{-1+d/2} = B\left(1 - \frac{d}{2}, \frac{d}{2}\right)$$

$$= \frac{\Gamma(1 - d/2)\Gamma(d/2)}{\Gamma(1)} = \frac{\pi}{\sin(\pi d/2)}. \tag{18.189}$$

Above, we employed the standard definition of the Beta function,

$$B(u, v) = \int_0^1 dy y^{u-1}(1 - y)^{v-1}, \tag{18.190}$$

and the identities

$$B(u, v) = \frac{\Gamma(u)\Gamma(v)}{\Gamma(u+v)}, \quad \Gamma(1-u)\Gamma(u) = \frac{\pi}{\sin(\pi u)}, \tag{18.191}$$

which are used above for the case $u = 1 - d/2$ and $v = d/2$. Substituting the result (18.188) with (18.189) into the equation (18.183), we obtain for $d < 2$

$$\Gamma(d/2) \left(\frac{T_{\text{quantum}}}{T} \right)^{d/2} = \epsilon^{-1+d/2} \frac{\pi}{\sin(\pi u)}. \tag{18.192}$$

Thus, for $d < 2$ and $T \ll T_{\text{quantum}}$, the quantity $\epsilon = Y - 1$ is given by the expression

$$Y - 1 \approx A(d) \left(\frac{T}{T_{\text{quantum}}} \right)^{d/(2-d)}$$

$$\Rightarrow \mu \approx -kTA(d) \left(\frac{T}{T_{\text{quantum}}} \right)^{d/(2-d)}, \tag{18.193}$$

which indeed demonstrates that $Y \to 1$ and $\mu \to 0$ only in the limit $T \to 0$. In the equation (18.193), the numerical prefactor $A(d)$ is given by

$$A(d) = \left[\frac{\pi}{\Gamma(d/2)\sin(\pi d/2)} \right]^{2/(2-d)}. \tag{18.194}$$

Thus, we have proved the result in the equation (18.154). It holds for any $d < 2$, in particular for the experimentally interesting case $d = 1$, for which,

$$A(1) = \left[\frac{\pi}{\Gamma(d/2)\sin(\pi d/2)} \right]^{2/(2-1)} = \left(\frac{\pi}{\sqrt{\pi}\sin(\pi/2)} \right)^2 = \pi. \tag{18.195}$$

Thus, by (18.193), for $d = 1$ and $T \ll T_{\text{quantum}}$, we have

$$Y - 1 \approx \pi \frac{T}{T_{\text{quantum}}} \quad \Rightarrow \mu \approx -kT\pi \frac{T}{T_{\text{quantum}}}. \tag{18.196}$$

Comment on the equation (18.172): For $T < T_C$, the total number of bosons in the excited states, with $\vec{\kappa} \neq \vec{0}$, can be written as follows:

$$N - N_0 = \sum_{\vec{\kappa} \neq 0} \bar{n}(\vec{\kappa}) = \sum_{\vec{\kappa} \neq 0} \frac{1}{\exp\left(\frac{E(|\vec{\kappa}|)}{kT} \right) - 1}$$

$$= \frac{V}{(2\pi)^d} S(d) \int_{\kappa_{\min}}^{\infty} \frac{\kappa^{d-1} d\kappa}{\exp\left(\frac{E(\kappa)}{kT} \right) - 1}. \tag{18.197}$$

In the above integral, the lower limit $\kappa_{min} \sim 1/L$ reflects the fact that the $\vec{\kappa} = \vec{0}$ state is excluded from the sum over the allowed values of κ (recall that the allowed values form a cubic lattice with the lattice cell size $\sim 1/L$). Since $\kappa_{min} \sim 1/L \to 0$ for macroscopic systems of interest here, it would be tempting to simply set $\kappa_{min} = 0$ in the integral of the equation (18.197) [as done in solving the item (b) of this Problem]. It is, however, important to note that this step is justified only for $d > 2$ [as was the case in the item (b)]. To see this, note that, for small energies, i.e., for small κ, we have

$$\exp\left[\frac{E(\kappa)}{kT}\right] - 1 \approx 1 + \frac{E(\kappa)}{kT} - 1 = \frac{E(\kappa)}{kT} \sim \kappa^2. \qquad (18.198)$$

Thus, at small κ-values, the integral in the equation (18.197) behaves as the integral

$$\int_{\kappa_{min}} \frac{\kappa^{d-1} d\kappa}{\kappa^2} = \int_{\kappa_{min}} \kappa^{d-3} d\kappa. \qquad (18.199)$$

From (18.199), by elementary integration rules, it is easy to see that one can indeed set $\kappa_{min} \sim 1/L \to 0$, but only for $d > 2$, when the integral (18.199) has a finite value in the limit $\kappa_{min} \sim 1/L \to 0$. On the other hand, for $d < 2$,

$$\int_{\kappa_{min}} \frac{\kappa^{d-1} d\kappa}{\kappa^2} = -\frac{(\kappa_{min})^{d-2}}{d-2} + \text{finite terms}, \qquad (18.200)$$

or, using $\kappa_{min} \sim 1/L$,

$$\int_{\kappa_{min}} \frac{\kappa^{d-1} d\kappa}{\kappa^2} \sim \frac{L^{2-d}}{2-d} + \text{finite terms} \to \infty, \qquad (18.201)$$

for $L \to \infty$. Likewise, for $d = 2$,

$$\int_{\kappa_{min}} \frac{\kappa^{2-1} d\kappa}{\kappa^2} = \int_{\kappa_{min}} \frac{d\kappa}{\kappa} = \ln(\frac{1}{\kappa_{min}}) + \text{finite terms}, \qquad (18.202)$$

or, using $\kappa_{min} \sim 1/L$,

$$\int_{\kappa_{min}} \frac{\kappa^{2-1} d\kappa}{\kappa^2} = \ln(L) + \text{finite terms} \to \infty, \qquad (18.203)$$

for $L \to \infty$. So for any $d < 2$ and $d = 2$, one is not allowed to set $\kappa_{min} \sim 1/L \to 0$ in the integral (18.199), i.e., in the integral (18.197). On the other hand, recall that, for solving the item (b), only the case $d > 2$ was relevant, since only then the critical temperature is positive (nonzero) and one has a real Bose–Einstein condensation. So, by the above discussion, we were allowed to set the $\kappa_{min} \sim 1/L \to 0$ in the item (b) calculations.

19 Special Topics

19.1 Ultrarelativistic Fermion Gas

In this section, we study a gas consisting of the so-called *ultrarelativistic fermions*, having very high energies $E \gg mc^2$, where m stands for the fermion rest mass. Such gases are encountered in some extreme physical conditions, e.g., during the early days of the present universe. When the fermion gas is compressed to a high density, the average energy of the constituent fermions may become very large, due to Pauli exclusion principle (see Chapter 17, "Photon Gas in Equilibrium"). When it becomes comparable or larger than mc^2, the relativistic effects become significant. In this discussion, we will assume that the energy of the fermions is large compared to mc^2. Thus, the relation of the energy and momentum of an ultrarelativistic fermion is given by

$$E = \sqrt{p^2c^2 + m^2c^4} \approx pc. \tag{19.1}$$

It should be noted that the equation (19.1) is exactly valid for massless fermions, like neutrinos. In such a case, by using the relation $E = pc$, we do not make any approximation. As for the nonrelativistic fermion gas, in the ground state $(T = 0)$, we have the following fermion distribution over energies,

$$dn_E = dg(E) \times \begin{Bmatrix} 1 & E < E_F \\ 0 & E > E_F \end{Bmatrix}. \tag{19.2}$$

In order to calculate $dg(E)$, we recall the result (15.46), i.e.,

$$dg(p) = g(\sigma)\frac{V}{h^3}4\pi p^2 dp. \tag{19.3}$$

For fermion gases consisting of ultrarelativistic fermions, we can use the approximation (19.1) to obtain

$$dg(E) = g(\sigma)\frac{V}{c^3 h^3}4\pi E^2 dE. \tag{19.4}$$

Using the results (19.2) and (19.4), we can calculate the total number of relativistic fermions in the form

$$N = \int\limits_{E < E_F} dg(E) = 4\pi g(\sigma)\frac{V}{c^3 h^3}\int\limits_{0}^{E_F} E^2 dE, \tag{19.5}$$

Introductory Statistical Thermodynamics
Copyright © 2011 Elsevier, Inc. All rights reserved.

or

$$N = \frac{4\pi}{3} g(\sigma) V \left(\frac{E_F}{ch}\right)^3. \tag{19.6}$$

The number density of the ultrarelativistic fermions is then obtained as follows:

$$n = \frac{N}{V} = \frac{4\pi}{3} g(\sigma) \left(\frac{E_F}{ch}\right)^3. \tag{19.7}$$

Thus, the Fermi energy for an ultrarelativistic relativistic fermion gas is given by

$$E_F = hc \left(\frac{3n}{4\pi g(\sigma)}\right)^{1/3}. \tag{19.8}$$

The ground-state internal energy is given by

$$U_0 = \int_{E<E_F} E dg(E) = 4\pi g(\sigma) \frac{V}{c^3 h^3} \int_0^{E_F} E^3 dE, \tag{19.9}$$

or

$$U_0 = \frac{4\pi}{3} g(\sigma) V \left(\frac{E_F}{ch}\right)^3 \frac{3}{4} E_F = \frac{3}{4} N E_F. \tag{19.10}$$

For the ultrarelativistic fermion gas at finite temperatures, the fermion distribution over energies is given by

$$dn_E = \frac{dg(E)}{\exp\left(\frac{E-\mu}{kT}\right) + 1}. \tag{19.11}$$

Substituting (19.4) into (19.11), we obtain

$$dn_E = 4\pi g(\sigma) \frac{V}{c^3 h^3} \frac{E^2 dE}{\exp\left(\frac{E-\mu}{kT}\right) + 1}. \tag{19.12}$$

The total number of particles in the ultrarelativistic fermion gas is now obtained as follows:

$$N = 4\pi g(\sigma) \frac{V}{c^3 h^3} \int_0^\infty \frac{E^2 dE}{\exp\left(\frac{E-\mu}{kT}\right) + 1}, \tag{19.13}$$

or

$$N = \frac{4\pi}{3}g(\sigma)V\left(\frac{E_F}{ch}\right)^3 \frac{3}{E_F^3}\int_0^\infty \frac{E^2 dE}{\exp\left(\frac{E-\mu}{kT}\right)+1}. \tag{19.14}$$

Substituting (19.6) into (19.14), we obtain

$$N = N\frac{3}{E_F^3}\int_0^\infty \frac{E^2 dE}{\exp\left(\frac{E-\mu}{kT}\right)+1}. \tag{19.15}$$

Thus,

$$\frac{3}{E_F^3}\int_0^\infty \frac{E^2 dE}{\exp\left(\frac{E-\mu}{kT}\right)+1} = 1. \tag{19.16}$$

The equation (19.16) is an implicit equation, which determines the chemical potential μ of the ultrarelativistic fermion gas. It can be rewritten as follows:

$$3\left(\frac{kT}{E_F}\right)^3 I_N = 1, \tag{19.17}$$

where

$$I_N = \frac{1}{(kT)^3}\int_0^\infty \frac{E^2 dE}{\exp\left(\frac{E-\mu}{kT}\right)+1}. \tag{19.18}$$

The internal energy for the ultrarelativistic fermion gas is given by

$$U = 4\pi g(\sigma)\frac{V}{c^3 h^3}\int_0^\infty \frac{E^3 dE}{\exp\left(\frac{E-\mu}{kT}\right)+1}, \tag{19.19}$$

or

$$U = \frac{4\pi}{3}g(\sigma)V\left(\frac{E_F}{ch}\right)^3 \frac{3}{E_F^3}\int_0^\infty \frac{E^3 dE}{\exp\left(\frac{E-\mu}{kT}\right)+1}. \tag{19.20}$$

Substituting (19.6) into (19.14), we obtain

$$U = \frac{3N}{E_F^3}\int_0^\infty \frac{E^3 dE}{\exp\left(\frac{E-\mu}{kT}\right)+1}, \tag{19.21}$$

or

$$U = 3NE_F \left(\frac{kT}{E_F}\right)^4 I_E, \tag{19.22}$$

where

$$I_E = \frac{1}{(kT)^4} \int\limits_0^\infty \frac{E^3 \, dE}{\exp\left(\frac{E-\mu}{kT}\right) + 1}. \tag{19.23}$$

In order to solve the integrals (19.18) and (19.23), we introduce a new variable w as follows:

$$E = \mu + kTw \Rightarrow dE = kT \, dw. \tag{19.24}$$

Thus, we obtain

$$I_N = \int\limits_{-\mu/kT}^\infty \frac{(w + \mu/kT)^2 \, dw}{e^w + 1}, \tag{19.25}$$

and

$$I_E = \int\limits_{-\mu/kT}^\infty \frac{(w + \mu/kT)^3 \, dw}{e^w + 1}. \tag{19.26}$$

The Helmholtz thermodynamic potential (15.55) for the ultrarelativistic fermion gas is given by

$$\Omega = -4\pi g(\sigma) \frac{VkT}{c^3 h^3} \int\limits_0^\infty E^2 \ln\left[1 + \exp\left(\frac{\mu - E}{kT}\right)\right] dE. \tag{19.27}$$

Integrating the right-hand side of the equation (19.27) by parts, in the same way as for the nonrelativistic quantum gas, we obtain

$$\Omega = -\frac{4}{3}\pi g(\sigma) \frac{V}{c^3 h^3} \int\limits_0^\infty \frac{E^3 \, dE}{\exp\left(\frac{E-\mu}{kT}\right) + 1}. \tag{19.28}$$

Comparing the results (19.28) and (19.19), we obtain the thermal-state equation of the relativistic fermion gas as follows:

$$\Omega = -pV = -\frac{U}{3} \Rightarrow pV = \frac{1}{3}U. \tag{19.29}$$

Thus, for an extremely relativistic fermion gas, the thermal-state equation (15.59) is replaced by the equation (19.29). This equation is identical with the equation (17.29) valid for extremely relativistic boson gas, i.e., the photon gas.

From the above results, we see that if we want to calculate the thermodynamic variables of the ultrarelativistic fermion gas, we need to calculate the integrals I_N and I_E defined by the equations (19.18) and (19.23), respectively.

19.1.1 Ultrarelativistic Fermion Gas for $T/T_F \ll 1$

In case of the ultrarelativistic fermion gas with $T/T_F \ll 1$, we can use the general result (16.31) to calculate the two integrals I_N and I_E, defined by the equations (19.18) and (19.23). The integral I_N, with $f(\lambda) = \lambda^2$, can be rewritten as follows:

$$I_N = \int_0^{\mu/kT} \lambda^2 d\lambda + \int_0^\infty \frac{[(w + \mu/kT)^2 - (-w + \mu/kT)^2]dw}{e^w + 1}. \tag{19.30}$$

On the other hand, the integral I_N, with $f(\lambda) = \lambda^3$, can be rewritten as follows:

$$I_E = \int_0^{\mu/kT} \lambda^3 d\lambda + \int_0^\infty \frac{[(w + \mu/kT)^3 - (-w + \mu/kT)^3]dw}{e^w + 1}. \tag{19.31}$$

Unlike their counterparts for the nonrelativistic fermion gas, the integrals (19.30) and (19.31) can be solved exactly. By expanding the binomial expressions in the numerators of these integrals, they are significantly simplified, and we obtain

$$I_N = \int_0^{\mu/kT} \lambda^2 d\lambda + 4 \frac{\mu}{kT} \int_0^\infty \frac{w\,dw}{e^w + 1}, \tag{19.32}$$

and

$$I_E = \int_0^{\mu/kT} \lambda^3 d\lambda + 6 \left(\frac{\mu}{kT}\right)^2 \int_0^\infty \frac{w\,dw}{e^w + 1} + 2 \int_0^\infty \frac{w^3\,dw}{e^w + 1}. \tag{19.33}$$

The values of the two numerical integrals left in the two equations (19.32) and (19.33) are given by

$$\int_0^\infty \frac{w\,dw}{e^w + 1} = \frac{\pi^2}{12}, \quad \int_0^\infty \frac{w^3\,dw}{e^w + 1} = \frac{7\pi^4}{120}. \tag{19.34}$$

Using the values (19.34), we obtain

$$I_N = \frac{1}{3}\left[\left(\frac{\mu}{kT}\right)^3 + \pi^2\frac{\mu}{kT}\right],$$ (19.35)

and

$$I_E = \frac{1}{4}\left(\frac{\mu}{kT}\right)^4 + \frac{\pi^2}{2}\left(\frac{\mu}{kT}\right)^2 + \frac{7\pi^4}{60}.$$ (19.36)

Substituting (19.35) and (19.36) into (19.17) and (19.22), respectively, we obtain

$$\left(\frac{\mu}{E_F}\right)^3 + \pi^2\left(\frac{kT}{E_F}\right)^2\frac{\mu}{E_F} = 1,$$ (19.37)

and

$$U = \frac{3}{4}NE_F\left[\left(\frac{\mu}{E_F}\right)^4 + 2\pi^2\left(\frac{kT}{E_F}\right)^2\left(\frac{\mu}{E_F}\right)^2 + \frac{7\pi^4}{15}\left(\frac{kT}{E_F}\right)^4\right].$$ (19.38)

Using here $E_F = kT_F$, we can also write

$$\left(\frac{\mu}{E_F}\right)^3 + \pi^2\left(\frac{T}{T_F}\right)^2\frac{\mu}{E_F} = 1,$$ (19.39)

and

$$U = \frac{3}{4}NE_F\left[\left(\frac{\mu}{E_F}\right)^4 + 2\pi^2\left(\frac{T}{T_F}\right)^2\left(\frac{\mu}{E_F}\right)^2 + \frac{7\pi^4}{15}\left(\frac{T}{T_F}\right)^4\right].$$ (19.40)

In this case, it is possible to find the exact solution of the equation (19.39) for the relative chemical potential $x = \mu/E_F$, and then substitute that solution into the caloric-state equation (19.40). The equation (19.39) has the form of the exactly solvable cubic equation

$$x^3 + a^2x - 1 = 0, \quad a = \pi\frac{T}{T_F} \ll 1.$$ (19.41)

This equation has only one real solution, and it is given by

$$x = \left[\frac{1}{2} + \left(\frac{a^6}{27} + \frac{1}{4}\right)^{1/2}\right]^{1/3} + \left[\frac{1}{2} - \left(\frac{a^6}{27} + \frac{1}{4}\right)^{1/2}\right]^{1/3}.$$ (19.42)

Thus, we can write

$$\frac{\mu}{E_F} = \left[\frac{1}{2} + \left(\frac{\pi^6}{27}\left(\frac{T}{T_F}\right)^6 + \frac{1}{4}\right)^{1/2}\right]^{1/3} + \left[\frac{1}{2} - \left(\frac{\pi^6}{27}\left(\frac{T}{T_F}\right)^6 + \frac{1}{4}\right)^{1/2}\right]^{1/3}.$$

(19.43)

Substituting (19.43) into (19.40), we can obtain the "almost exact" caloric-state equation. However, the solution (19.43) is quite complicated and not useful for practical calculations. It is included here primarily to illustrate the possibility of solving the implicit equation for the chemical potential (19.16). In order to obtain a simpler result for the chemical potential, we can use the fact that $T/T_F \ll 1$.
Using here the approximation

$$\sqrt{1+x} \approx 1 + \frac{x}{2}, \quad x \ll 1,$$

(19.44)

we can write

$$\left(\frac{\pi^6}{27}\left(\frac{T}{T_F}\right)^6 + \frac{1}{4}\right)^{1/2} \approx \frac{1}{2} + \frac{\pi^6}{27}\left(\frac{T}{T_F}\right)^6.$$

(19.45)

Substituting (19.45) into (19.43), we obtain

$$\frac{\mu}{E_F} = \left[1 + \frac{\pi^6}{27}\left(\frac{T}{T_F}\right)^6\right]^{1/3} + \left[-\frac{\pi^6}{27}\left(\frac{T}{T_F}\right)^6\right]^{1/3}.$$

(19.46)

Neglecting the sixth-order contribution in the first term, we finally obtain

$$\frac{\mu}{E_F} = 1 - \frac{\pi^2}{3}\left(\frac{kT}{E_F}\right)^2.$$

(19.47)

Substituting (19.47) into (19.40) and keeping only terms of second order in the small quantity $T/T_F \ll 1$, we obtain the approximate caloric-state equation

$$U = \frac{3}{4}NE_F\left[1 + \frac{2}{3}\pi^2\left(\frac{kT}{E_F}\right)^2\right].$$

(19.48)

The specific heat at constant volume is given by

$$mc_V = \left(\frac{\partial U}{\partial T}\right)_V = \pi^2 Nk\frac{kT}{E_F}.$$

(19.49)

Using (19.29), we obtain the thermal-state equation

$$pV = \frac{1}{4}NE_F\left[1 + \frac{2}{3}\pi^2\left(\frac{kT}{E_F}\right)^2\right].$$ (19.50)

Using here the definition of the fermion number density $n = N/V$, we obtain the pressure of the ultrarelativistic fermion gas

$$p = \frac{1}{4}nE_F\left[1 + \frac{2}{3}\pi^2\left(\frac{kT}{E_F}\right)^2\right].$$ (19.51)

The Helmholtz thermodynamic potential is given by

$$\Omega = -pV = -\frac{1}{4}NE_F\left[1 + \frac{2}{3}\pi^2\left(\frac{kT}{E_F}\right)^2\right].$$ (19.52)

19.1.2 Ultrarelativistic Fermion Gas for $T/T_F \gg 1$

This case is of special interest for studying the behavior of ultrarelativistic fermions in the early stages of the expansion of the universe. In this case, we have $\mu/kT \to 0$, and the integral I_E, defined by (19.26), can be approximated by

$$I_E = \int_{-\mu/kT}^{\infty} \frac{(w + \mu/kT)^3 dw}{e^w + 1} \approx \int_0^{\infty} \frac{w^3 dw}{e^w + 1} = \frac{7\pi^4}{120}.$$ (19.53)

Substituting (19.53) into (19.22), we obtain

$$U = 3NE_F\left(\frac{kT}{E_F}\right)^4\frac{7\pi^4}{120} = \frac{7}{8} \times \frac{\pi^2 Vk^4 T^4}{15c^3\hbar^3}.$$ (19.54)

Comparing the result (19.54) for the relativistic massless fermion gas with the result (17.31) for the relativistic massless boson gas (photon gas), we see that they are identical except for the ratio between the fermion and boson integrals, defined by the equation (15.81), i.e.,

$$\gamma_F = \left(1 - 2^{1-2j}\right) = 1 - 2^{-3} = \frac{7}{8}, \quad j = 2.$$ (19.55)

Thus, the massless relativistic fermion gas, for $T/T_F \gg 1$, has the same behavior as the black-body radiation, except for the factor of $\gamma_F = 7/8$ in calculating the thermodynamic variables like the internal energy or the specific heat.

19.2 Thermodynamics of the Expanding Universe

In the present universe, there are two types of relativistic particles, which can be considered as the relics from the early phases of the expansion of the universe. These are the $T = 2.75$ K microwave photons and the three cosmic seas of $T = 1.96$ K relic neutrinos. It is generally assumed that the early universe was in an approximate thermal equilibrium involving many other species of ultrarelativistic bosons and fermions. In the early stages of the universe, the quantum gas consisting of these bosons and fermions was highly relativistic, and the relativistic conditions $kT \gg mc^2$ and $kT \gg \mu$ were satisfied with a good margin. Thus, we can use the following approximation of the result (19.12) for the particles, distribution over energies

$$dn_E \approx \frac{g(\sigma)}{2\pi^2} \frac{V}{c^3 \hbar^3} \frac{E^2 dE}{\exp\left(\frac{E}{kT}\right) \mp 1}, \tag{19.56}$$

where, as before, the upper sign applies to bosons and the lower sign applies to fermions. For notational simplicity, let us now introduce a new constant, which includes both the spin factor $g(\sigma)$ and the constants c and $\hbar$ as follows:

$$g = \frac{g(\sigma)}{c^3 \hbar^3}. \tag{19.57}$$

Thus, we obtain

$$dn_E \approx \frac{g}{2\pi^2} V \frac{E^2 dE}{\exp\left(\frac{E}{kT}\right) \mp 1}. \tag{19.58}$$

The particle number density of an ultrarelativistic quantum gas is then given by

$$n = \frac{N}{V} \approx \frac{g}{2\pi^2} \int_0^\infty \frac{E^2 dE}{\exp\left(\frac{E}{kT}\right) \mp 1}. \tag{19.59}$$

The energy density of an ultrarelativistic quantum gas is given by

$$\rho = \frac{U}{V} \approx \frac{g}{2\pi^2} \int_0^\infty \frac{E^3 dE}{\exp\left(\frac{E}{kT}\right) \mp 1}. \tag{19.60}$$

Using the thermal-state equation (19.29), we obtain the pressure of the relativistic quantum gas

$$p = \frac{U}{3V} = \frac{\rho}{3}. \tag{19.61}$$

Introducing a new variable $w = E/kT$, we obtain

$$n = \frac{g}{2\pi^2}(kT)^3 \int_0^\infty \frac{w^2 dw}{e^w \mp 1},$$ (19.62)

and

$$\rho = \frac{g}{2\pi^2}(kT)^4 \int_0^\infty \frac{w^3 dw}{e^w \mp 1}.$$ (19.63)

Using the results for the integrals (15.80) and (15.81), we obtain the particle density

$$n = \begin{cases} \left(\zeta(3)/\pi^2\right) g(kT)^3 & \text{Bosons} \\ (3/4)\left(\zeta(3)/\pi^2\right) g(kT)^3 & \text{Fermions} \end{cases},$$ (19.64)

where $\zeta(3) = 1.20206\ldots$, and the energy density

$$\rho = \begin{cases} (\pi^2/30)g(kT)^4 & \text{Bosons} \\ (7/8)\left(\pi^2/30\right) g(kT)^4 & \text{Fermions} \end{cases}.$$ (19.65)

19.2.1 Internal Energy of Elementary Particle Species

Let us assume that the equilibrium state of the early universe consisted of a number of different ultrarelativistic particles, which we will call the ultrarelativistic species of the universe. The different species, labeled by i, can be in thermal equilibrium at temperatures T_i different from the temperature of the photons (T). In the conditions of the early expansion of the universe, the contribution from the nonrelativistic particles is to a good approximation negligible relative to the contribution from the elementary particles belonging to ultrarelativistic species. Thus, we can express the total energy density ρ_R and pressure $p_R = \rho_R/3$ of all the species in the thermal equilibrium, in terms of the photon gas temperature T. Thus, we obtain

$$\rho_R = \frac{\pi^2}{30} g_R (kT)^4, \quad p_R = \frac{\pi^2}{90} g_R (kT)^4,$$ (19.66)

where g_R is the effective constant, which measures the total number of relativistic (massless) species, with the appropriate factors to account for the different equilibrium temperatures for different species, as well as for the difference between bosons and fermions. The constant g_R is given by

$$g_R = \sum_{i=\text{bosons}} g_i \left(\frac{T_i}{T}\right)^4 + \frac{7}{8} \sum_{i=\text{fermions}} g_i \left(\frac{T_i}{T}\right)^4.$$ (19.67)

For lower energies ($kT \ll 1$ MeV), the only relativistic species are photon and the three neutrino species, in which case we have $g_R \approx 3.36\,c^{-3}\hbar^{-3}$. At intermediate energies (100 MeV $\leq kT \leq 1$ MeV), the additional relativistic species are the electron and positron, and we have $g_R \approx 10.75\,c^{-3}\hbar^{-3}$. For high energies ($kT \geq 300$ GeV), all the species in the standard model of the elementary particle physics are relativistic, and we have $g_R \approx 106.75\,c^{-3}\hbar^{-3}$.

19.2.2 Entropy per Volume Element

During the expansion of the universe, it is assumed that the local thermal equilibrium has been maintained for the particles belonging to ultrarelativistic species. There is no heat "added externally" to the universe, $dQ = TdS = 0$, so the expansion of ultrarelativistic species can be treated as an adiabatic process. Let us assume that a comoving (coexpanding) volume element has the (expanding) physical volume $V = R^3$. From the first law of thermodynamics, we obtain

$$T dS = dU + p dV = d(\rho V) + p dV, \tag{19.68}$$

or

$$T dS = d[(\rho + p)V] - V dp. \tag{19.69}$$

The differential of entropy is then given by

$$dS = \frac{1}{T} d[(\rho + p)V] - \frac{V}{T} dp. \tag{19.70}$$

Using the general result (9.77), we obtain the condition that the expression (19.70) is a total differential, in the following form

$$\frac{\partial}{\partial p}\left(\frac{1}{T}\right) = \frac{\partial}{\partial[(\rho + p)V]}\left(-\frac{V}{T}\right), \tag{19.71}$$

or

$$-\frac{1}{T^2}\frac{\partial T}{\partial p} = -\frac{1}{T}\frac{\partial V}{\partial[(\rho + p)V]}. \tag{19.72}$$

Thus, we obtain the relation

$$T\frac{dp}{dT} = \rho + p \Rightarrow dp = \frac{\rho + p}{T} dT. \tag{19.73}$$

Substituting (19.73) into (19.70), we obtain

$$dS = \frac{1}{T} d[(\rho + p)V] + [(\rho + p)V]\left(-\frac{dT}{T^2}\right), \tag{19.74}$$

or

$$dS = d\left[\frac{(\rho+p)V}{T} + S_0\right].$$

(19.75)

Thus, up to an additive constant S_0, the entropy of a comoving volume element is given by

$$S = \frac{(\rho+p)V}{T} = \frac{(\rho+p)R^3}{T} = \text{Constant.}$$

(19.76)

If we now define the entropy density $s_V = S/V$, we obtain

$$s_V = \frac{\rho+p}{T} = \frac{\rho+\rho/3}{T} = \frac{4}{3}\frac{\rho}{T}.$$

(19.77)

Substituting (19.66) into (19.77), we obtain

$$s_V = \frac{2\pi^2 k}{45} g_S (kT)^3,$$

(19.78)

where

$$g_S = \sum_{i=\text{bosons}} g_i\left(\frac{T_i}{T}\right)^3 + \frac{7}{8}\sum_{i=\text{fermions}} g_i\left(\frac{T_i}{T}\right)^3.$$

(19.79)

During a part of the expansion of the universe, the different ultrarelativistic species in thermal equilibrium have had the same temperature. Due to this, g_S was equal to g_R. The conservation of the entropy $S = s_V V$ implies that $s_V \sim R^{-3}$. Thus, the adiabatic process of the expansion of the universe has $VT^3 = \text{Constant}$. This is the same conclusion as the one we made in Chapter 18, "Other Examples of Boson Systems," for the adiabatic process in the photon gas. Thus, the quantity $R^3 T^3$ remains constant as the universe expands. This is an important physical conclusion, which shows that as the universe expands, its temperature decreases according to the following qualitative relation

$$T \sim g_S^{-1/3} R^{-1}.$$

(19.80)

If the quantity g_S would be a constant, we would obtain the simple result $T \sim R^{-1}$. The factor g_S is of significance when the energy of a species decreases and it eventually becomes nonrelativistic. Their contribution to the above thermodynamic variables disappears (their entropy is transferred to lower energy species forming a thermal plasma). This effect causes a decrease of g_S, and by the equation (19.80), it causes the temperature to decrease less slowly. This, however, does not apply to the massless particles that become decoupled from the thermal plasma, when the temperature drops below the mass threshold of the species. The temperature of such a massless species scales as $T \sim R^{-1}$.

19.3 Problems with Solutions

Problem 1

The present universe can be modeled as a spherical cavity of a radius $R_U \approx 10^{26}$ m and the temperature of $T_U \approx 3$ K.

(a) Calculate the total number of photons in the universe.
(b) Calculate the total energy of photons in the universe.

Solution

(a) The total number of photons in the universe is given by

$$N = \frac{V}{\pi^2 c^3} \int\limits_0^\infty \frac{\omega^2 d\omega}{\exp\left(\frac{\hbar\omega}{kT}\right) - 1}. \tag{19.81}$$

If we now introduce a new integration variable w as follows:

$$\frac{\hbar\omega}{kT} = w \Rightarrow \omega = \frac{kT}{\hbar} w \Rightarrow d\omega = \frac{kT}{\hbar} dw, \tag{19.82}$$

we obtain

$$N = \frac{Vk^3 T^3}{\pi^2 c^3 \hbar^3} \int\limits_0^\infty \frac{w^2 dw}{e^w - 1} = \frac{Vk^3 T^3}{\pi^2 c^3 \hbar^3} \Gamma(3)\zeta(3), \tag{19.83}$$

or finally

$$N = 2.4 \frac{Vk^3 T^3}{\pi^2 c^3 \hbar^3} = 2.3 \times 10^{87}, \tag{19.84}$$

where $V = 4\pi/3 R_U^3$ is the volume of the universe.

(b) The total energy of photons in the universe is given by

$$U = \frac{\pi^2 V k^4 T^4}{15 c^3 \hbar^3} = 2.6 \times 10^{65} \, \text{J}. \tag{19.85}$$

Problem 2

In a given volume of the expanding universe, the number of photons N is, on average, 10^9 times larger than the number of nucleons (protons or neutrons), which are assumed to comprise most of the rest mass of the universe. This ratio of the number of photons and the number of nucleons are constant during the expansion of the universe. In a certain phase of the expansion of the universe, the energy of the background electromagnetic radiation has dominated all the other energy types. Calculate the temperature at which the energy of the background radiation is equal to the energy $10^{-9}\,Nmc^2$ corresponding to the total rest mass of the universe, where $m \approx 1,67 \times 10^{-27}$ kg is the nucleon mass and $c = 2.9979 \times 10^8$ m/s is the speed of light.

Solution

Using the result (19.83), we know that the number of photons in the universe is

$$N = \frac{Vk^3T^3}{\pi^2 c^3 \hbar^3}\Gamma(3)\zeta(3).$$
(19.86)

The number of nucleons is then equal to $10^{-9} \times N$, and the rest energy of the nucleons is given by

$$E_N = 10^{-9} \times mc^2 \times \frac{Vk^3T^3}{\pi^2 c^3 \hbar^3}\Gamma(3)\zeta(3).$$
(19.87)

On the other hand, using the result (19.85), the total energy of the photons in the universe is

$$U = \frac{\pi^2 Vk^4T^4}{15 c^3 \hbar^3}.$$
(19.88)

If we now assume that $E_N = U$, we obtain

$$10^{-9} \times mc^2 \times \frac{Vk^3T^3}{\pi^2 c^3 \hbar^3}\Gamma(3)\zeta(3) = \frac{\pi^2 Vk^4T^4}{15 c^3 \hbar^3}.$$
(19.89)

Using $\zeta(3) = 1.20206\ldots$, we finally obtain

$$T = 10^{-9} \times \frac{30\zeta(3)}{\pi^4} \times \frac{mc^2}{k} = 4030\,\text{K}.$$
(19.90)

Problem 3

A dense white-dwarf star contains a gas of ultrarelativistic electrons of mass m_e at a temperature $T \ll T_F$, where T_F is the Fermi temperature of the electron gas.

(a) Calculate the density of the star, for which the Fermi momentum is equal to $p_F = \sqrt{3}/2m_e c$.

(b) Calculate the pressure of the star under such conditions.

Solution

(a) The Fermi energy for a relativistic electron gas, with $g(\sigma) = 2$, is given by the following expression

$$E_F = p_F c = hc \left(\frac{3n}{8\pi}\right)^{1/3}.$$ (19.91)

The Fermi momentum is then given by

$$p_F = h \left(\frac{3n}{8\pi}\right)^{1/3} = \frac{\sqrt{3}}{2} m_e c,$$ (19.92)

such that we obtain

$$\frac{3nh^3}{8\pi} = \frac{3\sqrt{3}}{8} m_e^3 c^3 \Rightarrow n = \pi \sqrt{3} \left(\frac{m_e c}{h}\right)^3.$$ (19.93)

Using the numerical values, we obtain

$$n = \pi \sqrt{3} \left(\frac{9.1 \times 10^{-31} \times 3 \times 10^8}{6.62 \times 10^{-34}}\right)^3 \approx 3.81 \times 10^{35} \frac{1}{\text{m}^3}.$$ (19.94)

(b) For $T/T_F \ll 1$, we have the approximate caloric-state equation

$$U = \frac{3}{4} N E_F \left[1 + \frac{2}{3}\pi^2 \left(\frac{kT}{E_F}\right)^2\right] \approx \frac{3}{4} N E_F = \frac{3}{4} n V E_F.$$ (19.95)

The pressure of the star is then given by

$$p = \frac{U}{3V} = \frac{n}{4} E_F = \frac{1}{4} n p_F c = \frac{\sqrt{3}}{8} n m_e c^2.$$ (19.96)

Using the numerical values, we obtain

$$p = \frac{\sqrt{3}}{8} 3.81 \times 10^{35} \times 9.1 \times 10^{-31} \times 9 \times 10^{16},$$ (19.97)

or finally

$$p = 6.75 \times 10^{21} \frac{\text{N}}{\text{m}^2}.$$ (19.98)

A Physical Constants

Atomic mass unit	a.m.u.	1.6605×10^{-27} kg
Avogadro number	N_{AV}	6.023×10^{23} mol^{-1}
Boltzmann constant	k	1.3807×10^{-23} J/K
Electron or proton charge	e	1.6022×10^{-19} C
Electron mass	m_e	9.1094×10^{-31} kg
Faraday constant	f	96488.5 C/mol
H-bar constant	$\hbar$	1.0546×10^{-34} J s
Planck constant	h	6.626068×10^{-34} J s
Speed of light	c	2.9979×10^8 m/s
Standard atmosphere	1 Atm	1.0133×10^5 Pa
Stefan–Boltzmann constant	σ_{SB}	5.667×10^{-8} kg/s^3 K^4
Universal gas constant	R	8.314 J/K mol

Bibliography

Balescu, R. (1975). *Equilibrium and nonequilibrium statistical mechanics*. New York: Wiley.

Bowley, R. M., & Sanchez, M. (1996). *Introductory statistical mechanics*. Oxford: Oxford University Press.

Callen, H. B. (1960). *Thermodynamics*. New York: Wiley.

Dalarsson, M., & Dalarsson, N. (2005). *Tensors, relativity and cosmology*. San Diego, CA: Elsevier Academic Press.

Fermi, E. (1957). *Thermodynamics*. New York: Dover Publications.

Feynman, R. P. (1972). *Statistical mechanics: A set of lectures*. Reading, MA: Addison-Wesley.

Hill, T. L. (1960). *An introduction to statistical thermodynamics*. Reading, MA: Addison-Wesley.

Hill, T. L. (1987). *Statistical mechanics: Principles and selected applications*. New York: Dover Publications.

Kittel, C. (1958). *Elementary statistical physics*. New York: Wiley.

Landau, L. D., & Lifshitz, E. M. (1963). *Statistical physics*. Reading, MA: Addison-Wesley.

Reif, F. (1965). *Fundamentals of statistical and thermal physics*. New York: Clarendon Press, McGraw-Hill.

Tolman, R. C. (1934). *Relativity, thermodynamics and cosmology*. Oxford: Clarendon Press.

Tolman, R. C. (1955). *The principles of statistical mechanics*. New York: Dover Publications.

Vasilyev, A. M. (1983). *An introduction to statistical physics*. Moscow: Mir Publishers.

Zemansky, M. W. (1957). *Heat and thermodynamics*. New York: McGraw-Hill.

Index

Printed and bound by CPI Group (UK) Ltd, Croydon, CR0 4YY

03/10/2024

01040416-0007